PLATE
TECTONICS

PLATE
TECTONICS

An Insider's History
of the
Modern Theory
of the Earth

NAOMI ORESKES
EDITOR

with Homer Le Grand

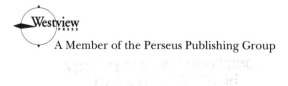

Westview
PRESS

A Member of the Perseus Publishing Group

Copyright © 2001 by Westview Press, A Member of the Perseus Books Group

Westview Press books are available at special discounts for bulk purchases in the United States by corporations, institutions, and other organizations. For more information, please contact the Special Markets Department at The Perseus Books Group, 11 Cambridge Center, Cambridge MA 02142, or call (617) 252-5298.

Published in 2001 in the United States of America by Westview Press, 5500 Central Avenue, Boulder, Colorado 80301–2877, and in the United Kingdom by Westview Press, 12 Hid's Copse Road, Cumnor Hill, Oxford OX2 9JJ

Find us on the World Wide Web at www.westviewpress.com

A CIP catalog record for this book is available from the Library of Congress.

ISBN 0–8133-3981–2

The paper used in this publication meets the requirements of the American National Standard for Permanence of Paper for Printed Library Materials Z39.48-1984.

10   9   8   7   6   5   4   3   2   1

To the late Chuck Drake,
who was always interested in people's stories

CONTENTS

Acknowledgments

The idea for this project emerged from the History Committee of the American Geophysical Union (AGU), then chaired by Ed Cliver. For some time, the AGU has been committed to the importance of history as a resource for the scientific community. With its commitment to publications, oral history, and historically oriented themes at annual meetings, the AGU has been a model of what history can do for a scientific community, bringing diverse specialists together in ways that the daily practice of science rarely does. Beyond expressing our thanks to the AGU as an organization, we are particularly indebted to Ed Cliver. The idea for this project was Ed's, and he convinced us to take it on. Moreover, he stayed involved every step of the way: helping to identify and track down contributors, calling to check progress, reading manuscripts, and providing moral support and good humor all along the way.

We also wish to express our appreciation for the 17 scientists whose essays appear in this volume. When we began this project, we did not know how our invitees would respond, and we had some concerns that they might be dismissive of history, as scientists sometimes are. This turned out not to be the case. On the contrary, our authors responded with enthusiasm and made time in their busy lives to take on this extra task. Each of them has been a pleasure to work with, and we have learned a great deal from our interactions with them. Nearly every one of our authors took additional time to help resolve discrepancies in the historical record; to discuss historical, philosophical, personal, and political questions raised by the history of plate tectonics; or to send us relevant reprints, photos, and interviews. To each and every one of them we are extremely grateful. We owe a special debt of gratitude to Xavier Le Pichon, Dan McKenzie, Peter Molnar, and Jack Oliver for extended conversations and email communications, and to local colleagues at the Scripps Institution of Oceanography (SIO) John Sclater and Bob Parker for their constant willingness to help clarify technical points. Jason Morgan and Ted Irving were unable to write essays for this volume, but generously gave of their time to talk about their contributions. We are grateful also to SIO archivist Deborah Day, whose professional title scarcely does justice to the myriad ways she supports and promotes historical understanding, not only at SIO, but throughout the earth sciences community.

Several colleagues read manuscripts and provided feedback at various stages in the project—Kenneth Belitz, Ron Doel, Gary Ernst, Norm Sleep, Ken Taylor, Gary Weir, and David van Keuren. Throughout the project, Duncan Agnew has been ever-ready and willing not only to read and comment on materials, but even to go to the library to find the reference that would resolve a question to which no one else knew the answer (or in some cases even where to look). The depth and breadth of his knowledge is nothing short of remarkable, and on more than one occasion he has saved us from embarrassing mistakes.

No work can be done without logistical support, and we wish to express our gratitude to SIO director Charles Kennel for his generosity in providing seed money for the larger project on the history of oceanography, of which this volume is a part, and ever-precious office space. Equally important is the ongoing support of colleagues in the Science Studies Program at the University of California, San Diego (UCSD), for creating a community interested in diverse approaches to understanding science and its role in all of our lives. I am particularly indebted to Robert Westman for deepening my appreciation of what it means to be a historian, my understanding of why history is so worth doing.

Several UCSD students worked on this project at various stages, and we need to thank Andrea Santos, Sue Kim, Brook Mangin, and Katie Atwood, and above all the inestimable James Peters, who simply did everything that needed to be done.

A book is only a pile of papers without an editor and a press, and here we have more than the usual debts to acknowledge. This project was originally signed at Columbia University Press, and we are grateful to Director William B. Strachan for releasing us from our contract and permitting us to move the project to Westview Press along with our editor and publisher, Holly Hodder. It is difficult to describe the work that Holly has done on this project without lapsing into platitudes. Suffice it to say that she has been there every step of the way: editing, talking, brainstorming, cajoling, supporting, encouraging. She has never wavered from her commitment to the importance, significance, and meaning of this project, never wavered from her belief that earth science is exciting, important, and worth knowing.

Naomi Oreskes
La Jolla, California, July 2001

Homer Le Grand,
Melbourne, Australia, 2001

PREFACE:
HISTORY AND MEMORY

> In reality, the interest of the past is that it illuminates the present.
> —Jacques Le Goff, *History and Memory,* p. xx.

HISTORIAN JACQUES LE GOFF HAS WRITTEN THAT THE JOB OF history is to correct memory.[1] If so, then this book is not a work of history. The essays presented here are works of memory, stories told by scientists whose work changed the way we think about the planet we live on. Before the 1960s, there was no generally accepted global theory to explain the major features of the earth: the continents and oceans, the mountains and valleys, the volcanoes and earthquakes. In the 1960s, a new theory emerged that explained all this and more as the result of the interactions of moving pieces of the earth's surface layer, henceforth to be known as *tectonic plates.* While the development of plate tectonics was a long time in coming – scientific evidence of continental mobility had been recognized since the early 20th century – its acceptance was rapid and nearly absolute. By the early 1970s, virtually all earth scientists accepted the new theory and textbooks were rewritten.[2]

It has been more than 30 years since the events retold here. When we invited the authors to write for this volume, we asked them to tell their stories as best as they could recall and to reflect on their significance with the benefit of hindsight. The authors were young when they did the work described here, and considerable time has elapsed. All of the authors are brilliant and creative people, and their scientific careers did not end with the contributions they made to plate tectonics. They all continued to work as scholars and teachers, some remaining in the specialties they began in, others shifting their focus. Our authors have had to reach back in time to write these essays, and the resulting stories are works of memory, not history. If memory is faulty – and we all know that it is – then why bother with it? One recent psychological study suggests that accurate memory of adolescent experience is no better than what might be expected by chance.[3] So why not simply write history, and correct memory?

There are at least three reasons why the essays in this volume are important contributions that complement histories written by professional historians.[4] The most obvious is that, while memory is often faulty, it is not always faulty. People do remember important and formative events in their lives, sometimes in extraordinary detail, and they often remember connections that are not recorded elsewhere. While historians prefer to rely on written documents, which are less readily subject to subsequent distortion or manipulation, contemporary documents are not always available. Even when they are, documentation is selective: many things are never written down, and most of what is written down is not saved.

The written record is most silent about the lives of ordinary people, and social historians have come to reply on oral accounts to capture the voices of people whose lives might otherwise go unheeded. However, the authors of these essays are not ordinary people – indeed, they are quite extraordinary – and their scientific work is amply documented in their published papers. Scientific research by its nature leaves an ample paper trail. But in their own way, scientific papers are as incomplete as any political or social records. While they recount the evidence and arguments at stake, they omit much of what is of human interest: how people came to their discoveries and insights and how they felt about them. Moreover, as historian Steven Brush pointed out some years ago, scientific papers are deliberately incomplete, if not downright misleading.[5] Scientific papers are written as if their authors knew from the start where they were heading and saw all along where the data were leading. The false starts, the misinterpretations, the wasted efforts, the failed experiments – these are almost always expunged from published reports. Philosopher Hans Reichenbach called this the "rational reconstruction of knowledge" – how it should have happened in a perfect world, if everything had been done right from the start.[6] The result is a picture of science and scientists as far more efficient than they really are. Because rational reconstruction is the norm for scientific reporting, many scientists follow this pattern even when speaking off the record, perpetuating the image of scientists as coldly rational, even robotic.

Beyond the cognitive cleansing that occurs in scientific publications, there is also an emotional cleansing: feelings are left out. Science is supposed to be about what we know, not about how we feel. Scientific papers are written as if the authors had no feelings about the matters under discussion (or anything else, for that matter).[7] They are written as if the authors didn't care about the outcome of their work.[8] Yet surely they do care, or why would they work so hard? Why would they call their colleagues – as Tanya Atwater vividly recounts – at 2 A.M. to discuss their latest idea?

If scientists' accounts of their work are drained of emotion, popular accounts often err in the opposite direction, painting scientific work as a steady stream of dramatic discovery. To anyone who has ever done scientific research, such accounts ring equally false as their reverse. The stories told here attempt to strike a realistic middle ground: to recount the genuine excitement their authors felt as they became involved in one of the great scientific developments of the 20th century, while conveying the frustrations and false starts as well. As important and true as Tanya Atwater's unbridled excitement is Xavier Le Pichon's poignant portrayal of the moment his world collapsed, as he realized that everything he had written in his just-finished Ph.D. dissertation was wrong.

A second reason for presenting these essays is to gather a multiplicity of perspectives in a single volume. Several of our authors have written about their work before, but never have their differing perspectives been presented together in one place. And their perspectives are indeed different. The 17 scientists who tell their stories here became scientists for different reasons, approached their work in different ways, and made important contributions by different means. As they look back now on their work, they come to various (and not always reconciliable) conclusions. One author – Gordon MacDonald – candidly recounts his objections to plate tectonics in the 1960s, which, he argues, have still not been adequately answered. As editors, we have not attempted to enforce a uniform style or to reconcile opposing views. We have done our best to correct errors on factual matters, but beyond that we have sought to preserve the diverse voices of our authors as an important part of the unique value of this volume.

While the value and legitimacy of multiple perspectives has become widely accepted in many fields – art, architecture, literature, history – in science we are still wedded to the notion of *the* right answer. While there is a single right answer to certain kinds of technical questions – How old is the earth? What is the composition of the sun? – many scientists extend the presumption of a single right answer to questions in philosophy of science: How does science advance? What is the correct scientific method? What makes a scientist great? (As if these were comparable questions to what the radius of the earth is!) The essays in this volume argue against a narrow answer to these kinds of questions. Some of the stories told here involve data-driven science, others involve conceptual or mathematical innovation, still others involve novel instruments and data analysis. Likewise, if one were to ask, "what kind of a personality does it take to succeed in science?" the answer provided by this volume would have to be multiple. The authors of these essays are unique and diverse individuals, and it would be no more possible to say what unites them than to say what unites all great artists or all wonderful mothers.

This leads to the third and most important reason for presenting these essays: the scientists writing in this volume speak with the voice of experience. Each of the authors has had time to consider his or her own scientific life and contributions. Each has, in some way, been forced into such consideration by the prominence of his or her contributions (or, in Lawrence Morley's case, by the poignancy of seeing someone else become famous for an idea that he also had, but saw rejected for publication at two leading scientific journals). By the nature of our sample – scientists writing about events 30 years later – these are people who made major contributions early in their scientific careers. All have had the opportunity to work on other things, to make contributions in other areas, and to reflect on what made the 1960s such a special time to be an earth scientist. Psychiatrist Daniel Offer and his colleagues call memories a form of "existential reconstruction" – a means by which people make sense of their lives.[9] The stories presented here are the sense that 17 distinguished scientists have made of their scientific lives. They may not be works of history, but they may well be works of wisdom.

Many Individuals, But Only a Few Institutions

Besides the insights from individual stories, there are patterns that emerge from the collective whole. Perhaps the most striking feature of the development of plate tectonics is the small number of institutions but large number of individuals involved. The bulk of the story told here takes place at only four institutions worldwide: Cambridge University, Columbia University's Lamont Geological Observatory, the University of California's Scripps Institution of Oceanography, and Princeton University.[10] A striking feature of the stories in this volume is how many of the players moved back and forth among Cambridge, Lamont, Princeton, and Scripps, and how data-sharing facilitated the rapid development of ideas, and idea-sharing facilitated the effective interpretation of data. Keith Runcorn brought the work of British paleomagnetism to the attention of scientists at Lamont; Dan McKenzie, Robert Parker, and John Sclater brought their physics-oriented Cambridge training to Scripps; Harry Hess brought his idea of sea floor spreading to the attention of Fred Vine at Cambridge. And so on. Research thrives where smart people can work together and share data and ideas.

The concentration of intellectual and material resources in these institutions was also self-perpetuating. Several of our authors had personal connections that helped them get to these places: a father who also studied at Cambridge, another father who was a physicist who knew geophysicists at Lamont. The importance of personal ties helps to explain why only a very few women, and no African Americans, appear in these stories: in the early 1960s women and African Americans were not admitted to graduate study at Princeton, and only begrudgingly at Scripps; the available evidence suggests the situation was similar at Cambridge and Lamont.[11] As Tanya Atwater makes clear in her essay, the women who made it to these places had to maintain their good humor despite numerous slights and petty obstacles. Atwater did her best to focus on the work she loved, ignoring the fact that many of the people around her considered her a "freak."

In contrast to the small number of institutions, the development of plate tectonics involved a large number of individuals. Seventeen of them tell their stories here; there could have been many more. Many key players have passed away: P. M. S. Blackett, Sir Edward Bullard, Drummond Matthews, and Keith Runcorn in Great Britain; Allan Cox, Robert Dietz, Bruce Heezen, Harry Hess, Bill Menard, and Tuzo Wilson in North America. As editors, we struggled with limitations of time and space and the need to balance contributions from scientists representing different institutions and specialties. Our solution, albeit an imperfect one, was that if a group of scientists worked together on a project, we generally asked only one of them to tell the story. Lamont alumni will therefore notice the absence of Jim Heirtzler, Lynn Sykes, Bryan Isacks, and Marie Tharp; their absence should by no means be read as a negative judgment on the importance of their work. Finally, some whom we invited (although only a very small number) declined to participate, being busy with other things. All in all, there are at least three dozen individuals who could easily be counted as major contributors to the development of plate tectonics, still more if we extend our view to include the recognition of its implications for continental geology and earth history.[12]

This raises a significant historical point. We tend to link scientific advance to scientific genius, which by definition is individual. When most of us think of the great advances in the history of science, we think of great names – Copernicus, Newton, Darwin, Einstein. An earlier generation of historians often labeled scientific advances by the names of the individuals credited with them: the Copernican Revolution, the Darwinian revolution. Certainly, simple labels are convenient. But when historians scratch the surface of scientific discovery, they usually find many scientists working around a topic. Often other individuals have either hit upon the same ideas or evidence as their more famous counterpart (think of Alfred Russell Wallace and Charles Darwin) or have been awfully close to it. So we might ask, is plate tectonics different than other major scientific advances in involving so many individuals? Or is it simply that time has yet to obscure the details?

Perhaps the large number of individuals involved in the development of plate tectonics is a function of the time when these events took place. The 1950s and 1960s were a period of unprecedented funding for scientific research, particularly in the United States, where much of the critical work of plate tectonics was accomplished. (Several of the British scientists whose work is discussed in this volume received funding from the U.S. Office of Naval Research.) As is well known, the expansive federal funding of American science in the 1950s coupled with the G.I. Bill, which greatly increased the numbers of individuals in higher education, dramatically boosted the number of American scientists.[13] Moreover, military funding of scientific research in aid of national security often involved large laboratories and team-oriented approaches.[14] This implies that, other things being equal, it is likely that discoveries in the 20th century will involve more people than discoveries in earlier centuries.

There is also something in the nature of these discoveries that helps to explain why so many people were involved. Plate tectonics is a global theory – the first global theory ever to be generally accepted in the entire history of earth science.[15] Putting it together was a work of synthesis, involving data of many kinds from many places. While Ron Mason and Walter Pitman were making paleomagnetic measurements of rocks on the sea floor, Lawrence Morley was making similar measurements on land, Jack Oliver and Bruce Bolt were analyzing seismic data from earthquakes, and John Sclater was measuring heat flow over the mid-ocean ridges. What made plate tectonics so compelling was the way it unified these different kinds of data from all parts of the earth. Unlike some kinds of theoretical arguments in physics or laboratory experiments in chemistry, which might conceivably be achieved by an individual, there is simply no way all this work could have been done by one person, or even a small handful of persons.

Moreover, many of the critical data of plate tectonics were collected on oceanographic expeditions, as the essays by Mason, Pitman, Opdyke,

Atwater, and others recount. Organizing these expeditions was a major undertaking, and every expedition involved many scientists, as well as technicians and crew. People also worked behind the scenes: beforehand to make the expeditions happen, afterward to compile, catalogue, and preserve the data. Frequently lurking in the background of our story is Maurice Ewing, the tireless director of the Lamont Geological Observatory, whose relentless pursuit of data – and the financing that kept his ships at sea almost continuously for two decades – made possible much of what is recounted here. Roger Revelle played a similar role at Scripps, but Revelle had diverse interests, and Scripps was far less systematic in its pursuit and cataloguing of data. This difference proved significant: when critical ideas were put forth, it was Lamont more than any other institution that was in a position to test them, to make sense of them, and to prove them right or wrong.

The importance of expeditions and their organization points to a second theme that emerges from these essays: the role of data – lots and lots of it. In history, philosophy, and sociology of science it has become routine to say that observations are "theory-laden," and that people can be resistant to information that fails to fit their cognitive frameworks. On many accounts, conceptual innovation is a prerequisite not merely to the reinterpretation of data, but even to its recognition. One popular rendition of this view, sometimes emblazoned on T-shirts, reads "If I hadn't believed it, I wouldn't have seen it." Philosophers of science in the mid- to late 20th century virtually abandoned the idea that observation could drive science, and focused instead on the role of hypotheses or theories in guiding observation, suggesting tests.

While conceptual innovation was an important part of the development of plate tectonics, the stories told here strongly suggest that the most powerful driving force was data. By his own account, Harry Hess was driven to develop the idea of sea floor spreading – that new ocean crust is generated at mid-ocean ridges, where the ocean floor splits apart, driving the motions of the continents – by the paleomagnetic data collected by British geophysicists Keith Runcorn, Ted Irving, P. M. S. Blackett, and their colleagues. These data showed that the continents had been moving, sometimes separately, sometimes together, throughout geological history. The evidence that this was so preceded the conceptual explanation of how it was so.

Paleomagnetic data also drove the further development of Hess' idea. As Ron Mason recounts in his essay, in the late 1950s and early 1960s, Arthur Raff, Victor Vacquier, and he were collecting magnetic data on rocks of the sea floor off the coast of California primarily because the U.S.

Navy was interested in paleomagnetism for its relevance to submarine detection; scientists, largely because it was there. When Mason and Raff discovered a distinctive pattern of "magnetic stripes" – zones whose magnetic polarity was the same as the present-day earth field, paralleled by zones whose polarity was opposite – they were frankly at a loss for how to explain them, and they said so in their published work. Others took up the challenge. As Lawrence Morley recounts, the "zebra pattern" was so peculiar, so *unexplained*, that it caused him to drop what he was doing to focus on interpreting its meaning. Fred Vine did the same. Dan McKenzie and John Sclater argue that it was precisely this, data that people acknowledged must be right, but could not be explained by available theory, which drove earth scientists toward a new explanation, however unlikely it had seemed at the outset of their investigations. In the development of plate tectonics, recalcitrant data drove conceptual innovation.

Geophysical data can't be gathered without geophysical instruments, and the stories told here also involve the development of new instruments and analytical techniques. As both Ron Mason and Lawrence Morley discuss, the invention of accurate magnetometers – motivated by both military concerns and commercial interests – was a prerequisite to the data collection that revealed the existence of sea floor magnetic stripes. Analytical techniques also played a critical role in permitting the interpretation of key seismic data. The accurate location of earthquakes and the understanding of slip mechanisms became major concerns of the U.S. government in the late 1950s and early 1960s, when people realized that underground nuclear tests produced seismic waves similar to but potentially distinguishable from earthquakes. So the U.S. government dramatically increased funding for seismology. With ratification of the Limited Test Ban Treaty in 1963, which forbade testing of nuclear weapons in the atmosphere or oceans, identification of underground nuclear tests became essential for treaty verification. Bruce Bolt, Jack Oliver, and Peter Molnar were among the beneficiaries of this largesse, and the data they produced proved crucial for developing the parameters of plate tectonics.

Advances in seismology were crucial to understanding the interactions at convergent plate boundaries, which were in turn critical to formulating an integrated global theory. The focus of an earthquake is the place where the rupture begins, and most earthquakes' foci are close to the earth's surface, in the upper 200 or 300 miles (300–450 kilometers). However, some occur as deep as 450 miles (720 kilometers) – these are known as "deep-focus" earthquakes – and they mostly occur on mountain chains on the edges of continents, for example, below the South

American Andes, or beneath island-arc chains within ocean basins, such as the Aleutian islands. In the 1950s, these deep-focus earthquakes were considered perplexing, because most seismologists thought the earth was too hot at depths to sustain the rigid motions that occur in quakes. Yet they did occur. Before the mid-1960s, most seismologists accepted the view of Canadian J. H. Hodgson that the motion on faults associated with deep-focus earthquakes was strike-slip, involving one block of rock slipping sideways past the other.[16] Yet if the plate tectonics model was correct, and pieces of the earth's crust were slipping down into the mantle, then one block of crust should be sliding under the other – what geologists call *thrust* (or *reverse*) *faults*. Hodgson's work showed otherwise.

Hodgson and his co-workers had been analyzing the slip directions in earthquakes around the world – so-called first motion studies – but they faced two substantial problems. First, their method involved making fault-plane solutions (graphical plots of zones of compression and zones of tension around the center of an earthquake) but fault-plane solutions are non-unique: for every set of data, there are two possible solutions. (This remains true today.) Second, their project involved compiling data from seismic stations around the globe, and the quality of these data was inconsistent and often poor. The first motions of earthquakes could be misinterpreted easily due to weak signals or incorrectly set-up galvanometer wires (in which case compression would look like tension, and vice versa). While Hodgson acknowledged these difficulties, others took his results as if they were established facts.[17] As late as 1963, Harry Hess was arguing that the structure beneath the trenches consisted of vertical downwarpings, or "tectogenes," in part because of Hodgson's work. As Bruce Bolt, Jack Oliver, and Robert Parker explain in their essays, it took the establishment of the world wide standard seismograph network (WWSSN), which solved the problems of inconsistent data, to resolve these ambiguities and demonstrate that the deep-focus earthquakes were coming from thrust faults, consistent with slabs of the earth's crust sliding down into the mantle.

The large number of people involved in the development of plate tectonics and the importance of data as a driving force for conceptual innovation helps to explain another striking feature of this story: the frequency of simultaneous discovery. Robert Dietz and Harry Hess both wrote seminal papers suggesting sea floor spreading; Lawrence Morley and Fred Vine and Drummond Matthews proposed independently that magnetic stripes could provide a test of the spreading theory; Jason Morgan and Dan McKenzie separately developed the quantitative analysis of crustal motions that now bears the name *plate tectonics*.

In the case of Hess and Dietz, it is widely known that Hess wrote his now-classic paper on sea floor spreading, "History of the Ocean Basins," in 1960, and circulated a pre-print in 1961. This was the same year that Dietz published his article, "Continent and Ocean Basin Evolution by Spreading of the Sea Floor," in the journal *Nature*. In 1962, Dietz acknowledged that Hess had priority, and later scientists have generally assumed that Dietz's work was derived from Hess'.[18] Perhaps it was, but Dietz's version of sea floor spreading was different from Hess', and by the standards of current knowledge closer to being correct.[19]

However Dietz came to sea floor spreading, there is no dispute that Lawrence Morley and Fred Vine independently developed the idea that sea floor magnetic stripes offered a test of it. Vine was a graduate student at Cambridge, working under Professor Dummond Matthews. Morley was a geophysicist working with the Geological Survey of Canada. The two men had never met, never corresponded, never read each other's work. But each had read the work of Mason and Raff, each had been struck by the remarkable pattern of magnetic anomalies off the coast of North America, and each realized that, if sea floor spreading was real, then combining it with global magnetic polarity reversals would lead to a symmetrical pattern of magnetic stripes on either side of the mid-ocean rift. That is, each time the earth's magnetic field shifted, the next batch of magmas that erupted at the mid-ocean ridges would have opposite polarity to the batch before.

However, as Vine and Morley recount here, their ideas did not receive identical treatment: Morley's was rejected by the editors at *Nature;* Vine and Matthews' was accepted. In retrospect, the Vine and Matthews' presentation was much more developed, including a sophisticated analysis of existing sea floor magnetic data. While many people believe that ideas are the key to science, the difference in the treatment of the two papers shows that good ideas alone are not enough; you need good data, too, and you need to show how the data fit with the idea. And perhaps Vine and Matthews' Cambridge credentials carried weight at *Nature* that Morley's Canadian ones did not.

A third major example of simultaneous discovery is the work of Dan McKenzie and Robert Parker at Scripps, and Jason Morgan at Princeton, which established the plate tectonic model: that crustal motions could be understood as rigid body rotations on a sphere. Both McKenzie and Parker and Morgan made use of a theorem developed by 18th-century Swiss mathematician Leonhard Euler, that the motion of a rigid body over a sphere can be described as a rotation about an axis through the center of the sphere. Imagine a satellite in orbit over Earth. Its motion can be described as a rotation about an axis. A second satellite with a different orbit can be described as a rotation about a different axis. The question is: what is the relative motion of the two satellites? As McKenzie explained, Euler's theorem states that these two rotations are "equivalent to a single rotation about a different axis, and therefore any relative motion of plates on the surface of a sphere is a rotation about some axis."[20]

If one assumed that the crustal segments were rigid, they could be treated as interlocking solid blocks and the motion of any one block calculated relative to another. Do this for each block in turn, and all the motions of the plates could be determined. Visually, one could imagine the plates like interlocking paving stones, albeit on a moving surface. McKenzie and Parker, working on the west coast of the United States at Scripps, based their analysis on seismic data in the North Pacific; Morgan, working on the east coast of the United States at Princeton, based his analysis primarily on the orientation of fracture zones and spreading rates from magnetic anomalies in the Atlantic.[21]

McKenzie and Parker's article was published first, in *Nature* in late December 1967; Morgan's was published in the *Journal of Geophysical Research* three months later, in March 1968. However, Morgan had already presented his results at a meeting of the American Geophysical Union the previous spring. McKenzie was present at that meeting, but he hadn't heard Morgan's talk. Morgan had originally intended to speak about gravity measurements in the Caribbean, and his title reflected this. Uninterested in gravity, McKenzie skipped the talk. But Morgan did not talk about gravity, he talked about rigid rotations of crustal blocks. So scientists who attended Morgan's talk, such as Xavier Le Pichon, have generally credited Morgan with priority for the idea.[22]

Did McKenzie somehow hear about Morgan's work? Perhaps, but Morgan and McKenzie were not the only ones thinking along these lines. In a paper presented at a symposium on continental drift at the Royal Society in London in 1964, and published in 1965, McKenzie's Cambridge mentor, Sir Edward ("Teddy") Bullard, had drawn on Euler's theorem to analyze the geometry of the fit of the continents across the Atlantic Ocean.[23] Also in 1964, Scripps scientist George Backus published an article in *Nature* also invoking Euler's theorem. Backus was on sabbatical leave at Cambridge, where he met Bullard, Vine, Matthews, and McKenzie, and suggested that the hypothesis of sea floor spreading could be tested by comparing the spacing of magnetic stripes in the North and South Atlantic. Since the latter was wider, the spreading rate must have been greater, and so the width of the stripes should be greater.

Moreover, he pointed out, Euler's theorem could be used to calculate just how wide those stripes should be as a distance from the pole of rotation. As he put it in nearly a throwaway line, "Of course, any rigid displacement of a continent on a spherical globe is a rotation about some point," and therefore the motion of any point could be calculated based on its position relative to the pole of rotation. (Think of a line of spinning ice skaters. The one at the center rotates slowly in place, the ones at the ends must skate much faster. If you knew the speed at which the center skater was rotating [and if the line were perfectly rigid], then you could calculate the speed of any skater along the line by knowing her or his distance from the center.) However, magnetic data for the South Atlantic were not available, so Backus presented his idea as a proposal for a test of sea floor spreading, rather than as a comprehensive theory of plate motions.[24]

More clearly influential was Tuzo Wilson's 1965 *Nature* paper, "A New Class of Faults and their Bearing on Continental Drift."[25] Wilson argued that if sea floor spreading was taking place, then the fracture zones that cut across the mid-ocean ridges might in fact be faults, along which crustal blocks were displaced. But they would be a different kind of fault than previously recognized. In the introduction to his paper, Wilson argued that the earth's mobile zones – the mountains and volcanoes along continental margins, and the rifts that ran through continents, like the East African Rift – were "connected into a continuous network of mobile belts about the Earth which divide the surface into several large rigid plates." Moreover, "any feature at its apparent termination may be transformed into another feature of one of the other two types."[26] In other words, a rift could be transformed into a strike-slip fault zone, a strike-slip fault could give way to an ocean trench, forming a continuous network of tectonic boundaries around the globe. Wilson coined the term *transform faults* to describe the junctions where horizontal shear faults terminate at both ends against other tectonic features, most notably the faults that displace segments of the mid-ocean ridges.

Later workers would focus on Wilson's definition of transform faults as a key element of the plate tectonic synthesis, but equally important at the time was Wilson's argument that the earth's mobile belts were continuous, dividing the world into a small number of segments, and that these segments could be thought of as rigid plates. If so, then the motions along transform faults could be described as rigid motions on a plane. (Think of the sliding plastic tiles in those number puzzles where you have to get the numbers back in order, but the tiles can only move up and down, right or left.) Both McKenzie and Parker and Morgan made the link between Wilson's rigid motions on a plane and Bullard's application of Euler's theorem for motions on a sphere: McKenzie and Parker described their work as an extension of Wilson's "concept of transform faults to motions on a sphere." Morgan called his work "an extension of the transform fault concept to a spherical surface."[27] Wilson laid out the idea of plates and rigid motions; McKenzie, Parker, and Morgan extended it to the spherical earth, quantified it, and showed that it was consistent with the geophysical data.

By the early to mid-1960s, many people had become convinced of continental mobility, and were thinking about how it might work. Like the British geologist Arthur Holmes before them (who first proposed an idea of sea floor spreading in the late 1920s), Hess and Dietz were pondering the mechanisms that generated ocean floor; Vine and Morley were trying to make sense of sea floor magnetic stripes in terms of sea floor mobility; and Bullard, Backus, Wilson, McKenzie, Parker, and Morgan were all thinking geometrically: how would the crustal pieces move and still fit together? Moreover, these are not the only examples of more than one scientist or group of scientists working on the same problem. As Neil Opdyke recounts, Jan Hospers had been working on the question of magnetic reversals in rocks since the 1950s; this was independently pursued by Allan Cox, Richard Doell, and Brent Dalrymple in the United States, and by Ian McDougall and Don Tarling in Australia. Xavier Le Pichon similarly notes that he recognized the similarity of the magnetic stripes over the Juan de Fuca and Reykjanes ridges at just about the same time as Fred Vine did.[28] Geophysicist Geoffrey Davies recently noted that Wilson's transform faults were proposed independently in 1965 by A. M. Coode.[29] These are examples of multiple discovery of good ideas; if one were to recount the history of multiple discovery of bad ideas, no doubt one would find still more.

If a problem is important, more than one person will be attracted to working on it, and sometimes more than one person will hit on the solution at more or less the same time. The more people working on the problem, the more they know and talk to each other, and the more public the relevant data, the more likely this is to be the case. The multiple cases of simultaneous discovery in plate tectonics emerge from a context in which many smart people were working closely in a small number of institutions where critical data were available.

A COMMON THREAD?

Throughout this introduction, I have emphasized the diversity of people, approaches, data, and ideas that contributed to the global synthesis of plate tectonics. Do our 17 authors have anything in common? One common thread is that they are all individuals who took advantage of opportunities and were willing to take chances. Fred Vine delivered an undergraduate geology club lecture on continental drift at Cambridge, knowing that many of his teachers opposed the idea. Neil Opdyke joined Keith Runcorn as a field assistant on only a few days' notice, having never before met the man. Tanya Atwater returned to the United States from Chile determined not to miss out on the revolution that she realized was taking place, and worried that she would get there too late. Peter Molnar refused to settle for a second-rate thesis topic, searching ardently for a problem he could fall in love with. Both John Dewey and Bill Dickinson were determined to show how this new geophysical theory could change geologists' worldview – Dewey by talking to anyone who would listen, Dickinson by organizing a conference in which he invited people he had never met to present their most outrageous ideas.

Among the authors who were graduate students or freshly minted academics at the time, what stands out most is their willingness to work hard, take chances, and not be afraid that their ideas might be scoffed at by their elders. Among the authors who were more senior at the time, what stands out is their willingness to admit that what they had previously believed was wrong, to rethink their commitments, regroup, recover, and keep working. Scientists who had been raised to believe that the earth's crust was fixed, that continental drift was impossible, and that Alfred Wegener was a crank, now contributed their own scientific brilliance to showing how the crust moved, what made it move, and that Wegener, in his overall insight if not in every detail, was right. This, perhaps, is the most important part of the story, a story of scientists who realized what they had previously believed was wrong, and set it right.

PART I
THE HISTORICAL BACKGROUND

The idea that continents move was first seriously considered in the early 20th century, but it took scientists 40 years to decide that it was true. Part I describes the historical background to this question: how scientists first pondered the question of crustal mobility, why they rejected the idea the first time around, and how they ultimately came back to it with new evidence, new ideas, and a global model of how it works.

CHAPTER 1

FROM CONTINENTAL DRIFT
TO PLATE TECTONICS

Naomi Oreskes

SINCE THE 16TH CENTURY, CARTOGRAPHERS HAVE NOTICED THE
jigsaw-puzzle fit of the continental edges.[1] Since the 19th century, geologists have known that some fossil plants and animals are extraordinar-
ily similar across the globe, and some sequences of rock formations in
distant continents are also strikingly alike. At the turn of the 20th century, Austrian geologist Eduard Suess proposed the theory of Gondwanaland to account for these similarities: that a giant supercontinent
had once covered much or all of Earth's surface before breaking apart
to form continents and ocean basins. A few years later, German meteorologist Alfred Wegener suggested an alternative explanation: continental drift. The paleontological patterns and jigsaw-puzzle fit could be
explained if the continents had migrated across the earth's surface,
sometimes joining together, sometimes breaking apart. Wegener argued
that for several hundred million years during the late Paleozoic and
Mesozoic eras (200 million to 300 million years ago), the continents
were united into a supercontinent that he labeled *Pangea*—all Earth.
Continental drift would also explain paleoclimate change, as continents
drifted through different climate zones and ocean circulation was
altered by the changing distribution of land and sea, while the interactions of rifting and drifting land masses provided a mechanism for the
origins of mountains, volcanoes, and earthquakes.

Continental drift was not accepted when first proposed, but in the
1960s it became a cornerstone of the new global theory of plate tectonics. The motion of land masses is now explained as a consequence of
moving "plates"—large fragments of the earth's surface layer in which
the continents are embedded. These plates comprise the upper 45 to 60

miles (80 to 100 kilometers) of the earth's surface (now called the lithosphere), and move at a rate of 1 to 4 inches (3 to 10 centimeters) per year. Earthquakes, volcanoes, and mountains are concentrated on plate margins where two plates collide, split apart, or slide past one another. Moreover, the global configuration of continents and oceans is constantly changing. As Wegener suggested, the breakup of Pangea produced the configuration of continents and oceans that we have today.

BEFORE CONTINENTAL DRIFT: VERSIONS OF CONTRACTION THEORY

One of the central scientific questions of 19th-century geology was the origin of mountains. How were they formed? What process squeezed and folded rocks like putty? What made the earth's surface move? Most theories invoked terrestrial contraction as a causal force. It was widely believed that Earth had formed as a hot, incandescent body, and had been steadily cooling since the beginning of geological time. Because most materials contract as they cool, it seemed logical to assume that Earth had been contracting as it cooled, too. As it did, its surface would have deformed, producing mountains.

In Europe, Austrian geologist Edward Suess (1831–1914) popularized the image of Earth as a drying apple: as the planet contracted, its surface wrinkled to accommodate the diminished surface area. Suess assumed that Earth's initial crust was continuous, but broke apart as the interior shrunk. The collapsed portions formed the ocean basins; the remaining elevated portions formed the continents. With continued cooling, the original continents became unstable and collapsed to form the next generation of ocean floor, and what had formerly been ocean now became dry land. Over the course of geological history, there would be a continual interchange of land and sea, a periodic rearrangement of the land masses.

The interchangeability of continents and oceans explained a number of other perplexing geological observations, such as the presence of marine fossils on land (which had long before puzzled Leonardo Da Vinci) and the extensive interleaving of marine and terrestrial sediments in the stratigraphic record. Suess' theory also explained the striking similarities of fossils in parts of India, Africa, and South America. Indeed, in some cases the fossils seemed to be identical, even though they were found thousands of miles apart. These similarities had been recognized since the mid-century, but they had been made newly problematic by Darwin's theory of evolution. If plants and animals had evolved independently in different places within diverse environments, then why did they look so similar? Suess explained this conundrum by attributing these similar species to an early geological age when the continents were contiguous in an ancient supercontinent called Gondwanaland.[2]

Suess' theory was widely discussed and to varying degrees accepted in Europe, but in North America geologist James Dwight Dana (1813–1895) had developed a different version of contraction theory. Dana suggested that the continents had formed early in earth history, when low-temperature minerals such as quartz and feldspar had solidified. Then the globe continued to cool and contract, until the high-temperature minerals such as olivine and pyroxene finally solidified: on the moon, to form the lunar craters; on Earth, to form the ocean basins. As contraction continued after Earth was solid, its surface began to deform. The boundaries between continents and oceans took up most of the pressure—like the seams on a dress—and so mountains began to form along continental margins. With continued contraction came continued deformation, but with the continents and oceans always in the same relative positions.[3] Although Dana's theory was a version of contraction, it came to be known as permanence theory, because it viewed continents and oceans as globally permanent features.

In North America, permanence theory was linked to the theory of geosynclines: subsiding sedimentary basins along continental margins. This idea was developed primarily by James Hall (1811–1889), state paleontologist of New York and the first president of the Geological Society of America (1889). Hall noted that, beneath the forest cover, the Appalachian mountains were built up of folded layers of shallow-water sedimentary rocks, thousands of feet thick. How did these sequences of shallow-water deposits form? How were they folded and uplifted into mountains? Hall suggested that materials eroded off the continents accumulated in the adjacent marginal basins, causing the basins to subside. Subsidence allowed more sediments to accumulate, causing more subsidence, until finally the weight of the pile caused the sediments to be heated, converted to rock, and then uplifted into mountains.[4] (The process of uplift, or mountain-building, is called *orogeny*.) Dana modified Hall's view by arguing that thick sedimentary piles were not the cause of subsidence but the result of it. Either way the theory provided a concise explanation of how thick sequences of shallow-water rocks could accumulate, but was vague on the question of how they were transformed into mountain belts.

CONTINENTAL DRIFT AS ALTERNATIVE
TO CONTRACTION THEORY

In the early 20th century, contraction theory was challenged by three independent lines of evidence. The first came from field mapping. Nineteenth-century geologists had worked in great detail to determine the structure of mountain belts, particularly the Swiss Alps and the North American Appalachians. When they mapped the folded sequences of rocks in these regions, they found the folds to be so extensive that if one could unfold them the rock layers would extend for hundreds of miles. Impossibly huge amounts of terrestrial contraction would have to be involved. Geologists began to doubt contraction theory as an explanation for the origins of mountains.

The second line of evidence came from geodesy—the science of the shape (or figure) of the earth. While field geologists were unraveling the structure of the Alps and Appalachians, cartographers with the Great Trignometrical Survey of India were making geodetic measurements to produce accurate maps of British colonial holdings.[5] In the early 1850s, Colonel (later Sir) George Everest, the surveyor-general of India, discovered a discrepancy in the measured distance between two stations, Kaliana and Kalianpur, 370 miles (600 kilometers) apart. When measured on the basis of surveyor's triangulations, the latitude difference was five seconds greater than when computed on the basis of astronomical observation. Everest thought the difference might be due to the gravitational attraction of the Himalayas on the surveyors' plumb bobs, and enlisted John Pratt (1809–1871), a Cambridge-trained mathematician and the archdeacon of Calcutta, to examine the problem. Pratt calculated the expected gravitational effect of the mountains, and discovered that the discrepancy was *less* than it should have been: it was as if part of the mountains were missing. Pratt proposed that the observed effects could be explained if the surface topography of the mountains were somehow compensated by a deficit of mass beneath them—an idea that came to be known as *isostasy*, or "equal standing." In the early 20th century, isostasy was confirmed by detailed geodetic and gravity measurements across the United States. John Hayford (1868–1925) and William Bowie (1872–1940), working at the U.S. Coast and Geodetic Survey, demonstrated that the distribution of gravity was most consistent with the assumption of isostasy, not just in mountain belts, but across the continents. Isostasy could be achieved either if the continents were less dense than the layers of rock beneath them, or if they had deep roots, like icebergs. Either way, they "floated" in the substrate beneath them, and therefore they could not sink to become ocean basins. Continents and oceans were not interchangeable.

Third, and most fundamental, physicists discovered radiogenic heat, which contradicted the basic assumption of contraction theory that the earth was steadily cooling. With contraction no longer assumed, earth scientists were motivated to search for other driving forces of deformation. By the 1920s, many considered the science to be in a state of crisis: with contraction theory discredited, how were geologists to account for the evidence of prior continental connections? How were they to reconcile the evidence from historical geology for the changing configuration of land masses with the apparent permanence of continents and oceans? This crisis was felt most acutely by European geologists who had accepted Suess' theory, but Americans also realized that they faced a dilemma. A number of scientists began to put forward alternative theories of continental fragmentation or migration. Alfred Wegener (1880–1930) is the most significant, for his theory was the most widely discussed at the time, and the one that was later vindicated.

A pioneering meteorologist and author of an early text on the thermodynamics of the atmosphere, Wegener realized that paleoclimate change could be explained if continents had migrated across climate zones and the reconfiguration of land masses altered Earth's climate patterns.[6] However, continental drift was more than just a theory of paleoclimate change. Wegener explicitly presented his theory as a means to reconcile historical geology with isostasy: on the one hand, paleontological evidence that the continents had once been connected; on the other, geodetic evidence that they could not be connected in the way European contractionists had supposed by now-sunken crust. Wegener's answer was to reconnect the continents by moving them laterally.

Wegener's theory was widely discussed in the 1920s and early 1930s. It was also hotly rejected, particularly by geologists in the United States, who labeled it bad science. The standard explanation for the rejection of continental drift is the lack of a causal mechanism, but this explanation is false. There was a spirited and rigorous international debate over the possible mechanisms of continental migration, which ultimately settled on the same explanation generally accepted today for plate tectonics: convection currents in the earth's mantle.

The debate over the mechanism of continental drift centered on the implications of isostasy. If continents floated in a denser substrate, then this substrate had to be either fluid or plastic, and continents could at least in principle move through it. There was good evidence that this was indeed the case: in Scandinavia, geologists had documented a progressive uplift of Finland and Scandinavia since the end of the Pleistocene epoch (10,000 years ago), which they called the *Fennoscandian rebound*. The accepted explanation for this phenomenon was that during the Pleistocene epoch, the region had been depressed under the weight of a thick sheet of glacial ice; as the ice gradually melted, the land surface gradually rebounded. This provided empirical evidence that continents could move through the substrate in which they were embedded, at least in the vertical direction and at least during the Pleistocene. However, in Scandinavia the cause of motion was generally agreed: first the weight of glacial ice, then the pressure release upon its removal. What force would cause horizontal movement? Would the substrate respond to horizontal movement as it did to vertical movement? Debate over the mechanisms of drift concentrated on the long-term behavior of the substrate and the forces that could cause continents to move laterally.

In the United States, the question was addressed by Harvard geology professor Reginald A. Daly (1871–1957), North America's strongest defender of continental drift. Daly argued that the key to tectonic problems was to be found in the earth's layered structure. Advances in seis-

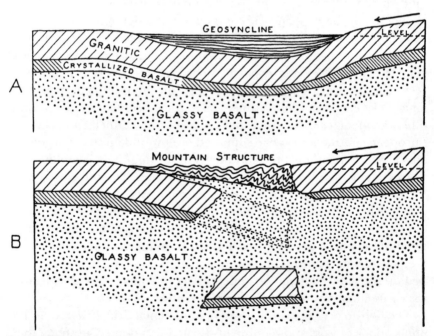

Reginald Daly's mechanism of continental drift by gravity sliding. Reprinted with permission of Scribner, a Division of Simon and Schuster, from *Our Mobile Earth* by Reginald A. Daly, copyright © 1926 by Charles Scribner's Sons, renewed 1954 by Reginald A. Daly, on p. 269.

mology suggested that the earth contained three major layers: crust, substrate (or mantle), and core. The substrate, he suggested, might be glassy, and therefore could flow in response to long-term stress just as old plates of glass gradually thicken at their lower edges and glassy lavas flow downhill. Continents might do the same. Building on the geosyncline concept of Dana and Hall, Daly suggested that sedimentation along the continental margins resulted in subtle elevation differences, which in turn produced gravitational instabilities. Eventually, the continent could rupture, sliding down over the glassy substrate under the force of gravity. The sliding fragment would then override the other half—an early suggestion of subduction—and, over time, the accumulation of small increments of sliding would result in global continental drift.[7]

Daly urged his American colleagues to take up the question of drift, but few did. Reaction in Europe was more favorable. Irish geologist John Joly (1857–1933) linked the question to discoveries in radioactivity. Trained as a physicist, Joly had demonstrated that the commonly observed dark rings in micas—so-called pleochroic haloes—were caused by radiation damage from tiny inclusions of uranium- and thorium-bearing minerals, such as apatite. Radioactive elements were therefore ubiquitous in rocks, suggesting that radiogenic heat was also ubiquitous. If it was, then it could be a force for geological change. Joly proposed that as radiogenic heat accumulated, the substrate would begin to melt. During these episodes of melting, the continents could move under the influence of small forces, such as minor gravitational effects, that would otherwise be ineffectual.[8] Periodic melting, associated with magmatic cycles caused by the build-up of radiogenic heat, would lead to the periods of global mountain-building that many geologists saw evidence of when they compared the geology of Europe and North America.

Joly's theory responded to a geophysical complaint against a plastic substrate, voiced most clearly by Cambridge geophysicist Harold Jeffreys (later Sir Harold), that the propagation of seismic waves indicated a fully solid and rigid Earth. Jeffreys argued on physical grounds that continental drift was impossible in a solid, rigid Earth; Joly noted that although Earth was solid now, it might not always have been. More widely credited was the suggestion of British geologist Arthur Holmes (1890-1965) that the substrate was partially molten or glassy—like magma. Underscoring arguments made by Wegener, Holmes emphasized that the substrate did not need to be *liquid,* only plastic, and that it might be rigid under high strain rates (during seismic events) yet still be ductile under the low strain rates that prevailed during orogeny (mountain-building). If it was plastic in response to long-term stress, then continents could move within it.

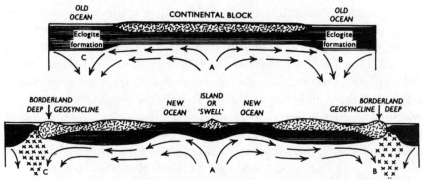

Arthur Holmes' model of continental drift driven by mantle convection currents, from Holmes (1929), Radioactivity and earth movements, *Transactions of the Geological Society of Glasgow* 18: 579 (1929), used by permission of the Geological Society of Glasgow.

Holmes' driving force was convection currents in the mantle. He argued that radiogenic heat would generate the convection: the mid-ocean ridges were fragments of continental crust left behind after continents had split apart above upwelling convection currents; the ocean deeps (geosynclines) were the sites of downwelling currents where continents deformed as the substrate descended. Between the ridges and the trenches, continents were dragged along in conveyor-belt fashion.[9]

THE REJECTION OF CONTINENTAL DRIFT

Arthur Holmes' papers were widely read and cited; many geologists thought he had found the cause of continental drift. However, opposition was nonetheless strong, particularly in the United States, where reaction to Wegener's theory was vitriolic.

Three main factors contributed to the American animosity to continental drift. First, Americans were widely committed to the method of multiple working hypotheses—the idea that scientific evidence should be weighed in light of competing (multiple) theoretical explanations, which one held provisionally until the weight of evidence was sufficient to compel assent. This provisional stage was thought to require a long time—certainly years, perhaps even decades. Most closely associated with the University of Chicago geologist T. C. Chamberlin (1843–1928), who had named it, the method of multiple working hypotheses reflected American ideals expressed since the 18th century linking good science to good government. Good science was anti-authoritarian, like democracy; good science was pluralistic, like a free society. Americans going back to Thomas Jefferson and Benjamin Franklin promoted the idea that good science provided an exemplar for good government; Jefferson advocated scientific study in large part for this very reason. And if good science was a model for a free society, then bad science implicitly threatened it.[10] Consistent with the methodology of multiple working hypotheses, Americans believed good scientific method was empirical, inductive, and modest, holding close to the objects of study and resisting the impulse to go further. Alfred Wegener's work was interpreted as violating these principles on several counts. It put the theory first and then sought evidence for it. It settled too quickly on a single interpretive framework. It was too large, too unifying, too ambitious. Features that were later viewed as virtues of plate tectonics were attacked as flaws of continental drift.[11]

Second, continental drift was incompatible with the version of isostasy to which Americans subscribed. While John Pratt had suggested that isostatic compensation could be achieved by subsurface density variations, British Astronomer Royal George Biddell Airy (1801–1892) had pointed out that the same surface effects could be produced by differences in crustal thickness. In Pratt's view, the mountains would be underlain by low-density crust, but the depth of isostatic compensation would be the same everywhere. In Airy's view, the depth of compensation would be variable, with the highest mountains underlain by the deepest roots. When Hayford and Bowie set out to investigate isostasy, they based their test on Pratt's model. By making the assumption of a uniform depth of compensation, they were able to predict the surface effects of isostasy very accurately throughout the United States—that is, to show that the data were consistent with the predictions of the model. Therefore, they concluded that the model was correct. Hayford and Bowie used Pratt's model because it was simpler and therefore easier to use. What began as a simplifying assumption evolved into a belief about the structure of the crust. This belief had consequences for the reception of the theory of drift, for if continental drift were true, then the large compressive forces involved would squeeze the crust to generate thickness differences, ultimately ending up with the Airy version of isostasy. Continental drift seemed to refute Pratt isostasy, which had worked for Americans so well. Rather then reject Pratt isostasy, they rejected continental drift.

Third, Americans rejected continental drift because of the legacy of uniformitarianism. Uniformitarianism was the principle, articulated most famously by British geologist Sir Charles Lyell (1797–1875), that the best way to understand the geological record was by reference to presently observable processes. To understand how sandstones formed, study beach processes. To understand volcanic rocks, study modern volcanoes. To understand fossils, study modern organisms in similar habitats. And so on. Lyell proposed uniformitarianism in part as an intellectual response to the difficulties of interpreting the rock record, and in part as a reaction against an earlier generation of natural historians who had looked to the Bible as a basis for interpreting earth history. So uniformitarianism was associated in many geologists' minds with the exclusion of religious arguments from geology and the consolidation of geology as a science.

Whether or not Lyell's arguments were correct, by the early 20th century the methodological principle of using the present to interpret the past was deeply embedded in the practice of historical geology. Historical geologists routinely used fossil assemblages to make inferences about climate zones. According to drift theory, however, continents in tropical latitudes did not necessarily have tropical faunas, because the reconfiguration of continents and oceans might change matters altogether. Wegener's theory raised the specter that the present was not the key to the past—that it was just a moment in earth history, no more or less characteristic than any other. This was not an idea that Americans were willing to accept.

In North America, the debate over continental drift was quelled by an alternative account of the faunal evidence. In 1933, geologists Charles Schuchert (1858–1942) and Bailey Willis (1857–1949) proposed that the continents had been intermittently connected by isthmian links, as the isthmus of Panama presently connects North America and South America and the Bering Land Bridge recently connected North America and Asia. The isthmuses had been raised up by orogenic forces, then subsided under the influence of isostasy. This explanation was patently ad hoc—there was no evidence of isthmian links other than the paleontological data they were designed to explain (away). Nevertheless, the idea was widely accepted, and it undercut a major line of evidence of continental drift. In 1937, South African geologist Alexander du Toit (1878–1948) published *Our Wandering Continents*, a comprehensive synthesis of the geological evidence of continental drift, but it had little impact in North America. Elsewhere, particularly in South Africa and Australia, some geologists continued to advocate drift and to use it to interpret their geological data, but these individuals were mostly isolated. The consensus of scientific opinion was against continental drift. There the matter rested for two decades, until the debate was reopened on the basis of entirely new evidence.

FROM LAND TO SEA:
GRAVITY ANOMALIES AND CRUSTAL MOTIONS

Schuchert and Willis' alternative theory satisfied most North American geologists that continental drift was no longer something they needed to worry about, but the issue did not quite stop there. In the 1920s, a group of American scientists led by William Bowie had begun a program in cooperation with the U.S. Navy to measure gravity at sea. Bowie and Hayford had demonstrated that isostasy applied over the continents, but did it also apply over the oceans? What *was* the structure of the crust under the ocean basins? What was the ocean floor made of? The answers to these fundamental questions were unknown, and one's view of the earth might change dramatically according to what the answers turned out to be.

Measuring gravity at sea was extremely difficult, because wind and waves disturbed the sensitive apparatus used. The world's expert on the subject was a Dutch geodesist, Felix Vening Meinesz (1887–1966), who had invented a novel gravimeter that was resistant to external disturbance. In 1923, he demonstrated its efficacy in a series of Dutch submarine expeditions to Indonesia, where he had discovered major gravity anomalies associated with the Java Trench. Supporters of Wegener had proposed that the Java Trench was the site of convergence of two giant crustal slabs, and Vening Meinesz became interested in the possible connection among gravity anomalies, ocean trenches, and crustal movements. In 1928, Bowie invited Vening Meinesz to the United States, and a series of gravity expeditions followed, focused on the Caribbean Sea and the Gulf of Mexico. Among the scientists who participated in these expeditions were two assistant professors, Harry H. Hess (1906–1969), a young petrologist at Princeton, and Maurice Ewing (1906–1974) a fledgling geophysicist at Lehigh who was rapidly becoming known for his pioneering work on refraction seismology (using explosives to send shock waves through the earth's crust to determine its structure). On the 1937 *Barracuda* expedition, they were joined by another rising star, British geophysicist Edward ("Teddy") Bullard (1907–1980).[12]

These expeditions confirmed Vening Meinesz's earlier discoveries: gravity measurements in the Caribbean and the Gulf of Mexico demonstrated an association between negative gravity anomalies (regions of lower than normal gravity) and regions where the ocean was particularly deep. Hess discussed these results with Vening Meinesz, and both agreed that they indicated some form of crustal disturbance or deformation. Apparently the ocean basins were not static, but actively deforming, at least in certain zones.

Teddy (later Sir Edward) Bullard, taking a break from gravity measurements somewhere in the Caribbean, ca. 1937. Photo courtesy of Robert
Parker, Scripps Institution of Oceanography.

Familiar with European arguments over continental drift, Vening Meinesz
proposed that convection currents might be dragging the crust downward
into the denser mantle below, explaining both the ocean deeps and the negative gravity anomalies associated with them.[13] Hess imagined vertical buckles in the crust, expressed on the surface as ocean trenches or deeps, and in
gravity measurements as negative anomalies. Borrowing a term from German geologist Erich Haarmann, he called these downwarpings *tectogenes*.[14]

   The tectogene concept received support from Vening Meinesz's

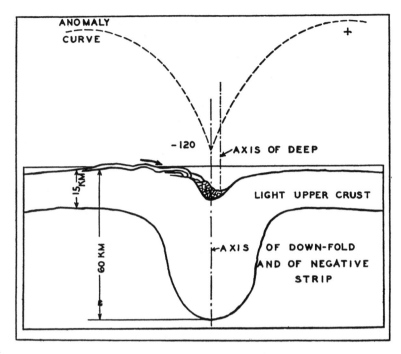

Harry Hess' tectogene concept explaining the origins of ocean deeps associated with negative gravity anomalies, from Hess (1933), Interpretation of geological and geophysical observations, in *The Navy–Princeton Gravity Expedition to the West Indies in 1932*, edited by R. M. Field. Washington, D.C., U.S. Government Printing Office, p. 30.

Dutch colleague, Philip Kuenen, who undertook a series of experiments to show that the idea was at least physically possible, and from University of California professor David Griggs, who created a laboratory model of continental drift using a layer of paraffin over a tank of oil, in which convection currents were simulated by the action of two rotating drums.[15] While his experimental apparatus was very small, Griggs argued that the scale of mantle convection currents could be very large, perhaps "covering the whole Pacific basin, comprising sinking peripheral currents localizing the circum-Pacific mountains and rising currents in the center."[16] He noted that seismologists such as Caltech's Beno Gutenberg and Charles Richter had noticed that the earthquakes around the edges of the Pacific basin were concentrated in zones that dipped about 45 degrees toward the continents; perhaps these quakes were "caused by slippage along the convection current surface."[17] Hess was excited by these suggestions, which helped to link his Caribbean work to global theory. In 1939 he began to put the pieces together, writing:

Recently an important new concept concerning the origins of the negative strip [of gravity anomalies] . . . has been set forward by David Griggs. It is based on model experiments in which . . . by means of horizontal rotating cylinders, convection currents were set up in a fluid layer beneath the "crust," and a convection cell was formed. A down-buckle in the crust, similar to that produced in Kuenen's experiments, was developed where two opposing currents meet and plunge downward. So long as the currents are in operation, the down-buckle is maintained. . . . The currents which Griggs suggested would have velocities [in nature] of one to ten centimeters [1/2 to 4 inches] per year.[18]

The year was 1939, and Griggs and Hess had hit upon what scientists would later affirm as the rate of plate motions. But before they could go any further, World War II broke out.

A New Age of Exploration

In the 1920s the Navy had been cautious about funding basic scientific research, concerned about the appropriate expenditure of Navy funds and doubtful that work such as gravity measurement was likely to be of operational use. World War II changed the situation, largely because of submarine warfare. Allied forces suffered heavy losses in the early part of the war from attack by German U-boats, and the U.S. Navy realized that geophysics and oceanography might provide means to detect or avoid submarines. Particularly salient were two lines of research: magnetics, which might provide direct means of submarine detection, and physical oceanography, which might guide evasive maneuvers.

In the early 1940s, the U.S. Navy was experiencing difficulties with its sonar equipment, which tended not to work well in the afternoon. Thinking that marine organisms were interfering with transmissions (or that operators were dozing off after lunch), the Navy asked Maurice Ewing, then working at the Woods Hole Oceanographic Institution, to investigate. Together with colleague J. Lamar (Joe) Worzel, Ewing discovered that temperature effects were bending the sound waves in such a way as to create a "shadow zone"—a region in which sonar transmissions went undetected. This discovery had enormous implications for submarine warfare: if a submarine commander could accurately locate the shadow zone, he could hide his ship within it. Moreover, Ewing and Worzel discovered that under certain conditions sound waves would be focused into a narrow region, in which they traveled for great distances.

They called this phenomenon *sound channeling,* and it became the basis for SOFAR (SOund Fixing and Ranging), which the Navy used during the war to locate downed airmen, and SOSUS (SOund SUrveillance System), the Navy's Cold War underwater acoustic array established to detect Soviet submarines.[19]

While Ewing worked on underwater sound in a civilian capacity, Hess joined the Naval Reserve and in 1941 was called to active duty. He became the captain of an assault transport, the USS *Cape Johnson,* and among her tasks was the echo-sounding of the Pacific basin. This was a project with both military and scientific significance: for the Navy, an accurate topographic map of the sea floor would provide captains with an independent check on their navigation; for scientists, understanding of the sea floor would be greatly enhanced by knowing its shape and structure. This latter hope was fulfilled by Hess' discovery of "guyots"— flat-topped mountains, which he named after Arnold Guyot, the first professor of geology at Princeton. Hess interpreted these mountains as ancient volcanoes whose tops had been eroded by wave action as they gradually sank on a subsiding ocean floor.[20] Guyots were strong evidence that the ocean basins were not fossils of an early stage of earth history, but were geologically active throughout time.

By war's end, the U.S. Navy was convinced of the value of geophysical research. Through the newly established Office of Naval Research (ONR), funds began to flow generously into American laboratories.[21] Three institutions particularly benefited from ONR support: Woods Hole, the Scripps Institution of Oceanography, and the newly created Lamont Geological Observatory at Columbia University, now directed by Ewing. Work at these institutions focused on physical oceanography for its relation to underwater sound, magnetics for its relevance to submarine detection, and bathymetry for mapping the sea floor. At Scripps and Lamont, seismology—the study of earthquakes and how shock waves travel through the earth—was also developed, first as means to investigate the structure of the sea floor and the nature of earthquakes; later to detect underground nuclear explosions.

The years 1945–1970 may well have been the most exciting time in the history of American earth science, as abundant funding led to a new age of scientific exploration—not to get across the oceans, but to spend time within and under them, and ultimately to understand them. Woods Hole, Scripps, and Lamont launched a series of major oceanographic expeditions, collecting an enormous quantity of diverse data on the bathymetry and structure of the sea floor, the physical and chemical properties of the water column, the air-sea interaction and generation of waves and currents, the sediments on the sea floor, and the magnetic and gravity signatures of the solid rocks at the bottom of the sea. More was learned about the oceans during these 25 years than in the entire previous history of science. But there was one downside: much of the data gathered was classified.

In the United Kingdom as in the United States, many scientists worked during the war on military-scientific problems, among them Teddy Bullard and P. M. S. Blackett (1897–1974). In the late 1920s, Bullard was a graduate student at the Cavendish Laboratory at Cambridge University, directed by Nobel Laureate Ernest Rutherford. Blackett was also a member of the lab and Bullard was assigned to work under Blackett on the scattering of electrons in gases; Bullard soon discovered diffraction patterns that supported recent theoretical advances in quantum mechanics.[22] Bullard's career was off to an outstanding start, but the year was 1931, the Depression was at its nadir, and there was no work to be had. Rutherford advised him to take whatever job he could find; that turned out to be teaching surveying under Cambridge geodesist Colonel Sir Lenox-Conyngham. Bullard became a demonstrator in the newly established Department of Geodesy and Geophysics—now consisting of two men.

Over the next eight years, Bullard worked on gravity measurements, including a 1937 trip to the United States where he met Hess and Ewing. Through Ewing, he also learned about refraction seismology, and began studies of the continental shelf on the British side of the Atlantic Ocean to parallel Ewing's studies of the North American side. Meanwhile Blackett was continuing work he had begun under Rutherford on the origin of cosmic rays, for which he would win the 1948 Nobel Prize in Physics.

In 1939 both Bullard and Blackett became involved in war work. Among other things, Bullard concentrated on magnetic minesweeping and demagnetizing ships. After the war, both Bullard and Blackett turned to questions of geomagnetism. For Blackett, the decision was a conscious move away from nuclear physics, with its connections to the atomic bomb.[23] In 1947, now working at the University of Manchester, Blackett proposed a theory to explain the earth's magnetic field: that magnetism arose as a fundamental property of rotating matter. When the planet rotated, it generated a magnetic field. To test his theory, Blackett designed an astatic magnetometer, a highly sensitive device in which he would rotate a massive object in an attempt to generate a detectable magnetic field. Drawing on rich political connections from his war work and a distinguished family background, Blackett arranged to borrow 37.4 pounds (17 kilograms) of pure gold from the Royal Mint, which he rotated at high speed to simulate the effects of the more massive earth moving at lower speed.[24] The experiment failed—no discernable field was generated.

Meanwhile, Bullard had become an advocate of an alternative view: that the earth's field resulted from transient factors such as convection currents in a liquid iron core—the so-called dynamo theory.[25] This led Bullard to conceive a test of the two theories. If Blackett were correct, and the magnetic field arose from the total mass of the earth (like gravity), then it would be a distributed property and the intensity of magnetism would decrease with depth (as does gravity). On the other hand, if Bullard were correct, the strength of the planetary magnetic field would be unaffected by depth. This suggestion was taken up by Blackett's Manchester colleague, S. K. (Keith) Runcorn (1922–1995), who began taking magnetometers down the shafts of coal mines. He found no depth effect, and by 1951 it was clear that Blackett's theory was wrong.

At this point, Runcorn and Blackett turned their attention to magnetism in rocks. If the magnetic field was transient, then the history of variations in the magnetic field might be recorded in rock remanent magnetism—the ancient magnetic signatures of rocks. In the early 20th century, Pierre Curie had discovered that rocks cooled in a magnetic field take on the polarity of that field (the temperature at which this occurs eventually became known as the Curie point). Therefore, if the magnetic field varied, these variations might be recorded in rocks, particularly volcanic rocks that began life as magmas at temperatures above the Curie point. There was evidence that this was so dating back to the early 20th century; more recently the idea had been revived by Jan Hospers, a Dutch graduate student who had entered the Ph.D. program at Cambridge in 1949 trying to use remanent magnetism to correlate lava flows in Iceland, and by John Graham, working in the United States at the Carnegie Institution of Washington.[26] Runcorn, now back at Cambridge, borrowed Blackett's magnetometer and began to develop a geomagnetic research group. He also hired a field assistant, a recent geology graduate named Edward (Ted) Irving. Runcorn and Irving began a program of collecting samples of rocks from different age strata (rock layers) in the United Kingdom.

In 1953, Blackett moved to Imperial College, London, where he set up his own remanent magnetism group. He also encountered geology professor H. H. Read, the man who inspired Arthur Holmes to make geology his professional focus. During the war years with few students to teach, Holmes had written a comprehensive textbook that had an extensive discussion of continental drift, including the evidence of it and the possible role of convection currents to drive it. Years later at Imperial College, it was said that when Blackett turned to Read to learn about rocks, Read sent Blackett to the library to read Holmes. Whatever the truth of the matter, by the mid-1950s both Blackett and his group at Imperial and Runcorn and his group at Cambridge were convinced that remanent magnetism held a record of the variations in the earth's magnetic field, and that these variations showed that rocks had not remained stationary relative to Earth's magnetic field over the course of geological history.[27]

There were two possible interpretations of their data: either the earth's poles had moved relative to the land masses (true polar wander), or the land masses had moved relative to the poles (continental drift). Runcorn realized this ambiguity could be resolved by comparing magnetic variations in rocks of the same age on different continents. By compiling remanent magnetism of rocks of varying ages, one could construct a record of how the poles had seemed to move over time, an "apparent polar-wandering path." If all the continents produced the same apparent polar wandering path, it would mean that the poles had moved. If they varied, it would indicate continental drift. Irving left Cambridge for the Australian National University, where he began to compare apparent polar-wandering paths for Australia, India, North America, and Europe. The result? The paths were distinctly different among the continents. By 1956, both Irving and Blackett's group—now working on rocks from India—were arguing for the paleomagnetic data as evidence for continental drift, and Runcorn soon accepted their views.[28] So did Teddy Bullard, and so did Harry Hess.

Inspired by these developments, Hess revisited the question he had set aside when he had gone off to war 20 years before: whether mantle convection currents might drive continental drift. In a paper written in 1960, although not published until 1962, Hess argued that the British paleomagnetic work had reopened the question, and the answer was drift. Moreover, heat flow measurements by Bullard, working with Scripps scientists Arthur Maxwell and Roger Revelle, showed that heat flow through the oceanic crust was greatest at the mid-ocean ridges, consistent with rising convection currents.[29] Hess therefore suggested that mantle convection might be driving the crust apart at mid-ocean ridges and downward at ocean trenches, forcing the continental migrations in their wake. "One may quibble over the details," he wrote, "but the general picture on paleomagnetism is sufficiently compelling that it is more reasonable to accept than to disregard it."[30] He interpreted the oceanic crust as an upper layer of the mantle that had been altered by interaction with sea water; Scripps geologist Robert Dietz (1914–1995) modified the hypothesis by arguing that the ocean crust was formed by submarine basalt eruptions, and gave it the name it holds today: *sea floor spreading*. Dietz's interpretation was later confirmed by direct examination of the sea floor.

Hess referred to his paper as an "essay in geopoetry," no doubt to deflect criticism from the many North Americans who were still hostile to continental drift.[31] While the British had generally viewed the outcome of the 1920s debate as a stalemate, and therefore open to reconsideration on the basis of new data, Americans generally believed that drift had been refuted.[32] It would take more work to convince North American scientists to reconsider. Moreover, while Hess grew convinced of continental drift on the basis of the apparent polar-wandering paths, others doubted the paleomagnetic data. While it was true that some rock sequences produced highly coherent patterns, others were less coherent, and some were *reversely* magnetized. That is, the polarity of the magnetic field recorded in the rock was opposite to Earth's magnetic field. Most people interpreted this as a sign that the data were unstable: some rocks accurately recorded the surrounding magnetic field, others didn't. Perhaps some minerals did not record the surrounding field, but somehow reversed the direction. Or perhaps the polarities were altered by later events.

Or perhaps Earth's magnetic field periodically reversed its polarity. Early in the 20th century, French physicists B. Brunhes and P. L. Mercanton had suggested this idea: that reversed remanant magnetism in rocks might be recording reversed polarity in the planetary field. But the origin of the earth's field was then unknown; to postulate reversals in a field of unknown origin was speculative in the extreme.[33] In the 1920s, Japanese geophysicist Motonari Matuyama undertook a detailed study of magnetism in volcanic rocks in Japan and found a very consistent pattern: recently erupted lavas were consistently polarized in line with the present field, but reversed rocks were all Pleistocene in age or older (more than 10,000 years). Matuyama argued for a Pleistocene field reversal: that sometime around 10,000 years ago, Earth's magnetic field reversed its polarity. But his work appears to have been largely ignored by European and American scientists.[34] Working in Iceland in the early 1950s, Jan Hospers found similar results: basalt flows there were alternately normally and reversely magnetized.[35]

The question was taken up by a group in the United States at the University of California at Berkeley: geophysics professor John Verhoogen, his postdoctoral fellow Ian McDougall, and graduate students Allan Cox

(1923–1987), Richard Doell, and Brent Dalyrymple. They wanted to determine whether reversals reflected the ambient magnetic field or were a consequence of the physical properties of the minerals involved. Cox began a project analyzing hundreds of samples from the Snake River basalts in the northwest United States, and found results that confirmed the work of Matuyama and Hospers: the patterns were coherent, and they appeared to depend upon the age of the basalt flows. To pin this down, Cox needed accurate ages for the flows.

At this point, a key instrumental development emerged. The radiometric uranium-lead (U-Pb) method for dating rocks had been around since the 1910s, but given the long half-life of uranium, it was accurate only for very old materials. However, Berkeley geochemists had developed the potassium-argon (K-Ar) dating technique to the point where it was accurate for very young rocks, including basalts that might be only a few hundred thousand years old. By this time, Cox, Doell, and Dalrymple had been hired as scientists at the U.S. Geological Survey, and McDougall had moved to the Australian National University, where he established a K-Ar laboratory with colleagues Don Tarling and François Chamalaun. The two groups were now working concurrently on the same problem: accurate K-Ar dating of the magnetic reversals in rocks to prove whether they recorded time-specific events in earth history, and, if so, when they had occurred. By 1963, the combined work of the two groups led to the establishment of a paleomagnetic timescale, with four clearly dated reversals extending over the past four million years. Scientists named the first two of these periods after Brunhes and Matuyama: we live in the Brunhes normal epoch, which was preceded, starting around 700,000 years ago, by the Matuyama reversed epoch.[36]

Meanwhile, throughout the 1950s, researchers at Scripps and Lamont had been collecting sea floor magnetic data, with funds and logistical support provided by the U.S. Navy. In 1961, Scripps scientists Ronald Mason and Arthur Raff published a widely read paper documenting a distinctive pattern of normal and reversely magnetized rocks off the northwest coast of the United States. The anomalies formed a series of stripes, roughly parallel to the shoreline. Published in black and white, they looked a bit like zebra stripes—slightly irregular, but stripes nonetheless. Magnetic reversals plus sea floor spreading added up to a testable hypothesis, proposed independently by Canadian geophysicist Lawrence Morley and Cambridge geophysicists Frederick Vine and Drummond Matthews (1931–1997). If the sea floor spreads while Earth's magnetic field reverses, then the basalts forming the ocean floor will record these events in the form of a series of parallel "stripes" of normal and reversely magnetized rocks. Both Morley and Vine and Matthews realized that Mason and Raff's zebra stripes might be the tangible evidence needed to convert Hess' geopoetry into geo-fact.

The group best situated to examine the evidence was at Lamont, led by James Heirtzler. Throughout the 1950s, Ewing had made sure that magnetometers were towed behind every ship, and that the data collected were catalogued systematically. For some years, Heirtzler and his students had been studying sea floor remanent magnetism, and they had inadvertently amassed the data needed to confirm or deny sea floor spreading. Very quickly they did.[37] In 1965, Heirtzler and Xavier Le Pichon published the first of several articles documenting the magnetic patterns of the Atlantic Ocean; by 1967–1968, Lamont scientists, including Walter Pitman, proved that the sea floor magnetic stripes were consistent with the predictions of the Vine and Matthews model.[38] Meanwhile Neil Opdyke, also working at Lamont, showed that marine sediments recorded the same magnetic events as terrestrial and sea floor basalts, linking the continents with the oceans.[39]

Another group at Lamont had focused on bathymetric data—measurements of the depth of the sea floor—primarily in the Atlantic. These data were highly classified, but Bruce Heezen (1924–1977) and Marie Tharp had found a creative means around security restrictions: a physiographic map, essentially an artist's rendition of what the sea floor would look like drained of water, based on quantitative measurements, but without actually revealing them. In one glance, a geologist could see the most important feature: a mountain chain running down the middle of the Atlantic Ocean floor, crosscut by an enormous series of east-west bearing fractures that dislocated the ridge all along its length. A fracture zone also ran down the middle of the mid-ocean ridge, and Tharp noted that the shape of this central fracture zone suggested it was a rift, a place where the ocean floor was being pulled apart. Heezen interpreted the medial rift as evidence in support of the expanding earth hypothesis, an idea that had been promoted in the mid-1950s by Australian geologist S. Warren Carey. But other Lamont scientists now saw it as strong evidence of Hess' theory. The sea floor was split down the middle, the two sides were moving apart, and the rocks on either side preserved a symmetrical pattern of the periodic reversals of Earth's magnetic field.

One more piece in the puzzle would help to bring the whole picture together: the recognition of transform faults by Canadian geologist J. Tuzo Wilson (1908–1993). An unusually creative and insightful scientist, Wilson had been studying Pacific oceanic islands, such as the Hawaiian

World Ocean Floor, Bruce C. Heezen and Marie Tharp, 1977. Copyright Marie Tharp, 1977.

chain, and found that the ages of the islands increased as one moved farther from the East Pacific Rise—a mountainous region on the eastern side of the Pacific. He realized this could be explained if the rise were a volcanic center above an upwelling convection current and the islands were moving progressively from that center by sea floor spreading.[40] The weight of geological data, together with the fit of the continents, revealed that the earth's surface was "divided into rigid blocks separated by zones of weakness," and that the "periodic break-up of continents and then their slow progression to a new pattern may have happened several times."[41] In 1965, Wilson visited Cambridge, where he spoke at length with Teddy Bullard, Fred Vine, Dan McKenzie, and others interested in continental mobility, including Harry Hess, who also visited Cambridge that year.[42]

Wilson now realized that the fracture zones that displaced the mid-Atlantic ridge—and similar fracture zones mapped by Scripps scientist W. H. (Bill) Menard (1920–1986) in the Pacific—provided a clear test of the idea that the two sides of the ridge were moving apart as solid blocks. Most people assumed that these fracture zones were strike-slip faults, because the ridges were displaced across them. But Wilson had a new idea. Normally, when geologists look at blocks of rock disturbed by an earthquake, they can determine which direction the land has moved based on the observable features that are displaced: a fence, a road, a bridge, or a distinctive rock layer. If the fault is a strike-slip (or transcurrent) fault, where two blocks slip alongside each other as they do along the San Andreas Fault, then geologists look across the fault to see which way things have moved: if objects have moved to the right, then it's a right-lateral fault; if they have moved to the left, then it's a left-lateral fault. But if the mid-ocean ridges were rifts, with the ocean floor splitting apart along them, then the slip directions on the faults that displaced the ridges—what Wilson now called *transform faults*—would be the opposite of what they would be along conventional strike-slip faults.[43]

This was a clear and unequivocal test, and developments in seismology, hastened by the U.S. government's funding of a world wide standard seismograph network (WWSSN), had recently made it possible to accurately determine the slip directions on faults. Once again, Lamont scientists were positioned to perform the test. In 1967, seismologist Lynn Sykes demonstrated that the slip directions on the fracture zones that cut across the mid-Atlantic ridge were consistent with Wilson's interpretation. The offsets were not transcurrent faults, but, in Wilson's new terminology, transform faults, where a mid-ocean rift was locally transformed into a zone of crustal sliding, and then back again into another

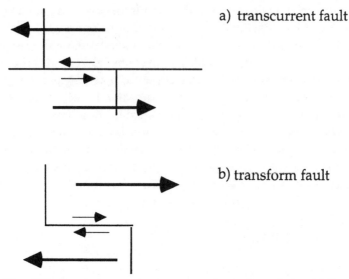

a) transcurrent fault

b) transform fault

The difference between transcurrent and transform faults. (a) In a transcurrent (or strike-slip) fault, the direction of movement can be determined from the offset of a feature intersecting the fault. If the feature is moved to the left, it is a left-lateral fault, as shown here. The north side of the fault has moved to the left (west), the south side of the fault has moved to the right (east), and the fault may continue indefinitely. (b) In a ridge-to-ridge transform fault, a section of the mid-ocean ridge is fractured perpendicular to its length. In this case, the right side of the ridge is moving to the right (east), the left side is moving to the left (west), and the sense of motion is opposite of that illustrated in (a). Note also that the fault does not extend indefinitely, but terminates against the north-south running ridge segments.

rifting ridge segment. There was no longer any doubt that the oceans were splitting apart.

Sykes and co-workers Jack Oliver and Bryan Isacks also examined the slip directions on earthquakes associated with the edges of ocean basins. These edges are characterized by zones of deep-focus earthquakes, either beneath volcanic island chains like the Aleutians on the northern edge of the Pacific, or beneath continental margin mountain belts such as the Andes on the eastern edge of the Pacific. Sykes, Oliver, and Isacks found that the slip directions were consistent with the overlap of one crustal plate onto another, with the lower one slipping downward; the zones of deep-focus earthquakes marked the position of the down-going slab.[44]

A global picture now emerged. Oceans split apart at their centers, where new ocean floor is created by submarine volcanic eruptions. The crust then moves laterally across the ocean basins. Ultimately, it collides with continents along their margins (edges), where the ocean crust sinks underneath, back into Earth's mantle. As it does, it compresses the continental margins, generating folded mountain belts and magmas that rise to the surface as volcanoes, and deep earthquakes as the cold, dense ocean slab sinks farther and farther back into the earth.[45]

In 1967–1968, this picture was integrated into a synthetic, quantitative theory. Working independently, Daniel P. McKenzie and Robert L. Parker at Scripps and Jason Morgan at Princeton established the plate tectonic model: that crustal motions could be understood as rigid body rotations on a sphere.[46] Building on Morgan's work, Xavier Le Pichon summarized the relevant data in a map of the world divided into plates, and calculated their rates of movement on the basis of paleomagnetic data.[47] The result became known as plate tectonics, and it was now the unifying theory of the earth sciences. By the early 1970s, geologists were working out its meaning for continental tectonics.[48] After nearly a century, scientists had finally answered the question of the origin of mountains: they form when plates collide.

This has been a very broad overview. We turn now to how these events looked at the time, to the people who made them happen.

PART II
THE EARLY WORK:
FROM PALEOMAGNETISM
TO SEA FLOOR SPREADING

When World War II broke out, arguments about crustal mobility were put on hold as earth scientists applied their special knowledge and skills to surf fore-casting, submarine navigation, anti-submarine warfare, and other pressing issues of the day. Afterward, a group of British geophysicists who had worked on magnetism and warfare (mine-sweeping and demagnetizing ships) turned their attention to rock magnetism. Initially, they hoped to answer questions about the origins of the earth's magnetic field. But they discovered something else entirely: rocks on land recorded evidence that the position of the land masses relative to the earth's poles had changed over the course of geological time. Some of them began to think again about continental drift. Yet these data did not immediately cause a stampede, for they were new and uncertain, and people doubted their reliability.

Meanwhile, American scientists had been measuring the magnetism of rocks on the sea floor, partly out of curiosity, partly because the U.S. Navy hoped these measurements might suggest new means to hide or detect sub-marines. The result surprised everyone: a distinctive pattern in which some rocks were magnetized in concert with the earth's current field and some in opposition to it. When plotted on paper in black and white, the pattern looked like zebra stripes. Scientists wondered what these magnetic stripes meant, and no one at first connected the pattern to continental drift. Then, another group of scientists proved that over the course of geological history the earth's mag-netic field had reversed its polarity many times. Suddenly the meaning of the stripes became clear: the sea floor was splitting apart, or "spreading," and new volcanic rocks were magnetized in alignment with the earth's field each time they erupted on the sea floor. Once the idea was in place, it took only a few years to demonstrate that it was right.

CHAPTER 2

STRIPES ON THE SEA FLOOR

Ron Mason

IN 1955, THROUGH A FORTUNATE SEQUENCE OF EVENTS, THE SCRIPPS Institution of Oceanography was in a position to make an accurately positioned survey of the earth's magnetic field covering a significant area of the northeast Pacific. This was the first survey of its kind, and it was to have a quite unexpected outcome: the discovery of a linear pattern of magnetism in the rocks of the sea floor not previously seen anywhere else. These linear magnetic patterns later came to be called sea floor magnetic "stripes" (because that's what they looked like when plotted on a map) and they pointed to apparent movements of the ocean floor in excess of 600 miles (1,000 kilometers). The lineations themselves became the first step in what eventually became a new global theory of the earth: plate tectonics.

It all started in 1952, but my interest in geophysics goes back to my student days. On completing my undergraduate course in physics at Imperial College, London, in the immediate prewar years, I was looking for an alternative to spending my working life in a laboratory when I discovered geophysics. It appealed to me as a developing, outward-looking subject with various interesting opportunities, so I opted to take the master's course in geophysics at Imperial. But before I could settle down to a steady career I found myself drawn into the war effort, where I gained experience that was to prove invaluable in later life. When the war ended I returned to Imperial as a lecturer in geophysics. And that is when life started to become interesting.

In 1951 I took a year's sabbatical, which I spent at the California Institute of Technology (Caltech). While there, in the spring of 1952, I attended the annual meeting of the University of California Institute of Geophysics, held that year in La Jolla. The location of the meeting, right by the ocean, and the several presentations on marine seismology, a branch of geophysics new to me, set me thinking about other geophysical

Ron Mason with piston core sample of sediments from the sea floor, on the *Spencer F. Baird,* 1954. (Photograph courtesy of the Scripps Institution of Oceanography, used by permission of the University of California.)

techniques that had been or might be used for studying the oceanic crust. Apart from seismology, very little seemed to have been done. Some important gravity work had been undertaken using instruments installed in submarines, thus avoiding the large ups and downs of surface ships, which would swamp the small gravity variations expected of sea floor structures, but I was unaware of any attempt to exploit the earth's magnetic field, other than Project Magnet.

Project Magnet was a joint effort of several bodies, including the Office of Naval Research (ONR), the U.S. Geological Survey (USGS), and the Naval Ordnance Laboratory (NOL).[1] The main purpose of its initial phase was to map the magnetic anomalies associated with volcanoes and other structures in the Aleutians, and with two atolls in the Marshall Islands, Bikini and Kwajalein, using a magnetometer installed in an aircraft. It was the first serious attempt to study the magnetic anomalies arising from oceanic structures. Talking casually to Scripps' seismologist Russ Raitt during the morning coffee break, I asked whether anyone had thought of investigating the magnetic anomalies associated with sea floor structures by towing a magnetometer behind a ship, an operation that could enable ships to obtain valuable data while engaged in other operations. "What's that?" came a deep voice from behind me. Roger

Revelle, director of Scripps, had overheard the conversation. After the briefest of explanation, Roger, in his characteristically direct way, asked, "Well, do *you* want to do it?" to which I promptly replied, "Yes," and I became Scripps' magnetometer man.

My first task was to look for a suitable magnetometer. While trawling the United States, I discovered that the Lamont Geological Observatory (now the Lamont-Doherty Earth Observatory) had towed a magnetometer across the Atlantic four years earlier. This fact was not known at Scripps; the results were not to be published until a year later.[2] However, my visit to Lamont had one favorable outcome: Lamont offered to loan us their magnetometer for Scripps' upcoming *Capricorn* expedition to the southwest Pacific (September 1952–February 1953). This presented a great opportunity for us to familiarize ourselves with the problems associated with operating a ship-towed magnetometer, and we gratefully accepted. After a scramble to get it to the west coast in time, we towed it successfully over more than 8,000 miles (12,500 kilometers) of ship's tracks, during which we recorded magnetic anomalies associated with seamounts, atolls, scarps, and other features of the sea floor. Although the results had limited quantitative value, it was clear that there was a future in ship-towed magnetometry. We just *had* to acquire a magnetometer of our own.

THE SCRIPPS MAGNETOMETER

The heart of the Lamont magnetometer was a military ASQ-3A magnetic airborne detector (MAD), originally designed for installation in aircraft for the detection of enemy submarines. It was an instrument known as a fluxgate magnetometer, in which the measuring element was mounted in a gimbals mechanism and automatically maintained in the direction of the earth's magnetic field. It therefore measured the strength of the field without being adversely affected by the motion of the aircraft. Modified by NOL for geophysical investigations (and used in Project Magnet), it was further modified by Lamont, where the fluxgate unit was installed in a streamlined "fish," a container for towing it behind a ship.

From our experience on *Capricorn*, we felt that we could do no better than to base our magnetometer on the Lamont instrument. A particular advantage of doing this was that its highly developed electronic and mechanical components were immediately available as surplus from the military. However, although the Lamont instrument was quite adequate for making qualitative surveys of sea floor structures, its system was prone to unpredictable drift, which made it unsuitable for exacting geophysical tasks. We therefore implemented a development program aimed at producing an instrument with a more accurate and stable measuring system, for which my wartime experience stood me in good stead. We also experimented with the design of the fish and its towing arrangements, and with the length of the tow cable, so that the fish would ride as smoothly as possible, minimizing magnetic "noise" caused by erratic movements.

Through the generosity of Varian Associates in Palo Alto, we were able to study the short-term and long-term stabilities of our final instrument by comparing it with their newly developed proton-precession magnetometer, an instrument based on atomic principles. By contrast with the fluxgate instrument, which measures relative values of the magnetic field and needs to be calibrated, the proton magnetometer is an absolute instrument, whose output gives the true value of the magnetic field. The comparisons showed that our instrument was highly stable over periods of a few hours and had a steady long-term drift that could easily be corrected.

THE *PIONEER* SURVEY

Early in 1955 I learned from Scripps' marine geologist Bill Menard that the U.S. Coast and Geodetic Survey (USCGS) ship *Pioneer* was about to commence a detailed bathymetric survey off the west coast of the United States. The object was to produce a map of the sea floor topography by recording the depth along a grid of long parallel east-west lines about 5 miles (8 kilometers) apart, using a continuously operating echo sounder. Joining points of equal depth on adjacent lines would enable surveyors to construct a contour map of the sea floor. The area to be surveyed extended from the foot of the continental shelf outward for between 250 and 300 miles (400 and 500 kilometers), and from the Mexican border in the south to the southern end of Queen Charlotte Islands in the north. A radio navigation system with fixed beacons ashore would enable the position of the ship's tracks to be accurately determined. The probable error of a position would vary with time of day and with distance from the beacons, but was expected to be on the order of 300 feet (100 meters). The survey would occupy the best part of two years.

Scripps immediately sought permission from the U.S. Navy Hydrographic Office, the sponsors of the survey, to tow its magnetometer behind the *Pioneer*. Unfortunately, this was not immediately forthcoming, because the Hydrographic Office was concerned that towing the fish and handling it overboard might slow down their operation! So we missed the first few monthly cruises. Eventually, through the persuasive efforts of Roger Revelle, these fears were allayed, and my assistant Art

Raff and I joined the ship in August 1955. This was to be the first of 12 monthly cruises, the last of which took place in October 1956. By this time we had covered an area extending from 32°N to 52°N, a distance of more than 1,250 miles (2,000 kilometers).

The August 1955 cruise was in the nature of a trial run. We had no idea what to expect. Our hope was that we would be able to produce a meaningful contour map of the magnetic field by following the procedure used in making the bathymetric map, that is, by joining points of equal field value on adjacent tracks. There were a number of uncertainties: would the magnetometer prove sufficiently stable during the rigors of several weeks at sea, and would the spacing of the ship's tracks prove sufficiently close to enable us to contour the results? The first was answered by our calibrations of the magnetometer at Palo Alto, both before and after the cruise. These showed that the magnetometer was more than adequately stable. They also enabled us to adjust our readings so that they represented the true value of the earth's magnetic field. As it turned out, the spacing of the ship's east-west tracks was not a problem in plotting the results, because the dominant trend of the contours was north-south.

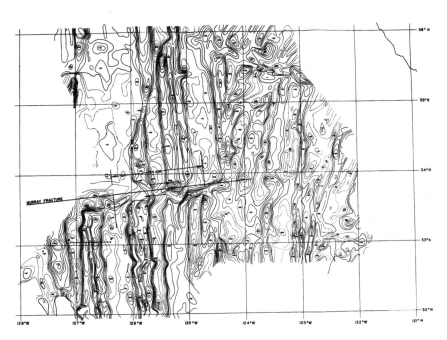

Initial results of the *Pioneer* survey of rock magnetism off the west coast of North America, after three months' operation. The distinctly linear pattern in the south-west corner of the map area, where we started, persuaded us to continue with the survey. (From Mason, 1958, Figure 2.)

Fortunately, the positions of the ship's tracks were made available to us in real time, so we were able to plot the results on a chart and build up a picture of the field as we went. The initial results were discouraging; they were apparently so erratic as to make it virtually impossible to contour them. At this stage we might easily have abandoned the whole operation. But as the data accumulated, the nature of the field began to emerge: it was dominated by bands of approximately north-south trending contours that extended the full 100 miles (160 kilometers) north-south extent of the August cruise.[3] This was a period of great excitement, because nothing like it had ever been observed before, on land or at sea. There was no longer any question of abandoning the survey.

After a further 11 monthly cruises, the survey was completed in October 1956, and the rather tedious task of plotting the results was completed in the first half of 1957. The outcome was a map dominated by contours trending mainly between north-south and northeast-southwest. However, before the results could be properly appraised it was necessary to separate out the geologically related magnetic anomalies from the earth's background field. We did this by overlaying our map on the map of the earth's magnetic field published by the Hydrographic Office, and subtracting the one from the other graphically. This rather tedious procedure greatly simplified the original map.

The final map showed that the bands of dominantly north-south contours bounded strips of positive or negative anomaly.[4] These lineated anomalies, up to one percent of the earth's background field in amplitude and a few tens of miles in width, covered most of the 1,250 miles (2,000 kilometers) north-south extent of the survey, with interruptions in places and some changes of direction. In particular, it was interrupted as it crossed two of the great east-west faults of the northeast Pacific, the Mendocino and the Murray, and a previously unrecognized fault, subsequently named the Pioneer Fault. At the Mendocino and Pioneer Faults there appeared to be no relation between the patterns on opposite sides, but at the Murray Fault they could be matched in such a way as to suggest that since its inception the fault had been offset by about 100 miles (160 kilometers) in a right-lateral sense (that is, crossing the fault, the pattern on the far side would be offset to the right). The absence of a match across the two other faults raised an intriguing question: could it be that displacements across them were so great as to exceed the width of the map? This was a matter that could only be settled by extending the survey in their neighborhoods, but this would have to wait for another day.

Late in 1956 I was diverted to head an International Geophysical Year (IGY) project in the equatorial Pacific. This was one of Scripps' contributions to the international program. It involved the setting up and oper-

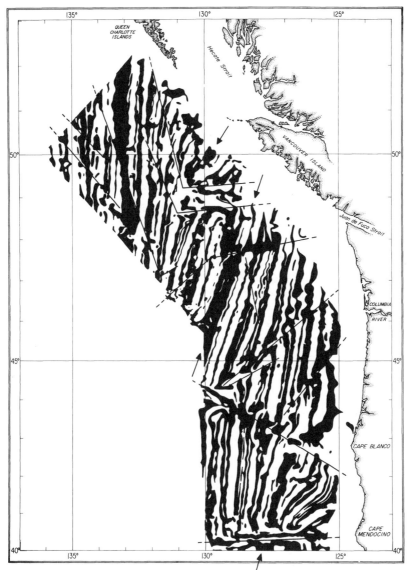

Summary map of the magnetic pattern in the area north of the Mendocino Fault, off the northwest coast of North America. The areas of positive magnetic anomaly are shown in black. (From Raff and Mason, 1961, Figure 1, reproduced courtesy of the Geological Society of America).

ation of temporary magnetic observatories on several islands spanning the magnetic equator. Its purpose was to study certain natural phenomena arising in the ionosphere, and we were certainly not prepared for the spectacular effects of the British nuclear tests on nearby Christmas Island, reflected in our magnetic records.[5] To take care of the logistics of this

operation Scripps chartered a schooner, with its owner, University of
Hawaii mathematician Martin Vitousek, while I took care of the science
and provided a link between Scripps and the field operation. This kept
me away from Scripps for much of the time, and Art Raff took care of the
remainder of the *Pioneer*'s 1956 field season and worked up the 1956 data.
In the meantime, Vic Vacquier came to Scripps to develop its ship-towed
magnetometry program.

During the following two years Vic's priority was to extend the survey,
which he did by running east-west lines, more than 1,250 miles (2,000
kilometers) long on both sides of all three faults, using Scripps' research
ships.[6] These were sufficiently long that in all cases he was able to obtain
matches between the patterns on opposite sides. On the Mendocino and
Pioneer Faults the patterns were displaced in a left-lateral sense by 710
and 160 miles (1,140 and 260 kilometers) respectively, and on the Murray Fault the displacements ranged between 95 and 425 miles (150 and
680 kilometers) in a right-lateral sense.[7] At the time, these results were
taken to imply transcurrent (strike-slip) displacements of the ocean
floor of the same order as the largest of those observed on the conti-

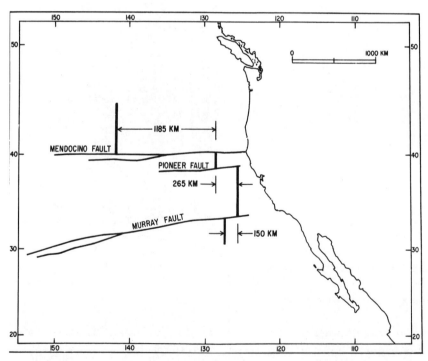

Summary of Victor Vacquier's lateral displacements across the Mendocino, Pioneer,
and Murray Faults. (After Menard, 1960, Figure 6.)

nents. This was a quite unexpected discovery; that such large movements of blocks of the oceanic crust could take place with so little distortion of the magnetic pattern at their margins was taken as evidence for the rigidity of the upper part of the oceanic crust. However, Tuzo Wilson of the University of Toronto was shortly to publish his ideas about a new class of oceanic faults connecting offsets of ocean ridges, to which he gave the name *transform faults*.[8] This would throw doubt on the reality of our proposed displacements.

Wilson's idea invoked the concept of sea floor spreading proposed by Harry Hess and Bob Dietz: that new crust is formed by magmatic intrusion along the crests of mid-ocean ridges, and then drifts steadily away from those crests.[9] Ocean ridges typically suffer numerous lateral offsets, and most people assumed the two segments were initially aligned, and were displaced to their present positions by movement along transcurrent (strike-slip) faults. Wilson proposed that, on the contrary, the offsets were primary features, having always been in their present positions. As a consequence, not only would relative displacements between opposite sides of the fault be confined to that section between the two ridge segments, but the two sides would move in opposite directions away from their respective ridges; the patterns on opposite sides would be the reverse of one another, and no simple match would be possible. Confirmation of Wilson's transform faults came from the work of Lamont's Lynn Sykes. Studying the first motions of earthquakes occurring on offsets of the mid-Atlantic ridge, he showed that movements on their two sides were in the directions predicted by Wilson.[10] Outside of this ridge-to-ridge section, no relative movement between the two sides takes place, but the fault trace continues, because the spreading process has brought crusts of different ages and characters into contact with one another. Hence, corresponding features on the two sides are displaced by an amount equal to the offset of the ridge. If the faults studied by Vacquier were transform faults, related to segments of the East Pacific Rise now buried under the North American continent, then no relative displacement between opposite sides would have occurred. Instead of the hundreds or thousands of miles of offset, there might not be any.

THE MAGNETIC LINEATIONS

The cause of the magnetic lineations – or stripes – led to much speculation. I was able to show that they could be explained by shallow slablike structures, immediately underlying the positive stripes and more highly

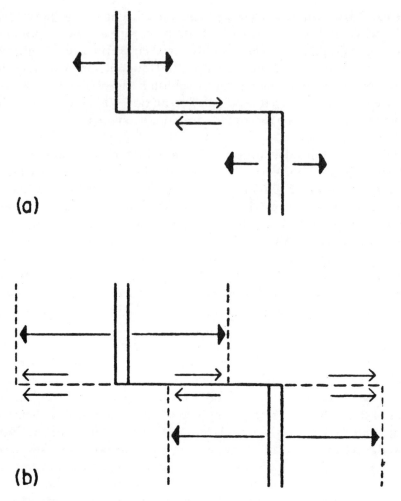

(a)

(b)

Evolution of a transform fault. (a) Two expanding ridges, connected by a transform fault. (b) The same after a lapse of time. The offset of the ridge has not changed, and movement across the fault is confined to the ridge-ridge section. (After Wilson, 1965, Figure 1.)

magnetized than the surrounding crust, but there was no plausible geological model to support such structures. The situation remained unexplained for five years, until Fred Vine and Drummond Matthews, studying two comparatively isolated volcano-like submarine structures in the northwest Indian Ocean, observed that whereas one appeared to be magnetized in the present direction of the earth's magnetic field (where a compass points toward the north magnetic pole), the other was magnetized in the opposite direction.[11] This, and other observations in their field area, suggested to them that perhaps half the oceanic crust was reversely magnetized. This led them to propose a new model to account for the magnetic patterns observed over mid-ocean ridges, bringing together ideas current at the time about sea floor spreading and periodic reversals of the earth's magnetic field.[12] New crust is formed as a result of magmatic activity along the crests of mid-ocean ridges, and as it cools it acquires permanent magnetization in the prevailing direction of the earth's field. If the earth's field then reverses its polarity, the next batch of crust would be magnetized in the opposite direction. Vine and Matthews suggested that these two processes – sea floor spreading and field reversals – would lead to successive strips of alternately normally and reversely magnetized crust drifting away from the axis of the ridge.

Vine and Matthews' hypothesis offered an elegant explanation of how the magnetic lineations of the northeast Pacific could have come about, although in this case there was no obvious connection with an ocean ridge. I was familiar with ideas about sea floor spreading and reversals of the earth's magnetic field, and I could have kicked myself for not thinking of the idea, particularly because, had I looked more carefully at our map, I would have realized that some of the seamounts might be reversely magnetized, and this might have headed my thoughts in the right direction. I had absolutely no doubt as to the correctness of their hypothesis. But to my surprise the idea was not universally accepted. One reason for doubt was that no detailed magnetic survey spanning an oceanic ridge was thought to be available. Perhaps more important, in 1963 neither sea floor spreading nor geomagnetic field reversals were universally accepted. In that regard, the Vine–Matthews hypothesis was built upon two other hypotheses, about which many people still had doubts.

However, it turned out that support for the Vine–Matthews hypothesis had been staring us in the face. In searching for examples of transform faults in the northern part of the magnetic map, Wilson had identified the Juan de Fuca and Gorda Ridges as short sections of young, active oceanic ridges, connected by a conjectured submarine extension of the San Andreas Fault, which he interpreted as a transform fault.[13] Both ridges were marked by seismic activity and symmetry of the magnetic pattern on opposite sides, as would be predicted by the Vine–Matthews hypothesis, extending in the case of the Juan de Fuca Ridge to 110 miles (175 kilometers) on each side. Shortly afterward, Jim Heirtzler and colleagues at Lamont published the results of an aeromagnetic survey spanning the Reykjanes Ridge, the most northerly part of the mid-Atlantic ridge, south of Iceland.[14] This showed magnetic lineations parallel to the ridge and symmetrical about it, extending to about 100 miles (160 kilometers) on

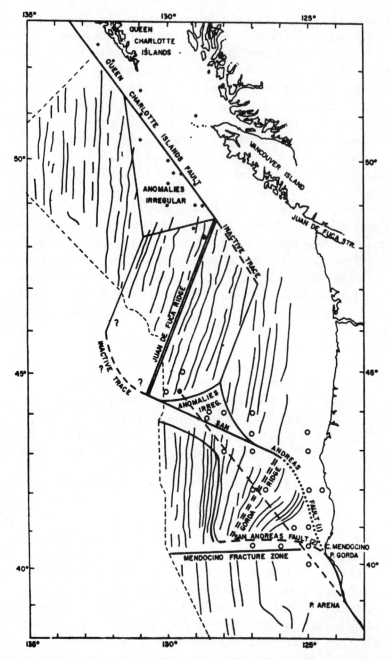

Sketch map showing the relation of the magnetic pattern and earthquake epicenters (dots and open circles) to the Juan de Fuca and Gorda Ridges. Two possible paths of a submarine extension of the San Andreas Fault are also shown. (After Wilson, 1965, Figure 3.)

each side. To my mind, these discoveries should have clinched the Vine–Matthews hypothesis. However, doubts were still being expressed about the universal association of magnetic lineations with spreading axes. It was not until 1966 that the Lamont group, who had been among the most influential doubters, finally came around to accepting it.

MAGNETIC STRIPES AND THE AGE OF THE SEA FLOOR

One consequence of the Vine–Matthews hypothesis is that it offered the possibility of assigning ages to individual stripes, and hence to the underlying sea floor. The method depends on matching the pattern of successive magnetic stripes to the timescale of reversals of the earth's magnetic field. Because field reversals occur at irregular intervals, it follows that, assuming a constant rate of spreading, the widths of successive stripes will follow a similar pattern; under favorable circumstances the two patterns can be matched, and ages of particular stripes determined. In 1968 Fred Vine did this for the northern part of the magnetic map north of the Mendocino Fault.[15] The timescale he used, covering the past 3.5 million years, is based on measurements of the direction of magnetization of rock samples and determination of their ages using radiometric methods. This timescale could be extended by taking advantage of the fact that the patterns of magnetic stripes about mid-ocean ridges are remarkably similar wherever they occur. From a very long profile of the magnetic field spanning the Pacific-Antarctic Ridge, and assuming a constant rate of sea floor spreading, Pitman and Heirtzler extrapolated the timescale back to 11 million years.[16] As expected, the youngest parts of the Pacific sea floor are the actively spreading Juan de Fuca and Gorda Ridges; the oldest is in the northwest corner of the area, with an age in excess of 10 million years. The process has since been applied to most parts of the ocean floor where magnetic stripes have been identified.

SERENDIPITY AND SCIENTIFIC DISCOVERY

I started the project described here with no thought of its possible relevance to continental drift. The discovery of sea floor magnetic stripes was serendipitous: we were not looking for them, nor could we have been, because no one knew they existed! Now, so long after the event, it is difficult to assess what influence their discovery had on the development of plate tectonics, but it is clear that they helped to set events in motion.

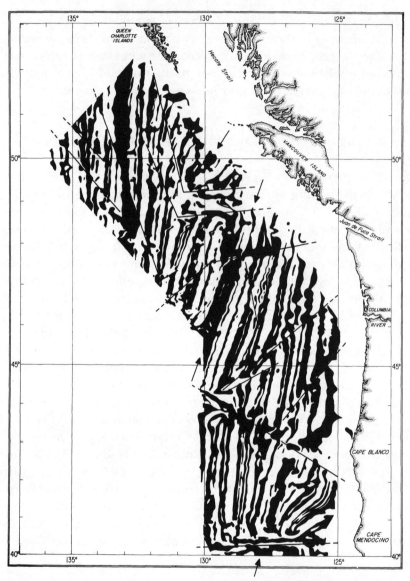

Summary of the magnetic anomalies southwest of Vancouver Island. Areas of positive anomaly, assumed to represent normal magnetization of the underlying crust, are colored to match the radiometrically derived reversal timescale for the past 4 million years, extended to 11 million years on the basis of a long magnetic profile spanning the Pacific-Antarctic Ridge. The straight lines mark geological faults. (From Raff and Mason, 1961, Figure 1, reproduced courtesy of the Geological Society of America.)

The recognition of sea floor magnetic stripes led immediately to the hypothesis of large displacements of the ocean floor which, although later thrown in doubt by Wilson's transform fault hypothesis, stimulated interest in the mobility of the oceanic crust, and it raised interesting questions. The Vine–Matthews hypothesis provided a speculative answer to one of those questions, and the work of Tuzo Wilson, followed shortly by Lamont's aeromagnetic survey of the Reykjanes Ridge, proved it to be correct. For most people these results placed beyond doubt the existence of ocean floor spreading, a factor fundamental to the concept of plate tectonics.

As to my own views on the subject, at no time had I any doubt about the reality of continental drift or the validity of plate tectonics. This is principally because I came to the subject at a time when favorable data were rapidly accumulating and, in contrast to the attitude toward continental drift on the American side of the Atlantic, opposition in the United Kingdom was muted. I had frequent contacts with Teddy Bullard, Keith Runcorn, and other British workers in the field and I cannot recall any outspoken opposition. Bullard had officially sat on the fence for many years, finally coming out in print in support of continental drift in 1964.[17] But, in fact, he had been fully convinced four years earlier by the palcomagnetic data brought together by P. M. S. Blackett and colleagues.[18] In 1960, they had presented their data in such a way that, as he wrote, "it clearly indicated the reality of continental drift."

ACKNOWLEDGMENTS

The sea floor magnetism project owed its existence to the enthusiasm and support of Roger Revelle, then director of the Scripps Institution of Oceanography. Various members of Scripps, in particular Jim Snodgrass and Jeff Frautschy, made valuable contributions to the design and construction of the magnetometer's hardware. Art Raff was a tower of strength throughout.

We owe our thanks to Captain Pearce and the officers and men of the *Pioneer* for putting up with us, to the numerous students and others who maintained shipboard watches, sometimes under difficult conditions, and to Martin Packard of Varian Associates for making calibration facilities available between cruises. The work was supported by the U.S. Office of Naval Research and the U.S. Bureau of Ships.

CHAPTER 3

REVERSALS OF FORTUNE

Frederick J. Vine

MY CONTRIBUTION TO THE NEW UNDERSTANDING OF GLOBAL tectonics predated the recognition of "plates," and the formulation of the plate tectonic paradigm, in 1967. In 1963, Drummond Matthews and I added a rider to the concept of sea floor spreading, which had been proposed by Harry Hess of Princeton University.[1] According to Hess, "conveyor belts" of crust and upper mantle move symmetrically away from mid-ocean ridges and passively drift continents apart.[2] We proposed that the conveyor belts might also act as tape recorders that record reversals in the polarity of the earth's magnetic field in the 'fossil' (i.e., remanent) magnetism of the oceanic crust. This record can be played back by measuring the changes in the intensity of the earth's magnetic field at or above sea level from ships or aircraft.

The validity of our idea depended on the reality of three phenomena: sea floor spreading, reversals of the earth's magnetic field, and the importance of remanent magnetism in the oceanic crust. Not one of these was widely accepted at the time, even by experts in the respective fields. In addition, from the limited magnetic data available to us, largely from the North Atlantic and the northwest Indian Oceans, it appeared that the record was not very clearly written. For example, there was no obvious symmetry of the observed anomalies in the earth's magnetic field about the ridge crests in these areas as predicted by the simple model. This seemed to imply a rather diffuse zone of formation of oceanic crust at a ridge crest, presumably by a process of extrusion and intrusion of basaltic magma. It was not surprising, therefore, that in 1963 the idea was, in general, rather poorly received.

However, in 1965, following the recognition of the Juan de Fuca Ridge southwest of Vancouver Island and the definition of the Jaramillo event within the geomagnetic reversal timescale for the past 3.5 million years,

Fred Vine, at Cambridge in 1967, examining magnetic profiles. (Photo courtesy of Fred Vine.)

it became clear that over this ridge the magnetic field anomalies are not only symmetrical about the ridge crest, but also reflect the reversal history.[3] This, in turn, implies an essentially constant rate of spreading on the Juan de Fuca Ridge, of just over an inch (2.4 centimeters) per year per ridge flank for the past 3.5 million years. In 1966, the acquisition of new magnetic anomaly data for the Pacific-Antarctic Ridge, in the South Pacific, and the Reykjanes Ridge, south of Iceland, confirmed the Juan de Fuca Ridge result.[4] This enabled Jim Heirtzler, of Lamont Geological Observatory, Columbia University, and me to make a convincing case for the validity of the sea floor spreading hypothesis at a meeting held at the Goddard Institute for Space Sciences, New York, in November of that year.[5] I also presented the case at the annual meeting of the Geological Society of America in San Francisco in November 1966. It is probably true to say that these presentations, and the publications by Jim Heirtzler and Walter Pitman also of Lamont, and myself, in December 1966, were instrumental in finally convincing most earth scientists of the validity of sea floor spreading, and hence of continental drift.[6]

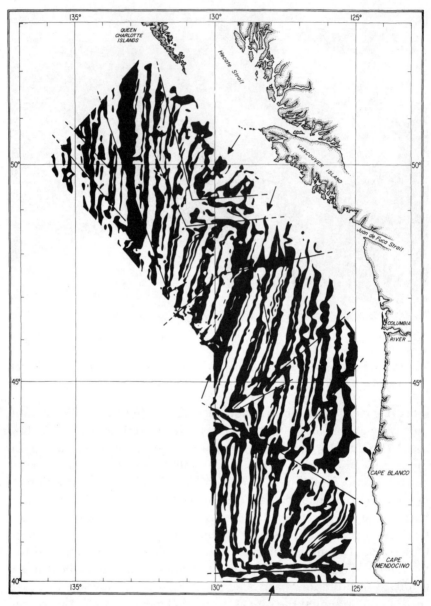

Summary of anomalies in the earth's magnetic field measured at sea level off British Columbia, Washington, and Oregon. Areas of anomalously high field strength are shown in black. Straight lines indicate faults offsetting the anomaly pattern; arrows, the axes of three short ridge lengths in the area – from north to south, the Explorer, Juan de Fuca, and Gorda Ridges. Reproduced courtesy of the Geological Society of America. (Raff and Mason, 1961, note 11.)

EARLY INTEREST IN CONTINENTAL DRIFT AND GEOPHYSICS

In April 1955, at the age of 15, I was a student at Latymer Upper School, West London, and studying for my 'O' (Ordinary) level examinations, which were just two months away. It was the Easter holiday and in a rather desultory attempt to study for geography, I opened a physical geography text, probably for the first time, for as I recall, on the first page of the first chapter, there was a diagram illustrating the approximate fit of the Atlantic coastlines of South America and Africa. In the text, it stated that although it had been suggested on the basis of this fit that these continents were once part of a supercontinent that subsequently split and drifted apart to form the South Atlantic Ocean, geologists had no idea whether there was any truth in this hypothesis. I was struck at once both by the boldness of the idea that seemingly stable continents might have drifted across the face of the earth in the past, and by the fact that we did not know whether this had occurred. It seemed to me that one could hardly conduct any meaningful study of the history of the earth until one had resolved this issue. Surely there must be some way of proving or disproving the concept of continental drift.

In my 'O' level examinations I did well in mathematics, physics, and geography, but only just managed to pass English, Latin, and French. I was not an "all-rounder" academically. However, my limited abilities were well-suited to the English 'sixth-form' system whereby, between the ages of 16 and 18, one studied just three or four subjects to 'A' (Advanced) level. My own preference would have been to take 'A' levels in mathematics, physics, and geography, but I was persuaded to take the then more conventional combination of pure and applied mathematics, physics, and chemistry, which could lead on to a wider spectrum of possible careers in the physical sciences or engineering. I retained my interest in geography, however, and continued to be an active member of the school's Geographical Society. As a sixth-former I had a great interest in the physical environment and read everything I could find, for example, about the International Geophysical Year (IGY), which lasted for 18 months (!) during 1957 and 1958. My performance in the 'A' level examinations earned me a state scholarship and a place at St. John's College, Cambridge, to study natural sciences.

UNDERGRADUATE YEARS AT CAMBRIDGE, 1959–1962

Only subsequently have I realized just how unusual the natural sciences course at Cambridge was at the time, given its large range of options. By now I was determined to drop chemistry; I had also decided that I wanted

to be a schoolteacher and would probably teach physics and/or mathematics. I had spent four terms, after 'A' levels and before entering Cambridge, teaching mathematics at Latymer, an experience I had not only
survived but enjoyed. My director of studies respected my wishes, but
pointed out that we needed to identify one or more options in addition
to mathematics and physics in order to fulfill the degree requirements.
He ran down the list and alighted on geology. He said, "How about geology?" There was a slight pause, and he looked up. "Do you like the open
air? Would you enjoy the fieldwork?" "Why, yes," I said, "of course." "Well
that's done then, geology it is."

At the time, it was not possible to do a degree course in geophysics,
and there was very little geophysics in other, related courses such as
geology. For the first two years of geology I had to content myself with
trying to identify anything in what I read or was taught that was incompatible with the concept of continental drift. I found nothing. In the
third and final year of the degree (so-called Part II), one studied just a
single discipline, in my case geology with emphasis on mineralogy and
petrology. There was also a short, introductory course on solid-earth geophysics given by Sir Edward Bullard, head of the Department of Geodesy
and Geophysics at Cambridge, which was essentially a small graduate
research school. Sir Edward (universally known as "Teddy") was an enter-
taining and inspiring lecturer. He certainly conveyed the excitement in
the field at the time, not least it seemed to me in the area of marine geophysics, where the development of new techniques during the previous
decade had yielded some surprising and fascinating results. Teddy's
department included a marine geophysics section (the only one in the
country) headed by Dr. Maurice Hill, and funded largely by the U.S.
Office of Naval Research. Teddy himself had been involved with the
development of a new technique to measure heat flow through the floor
of the deep ocean basins; Hill had pioneered the use of the proton precession magnetometer to make underwater measurements of the earth's
magnetic field.[7] Teddy's short "taster" course was followed by a longer,
more conventional course on geophysical methods given by Hill. These
two courses confirmed and strengthened my belief that this was the subject that interested me most, and that if I were to undertake postgraduate research it would be in this area, ideally in the subdiscipline of
marine geology and geophysics. At about this time, however, I also contacted the university's Department of Education with a view to doing a
postgraduate Certificate of Education the following year.

To have been taught geophysics by two of the leading exponents of
the field in my final year as an undergraduate was good fortune enough,

but there was more. In January 1962, Cambridge was host to the 10th Inter-University Geological Congress, an annual three-day meeting, organized primarily by undergraduates, the venue for which rotated around British universities that offered degree programs in geology. The theme for this particular meeting was "The Evolution of the North Atlantic," and the lead, guest speaker was Professor Harry Hess of Princeton University. As a student of geology, and mineralogy and petrology in particular, I was already familiar with and an admirer of Hess' work, which ranged over mineralogy, petrology, tectonics, geophysics, and marine geology. His talk, entitled "Impermanence of the Ocean Floor," was essentially equivalent to his paper published subsequently under the title "History of Ocean Basins." The first part was a summary of the geological and geophysical characteristics of the deep ocean floor and the second part an explanation of them in terms of mantle convection and what soon became known as "sea floor spreading," thanks to Bob Dietz, of the U.S. Navy Electronics Laboratory and the Scripps Institution of Oceanography, San Diego.[8] This radical idea was proposed in part to explain what was known of the ocean floor and mid-ocean ridges in general, but also to explain the distinctive characteristics of ridge crests in particular. From Hess' personal point of view it had the additional merit of providing what he regarded as the most plausible and satisfactory explanation for the subsidence of the flat-topped seamounts that he had discovered while captain of the USS *Cape Johnson* and on convoy escort duty across the Pacific during the latter part of the Second World War.[9]

To me this was an inspiring and exciting synthesis and explanation; above all, it was a testable hypothesis and a potential explanation of continental drift. The whole tenor of this meeting was, if only by implication, favorable toward the concept of continental drift. This was in great contrast, I suspect, to the climate of opinion in North America at the time, where such ideas would have been regarded as verging on the heretical. It is probably true to say that throughout the first half of the 20th century there was not the same degree of opposition to the concept of continental drift in Britain and the British Commonwealth countries as there was in North America.[10] In particular, in the late 1950s and early 1960s, there was renewed interest in the idea following the development of paleomagnetic techniques to determine paleolatitudes. In March 1962, for example, I attended a meeting of the Natural Sciences Club at Cambridge addressed by P. M. S. Blackett in which he provided an elegant summary of the results of such paleomagnetic studies; continental drift was assumed to be axiomatic. In contrast, in North America up to that time, the use of the paleomagnetic method to determine paleolatitudes had been largely neglected, if not shunned, perhaps because of the underlying assumptions.

In April 1962 I was awarded a three-year Shell Studentship to undertake a Ph.D. in the marine geophysics section of the Department of Geodesy and Geophysics at Cambridge. This was quite surprising in that the department consisted almost entirely of theoreticians (mathematical physicists) and applied physicists, who built new instruments. The one exception was Dr. Drummond H. Matthews, a geologist with a background similar to my own, who had entered the department as a graduate student in January 1958. It was decided that I should be Drum Matthews' first research student, and that I should work on the interpretation of magnetic data. At any one time, there was at least one graduate student working on each of the main geophysical techniques, such as gravity, heat flow, refraction seismics, magnetics, and so on. There was a 'vacancy' in the magnetic area and it was entirely appropriate that Drum Matthews should supervise me, not only because of our similar backgrounds, but also because he was involved in the acquisition of magnetic data at the time and had measured the magnetic properties of some basaltic rocks dredged from the ocean floor as part of his Ph.D. thesis. By this time, surface, deep-towed, buoy, and differential proton magnetometers had been built in the department, and the main requirement was to develop techniques and ideas for interpreting the magnetic field data, which were accumulating at an ever increasing rate, but were largely uninterpreted.

Presidents of the Geological Society at Cambridge – the Sedgwick Club – were traditionally drawn from the final year undergraduates, and held office for one term. During this term they were expected (or was it required?) to give a presidential address. In May 1962, as president for the summer term, I gave a talk entitled "HypotHESSes" at the 870th meeting of the club. This topic was a natural choice for me, in that Harry Hess' range of interests closely paralleled my own, and I had been inspired both by his papers and by his talk a few months earlier. The address was therefore something that I enjoyed preparing. In addition, it was useful review for my finals, which by then were very imminent. In the talk I summarized Hess' work and ideas on layered igneous intrusions, on the mineralogy and crystallography of pyroxenes, on the alteration of ultrabasic rocks and in marine geology, emphasizing the connections between them. I assumed that most or all of my audience had been present at the meeting in January, and the talk was intended therefore to provide the background to the development of Hess' current ideas. As a consequence I only made brief mention of the substance of his January talk.

Drum Matthews and Tony Laughton, from the National Institute of Oceanography, Wormley, were present, and it soon became clear during the discussion that followed that they had not been present at Hess' talk in January. This was their first encounter with the concept of sea floor spreading. Someone, quite possibly Drum or Tony, asked whether I thought that the north-south 'grain' of linear magnetic anomalies recently discovered in the northeast Pacific might be related to sea floor spreading.[11] (The lack of any reference to these anomalies, and to the central magnetic anomaly observed over ridge crests, was a notable omission from Hess' talk and subsequent paper.) I replied that I felt that they must in some way be an expression of mantle convection as envisaged by Hess, but I had no idea how this effect was produced.

BACKGROUND TO THE VINE–MATTHEWS HYPOTHESIS

When I joined the Department of Geodesy and Geophysics at Madingley Rise, Cambridge, in October 1962, Drum Matthews was at sea in the northwest Indian Ocean. At the time he was coordinator of the U.K. contribution to the International Indian Ocean Expedition. Initially I was put under the wing of Maurice Hill. My main assignment was clear, however: to review published magnetic surveys and traverses at sea, the methods that had been used in interpreting them, and current lines of approach.

The department was a very friendly and happy place, not least because of the lead provided by the senior staff. Coffee and tea breaks were something of an institution. Technicians, academic staff, students, and visitors (typically from North America, because of Teddy Bullard's reputation and contacts there) almost literally rubbed shoulders seated at long tables. Conversations ranged from serious science to whether Teddy's new car befitted his status. It clearly didn't but he would be the last person to be bothered by this. There was an air that doing science was fun, and that there had never been a better time to do geophysics. Certainly, as far as marine geophysics was concerned, the ship time and resources that had been made available since World War II had led to major developments in instrumentation and techniques that were yielding new data at an accelerating rate. The general philosophy in the department was that one might well waste one's first year as a research student investigating a difficult problem that ultimately proved to be intractable, but that it was only by taking such a risk that you increased your chances of doing significant and original research.

I had had little opportunity to discuss the project with Drum, but I had gotten the impression that he had in mind constructing analogue models with iron filings and putty, to simulate the volcanic topography at ridge crests, and then measuring the disturbances in an applied field caused by the induced magnetization in the model topography. Even if the simulated anomalies bore no relation to the observed anomalies, one might still be able to make a correction for the induced magnetization of the volcanic topography. Everyone was mystified by the fact that despite the relatively strong magnetization of the volcanic rocks of the ocean floor there was no systematic correlation between topography and the magnetic anomalies developed over it. In many ways this was the crux of the problem.

My completed literature review, dated January 1963, included the following statements: "Seamounts and volcanic islands give rise to large and obvious anomalies but these can rarely be explained by models assuming uniform magnetization throughout and directed parallel to the present earth's field. This discrepancy suggests that there may be a large thermoremanent component of magnetization, probably often reversed relative to the present earth's field."[12] Ron Girdler and George Peter, both working at Lamont at the time, considered it essential "to assume reversed magnetization in order to interpret a linear anomaly in the Gulf of Aden and support this by convincing calculations."[13] However, they favored a mineralogical self-reversal mechanism to explain the reversed magnetization rather than a reversal of the earth's magnetic field. I continued: "This does not strike one as being a necessary corollary. If current theories regarding impermanence of the ocean floor are correct, paleomagnetic evidence would suggest that the thermoremanent component of oceanic basalts should, in most cases, be approximately normal or reversed.[14] All too little is known about the magnetic properties of oceanic basalts. Work on dredged samples suggests that they are not essentially different from exposed basalts but would indicate that they invariably have a very strong remanent component, such that the remanence is very much greater than the induced intensity.[15] Values of susceptibility (that determine the induced magnetization) assumed in models simulating magnetic anomalies, often, necessarily, have to be high, higher than is reasonable in the light of existing measurements (of susceptibility) on basalt samples."[16]

I concluded that it seemed "highly desirable that any interpretation technique should be able to take account of remanent magnetic intensity even if unknown. Certain computer programs would appear to be capable of doing this. Possibly (analogue) model studies could also, but there is no evidence for this. Such studies would also appear to necessitate very elaborate, cumbersome and, presumably, costly apparatus." "The use of computers in more recent years to simulate anomalies, and conversely, magnetic bodies from anomalies, may herald a breakthrough in interpretation methods. Computer techniques are probably more potent than model studies and easier to handle, judging by the dearth of model studies in the literature."

On reading these conclusions Drum's face visibly fell. Quite apart from the fact that he was, I suspect, looking forward to playing around with physical models, he had not had the time to keep abreast of the rapid developments in scientific computing at that time. However, plenty of help was at hand. Several other research students in the department were using computer methods, albeit in other contexts, and Teddy himself was not only at the forefront of developments in computing in relation to geophysics, but had also written a program to compute a magnetic anomaly profile across a two-dimensional model. However, this program was written in machine code and could only assume induced magnetization in the body producing the anomaly. I had taken a course in computer programming during the previous term and, although I had learned machine code, it was clear that all future programming would be in higher-level autocode. I therefore set about writing my own two-dimensional program for the interpretation of profiles, using mathematical expressions published in several places in the literature, and allowing for any direction of resultant magnetization, that is, remanent plus induced magnetization.[17]

Drum Matthews returned from the Indian Ocean with a large quantity of magnetic data, including a detailed survey of the crest of the Carlsberg Ridge in the northwest Indian Ocean at 5°N. This survey, measuring 51 by 39 nautical miles and known as Area 4A, was the largest and most detailed survey of a known mid-ocean ridge crest at that time.[18] Clearly if I was to interpret this quantitatively, I would need a program capable of calculating the anomaly over three-dimensional features so that I could carry out the correction for induced magnetization as envisaged by Drum.

At about this time, I visited Imperial College, London, where a mathematician, Dr. K. Kunaratnam, had just completed his Ph.D. under Professor J. M. Bruckshaw and Dr. R. G. Mason.[19] He had developed a variety of techniques for interpreting both gravity and magnetic anomalies in either profile or survey form, assuming two- or three-dimensional models respectively. His program for a three-dimensional source region used a particularly elegant method both to approximate the body and to calculate the anomaly in the earth's magnetic field developed over it.

This was an important consideration at the time in view of the slow speed and small storage capacity of computers. The program could also deduce the direction and intensity of magnetization of a specific topographic feature, given details of the topography and of the anomaly observed over it. Such a program was ideal for interpreting the anomalies measured over isolated seamounts. Kunaratnam had developed it in order to interpret the anomalies associated with seamounts within the area surveyed off the west coast of North America in the mid- to late 1950s by Mason and Arthur Raff.[20] In a similar vein, part of my Ph.D. project was to interpret earlier magnetic surveys acquired by the department, and these were typically of isolated seamounts. Dr. Kunaratnam gave me a draft copy of his thesis and said that I was welcome to use any of the mathematical formulations he had developed. I therefore wrote the equivalent of his three-dimensional program for use on the Cambridge (Mathematical Laboratory) computer, EDSAC 2. Subsequently, in order to interpret larger surveys of irregular topography, such as the whole of Area 4A, I would need to utilize a more conventional, if less efficient, formulation in which the bathymetry (ocean floor topography) is approximated by a grid of vertical prisms.

In many ways the results of Drum's magnetic survey of Area 4A were so spectacular that they did not need quantitative interpretation. While making the survey he was concerned, on the basis of the bathymetry, that they were not over the ridge crest. With hindsight this was understandable because approximately one-third of the survey area is occupied by a transverse fracture zone (including what we would now call a transform fault), and away from this, the central valley is not well, or continuously, developed. At one point the central valley appears to be blocked, presumably by volcanism. In contrast to the bathymetry, away from the fracture zone the magnetic anomalies generally form areas of positive and negative anomalies separated by steep anomaly gradients that parallel the trend of the ridge and largely disregard the bathymetry beneath them. Within the areas of positive anomaly, there are some positive correlations with bathymetry, and within areas of negative anomaly some negative correlations. This is what one would expect for reversely and normally magnetized features respectively at this latitude. Thus qualitative inspection alone indicated that, away from the fracture zone, the area is underlain by blocks or avenues of normally and reversely magnetized crust that parallels the trend of the ridge crest. Moreover, the center or crest of the ridge is more reliably identified by a large amplitude negative magnetic anomaly, implying that it is underlain by normally magnetized crust, than by the median valley, which is less continuous.

Once I had seen the Area 4A magnetic survey and made this preliminary assessment of the results, it was a very small step to the formulation of what became known as the Vine–Matthews hypothesis, particularly in view of my prejudices regarding sea floor spreading and continental drift. (The latter, incidentally, amounted to no more than testing a hypothesis, which I had been taught was the essence of the scientific method.) What was now clear, however, was that Drum's decision to devote ship time to such a detailed and time-consuming survey was an inspired one.

I am unable to pinpoint exactly when the idea first came to me. It is clear from the quotations given above that I was quite close to it when writing my literature review in January 1963. It seems probable that it occurred in February or March 1963. Unbeknown to me, at precisely this time, Lawrence Morley, of the Canadian Geological Survey, penned a letter to *Nature* proposing exactly the same idea. He was unable, however, to draw on a survey of a known ridge crest and had to make the case with reference to the linear anomalies mapped in the northeast Pacific, which were not obviously related to a mid-ocean ridge. Morley's paper was rejected by *Nature,* and subsequently by the *Journal of Geophysical Research,* for being too radical and speculative. It was four years before I became aware of this remarkable coincidence and 18 before Morley's 'letter' was reproduced in print in full.[21]

Meanwhile, Drum Matthews and I had decided that however convincing the case might be to those well-versed in the interpretation of magnetic anomalies (which turned out not to be true) there were very few people in this category. In the hope of convincing a wider audience, therefore, we decided that I should undertake some computer-based interpretation before writing up the idea for publication. Thus it was May before I sat down and wrote the first draft of the Vine and Matthews paper; Drum was on his honeymoon at the time. It differed from the published paper in that it did not include the first two paragraphs and the penultimate paragraph (excluding the acknowledgments). It did, however, include more details of the acquisition and reduction of the Area 4A survey. It was reviewed internally by Maurice Hill, Teddy Bullard, and ultimately Drum. I cannot be certain, but I suspect that Hill was very unhappy with it, that Teddy was quite excited about it (recognizing the tremendous implications if it turned out to be correct), and that Drum, caught in the middle, did not know what to think, except perhaps that having a research student was something of a mixed blessing. All agreed that the full details of the acquisition and reduction of the Area 4A survey were inappropriate to a letter to *Nature* and so this section was removed. The problem then was that it became rather long on interpretation and speculation and short on original data.

In order to solve this problem, as I think he in particular saw it, Hill gave us permission to include two long, unpublished magnetic profiles across the crests and flanks of the North Atlantic and northwest Indian Ocean ridges acquired by the group in 1960 and 1962 respectively. The title was changed from "Magnetic Anomalies over the Oceans" to "Magnetic Anomalies over Oceanic Ridges," and the introductory paragraphs were added to set the scene and incorporate the ridge profiles. Knowing now the difficulty that Larry Morley had in getting his article published, this could have been a very significant addition, and I suspect that Maurice Hill had a very shrewd idea as to what would be acceptable to *Nature*. I think that the paper was submitted to *Nature* in late June or early July, probably by Drum or Maurice, for I have no record of it. It appeared in *Nature* for September 7, 1963. By this time the three of us were at sea on the RRS *Discovery* on a four-month expedition in the northwest Indian Ocean, returning to the United Kingdom in December.

A POOR RECEPTION

Initial reaction to the paper was, to say the least, muted. In particular, those most familiar with the interpretation of magnetic anomalies were less than impressed. At the Royal Society Discussion Meeting on Continental Drift held in London on March 19–20, 1964, Vic Vacquier, the only speaker to mention the hypothesis, said that this "attractive mechanism is probably not adequate to account for all the facts of observation. A theory consistent with the facts is still needed to account for the existence of the north-south magnetic lineations in the north-eastern Pacific. Where the East Pacific Rise can actually be seen, no lineated magnetic pattern was found."[22] Manik Talwani, of Lamont, in writing a review of marine geophysics, referred to our idea as "improbable, startling."[23] In papers published in 1965 George Peter and Harry Stewart, of the U.S. Coast and Geodetic Survey, Washington, and Jim Heirtzler and Xavier Le Pichon, at Lamont, failed to invoke or mention the possibility of reversely magnetized crust in modeling oceanic magnetic anomalies.[24] On the other hand, in letters to *Nature*, Harry Hess referred to the "fruitful Vine and Matthews hypothesis" (a pun of which he was quite proud), and George Backus, of the Institute of Geophysics and Planetary Physics, San Diego, who had probably been converted by an enthusiastic exposition from Teddy Bullard, wrote constructively, suggesting a possible test for the hypothesis.[25] Backus pointed out that, if the idea is correct, the magnetic anomalies measured along east-west profiles across the South Atlantic Ocean should increase in width to the south, reflecting the fact that the separation of South America and Africa becomes greater as one moves south.

For me, 1964 was a fallow year as far as the hypothesis was concerned, and I concentrated on producing more substantial, or at least conventionally acceptable, material for my thesis, as well as helping to run a Scout troop. The most significant thing I did was to get married, in March, and probably the next most significant thing was to write to Harry Hess at Princeton, toward the end of the year, to ask him if there was any possibility of finding me a job there once I had completed my Ph.D., hopefully by September 1965. I had heard that Hess was to spend a sabbatical at Madingley Rise during the early part of 1965, and I did not want to confront him face to face with this question, but to give him time to think about it. I did not receive a written reply, which was unsurprising, but it meant that I was on tenterhooks when his arrival was imminent. Much to my delight, on his arrival the first thing he said to me was that he thought that our idea was great, and the second thing he said was that he thought he would be able to find a position for me at Princeton.

THE JUAN DE FUCA RIDGE

Tuzo Wilson, from the University of Toronto, was also on sabbatical at Madingley Rise during the early part of 1965, and it was during this time that he formulated the concept of transform faults.[26] In applying this idea to the worldwide system of ridges and trenches, he eventually arrived at the Gulf of California and recognized that the San Andreas must be a major transform fault system. Farther to the north, the Queen Charlotte Islands strike-slip fault system appeared to be another transform fault terminating in ('transforming' into) the Aleutian trench at its northern end. However, off the states of Washington and Oregon there is a gap and offset in the seismicity associated with these two fault systems. The logic of Wilson's hypothesis predicted that there should be a short length of ridge between the two faults in this area which he named the Juan de Fuca Ridge after the Strait of Juan de Fuca, which forms the boundary between the United States and Canada along this coastline. Tuzo was explaining this to Harry Hess and me when Harry suddenly interrupted him and said, "If you want to put a ridge there, that is one of the few oceanic areas for which there is a detailed magnetic survey, and if Fred is right, there should be a clear expression of the ridge in that survey." I dashed upstairs to the library to look at the volume of the *Bulletin of the Geological Society of America* containing the relevant article by Raff and Mason.[27]

I cannot remember whether I took a quick look at the summary map before rushing back to Tuzo's office and setting it before Tuzo and Harry. All three of us stared at it in amazement. Not only were there linear magnetic anomalies paralleling the trend of Tuzo's putative ridge, but there was also a symmetry to the pattern of anomalies about the ridge crest. Despite the fact that this diagram had been in the literature for four years, no one it seems had noticed this symmetry. The irony of the discovery of the Juan de Fuca Ridge, or rather its non-discovery at an earlier date, is that because the survey was undertaken for military purposes during the Cold War – detailed maps of the bathymetry and gravity field were required for the nuclear submarine deterrent – only the full details of the magnetic data were declassified. Although the topographic expression of the Juan de Fuca Ridge is obscured by sediment from the Columbia River fan, which spills over it and infills much of the topography very rapidly, there is still enough on a detailed survey to reveal the location of the ridge. Had the detailed bathymetry been released at the same time as the magnetics, this would almost certainly be a very different story. In fact, Bill Menard, of the Scripps Institution of Oceanography, San Diego, was aware of the Juan de Fuca Ridge and the shorter Gorda Ridge to the south of it, having discovered the latter in 1952, and having seen the classified data. He was quite convinced however, that these ridges were not equivalent to mid-ocean ridges.[28]

Tuzo was planning to write up his discovery of the Juan de Fuca Ridge as an actively spreading ridge as an article for the journal *Science.* He proposed that he and I should write a second paper for *Science* on the interpretation of the magnetic anomalies over the ridge. He also persuaded me (and presumably Harry as well) that we should present these two papers at the annual meeting of the Geological Society of America to be held in Kansas City in November. My work schedule for the coming few months was now clear, if a little ambitious: to write the paper for *Science,* to complete and be examined on my thesis, and to move to Princeton. Progress on the *Science* article was slow, mainly but not entirely because of the competing demands of the thesis, and Tuzo became a little impatient because he was keen to submit the two papers together. The paper was finally finished in June, the thesis in August, and on September 16 my wife and I set sail from Southampton for New York on the *United States.*

Although the symmetry of the magnetic anomalies about the crest of the Juan de Fuca Ridge provided stunning support for the idea that they might result from a combination of sea floor spreading and reversals of the earth's magnetic field, there was a problem with the more detailed interpretation. If one assumed the reversal timescale for the past few million years as defined then by Allan Cox, Richard Doell, and Brent Dalrymple, of the U.S. Geological Survey, Menlo Park, the pattern of anomalies implied major changes in spreading rate with time.[29] Although inelegant, and counterintuitive if spreading was related to large-scale convection in the mantle, within this pre-plate tectonic paradigm it was not at all clear whether one would expect spreading to be at a uniform rate or somewhat erratic. Consideration of this and of the ambiguity in the thickness of the source region for the anomalies meant that writing the *Science* paper on the magnetic anomalies was not as straightforward as one would have wished. The two papers appeared in *Science* on October 22, 1965, just a few weeks before the meeting of the Geological Society of America in Kansas City.[30]

THE JARAMILLO EVENT

Tuzo was right, of course. The meeting in Kansas City provided me with an excellent opportunity to publicize my ideas and to meet North American geologists. One such meeting was particularly memorable and significant; indeed, it provided me with the last piece of the jigsaw puzzle and enabled me to make a convincing and essentially unarguable case for the validity of the Vine–Matthews hypothesis. It was with Brent Dalrymple, from Menlo Park, who told me that they were increasingly confident that they had discovered an additional detail of the reversal timescale at around 0.9 million years before present. It was a period of normal polarity, of perhaps 100,000 years' duration, and they had named it the *Jaramillo event.* I realized at once, having pored over the problem for so long and so recently, that with this revised timescale it would be possible to interpret the Juan de Fuca anomaly sequence with an essentially constant rate of spreading. To me, at that instant, it was all over, bar the shouting. From the situation less than a year earlier, when I thought that I might spend my whole career trying to convince people of the validity of our idea, I could now make a compelling case that the sea floor not only

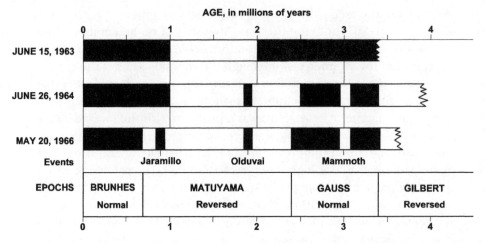

Successive refinements of the geomagnetic reversal timescale for the past 3.5 million years, obtained by Cox, Doell, and Dalrymple, of the U.S. Geological Survey, Menlo Park, during the 1963–1966 period. The dates are publication dates (Cox, A., R. R. Doell, and G. B. Dalrymple, 1963. Geomagnetic polarity epochs and Pleistocene geochronology. *Nature* 198: 1049–1051; Cox, Doell, and Dalrymple, 1964, note 29; Doell and Dalrymple, 1966, note 4.)

spreads symmetrically about mid-ocean ridges but at an essentially constant rate, and in doing so, it faithfully records the timescale of reversals of the earth's magnetic field. At once, the possibility of documenting the evolution of the present-day ocean basins and the geomagnetic reversal timescale for the past 150 million to 200 million years opened up.

INDEPENDENT CONFIRMATION

In February 1966 I visited the Lamont Geological Observatory (as it was then called), at the invitation of Neil Opdyke. Neil had moved there in 1963 to set up a paleomagnetic laboratory, and was housed in the same suite of offices as Jim Heirtzler and his Magnetics Department, whose work included the study of marine magnetic anomalies. When I arrived, Neil was working on a diagram on a light table. It transpired that it was the diagram that would become 'figure 1' of his paper on the paleomagnetism of Antarctic deep-sea cores that was published in *Science* later that year.[31] "Look at this Fred," he said. "Not only have we found the complete Cox, Doell, and Dalrymple timescale in these cores, but we have also discovered an additional, normal event at around 0.9 million years. We have called it the Emperor event." "Well, I am sorry to have to disappoint you," I said, "but the Menlo Park group have already resolved such an event and named it the Jaramillo." Neil's jaw dropped. On the wall of the same room were pinned up the now famous *Eltanin* magnetic and bathymetric profiles across the Pacific-Antarctic Ridge that his Lamont colleague Walter Pitman had recently reduced. By this time Walt had joined us. "Furthermore," I said, "one can see it on these magnetic profiles just as you can on the Juan de Fuca Ridge." Walt's jaw dropped.

With the addition of the work on deep-sea cores we now had three independent records of the geomagnetic reversal timescale for the past 3.5 million years, and I could, with some confidence, set about reviewing the status of the Vine–Matthews hypothesis in the light of new data, new ideas, and the revised reversal timescale. Either during this meeting at Lamont, or soon after, Jim Heirtzler let me have copies of the *Eltanin*-19 profile and details of the aeromagnetic survey of the Reykjanes Ridge south of Iceland, which was the second extensive survey to demonstrate the symmetry of the magnetic anomalies about the ridge crest. Presumably this had already been prepared for publication, in that it appeared in print a few months later in the journal *Deep Sea Research*.[32]

In the months that followed my memorable meeting with Neil, Walter, and Jim, I prepared my review article and made a further visit to Lamont to give a talk. In May or June, Walter and Jim visited Princeton and were surprised, I think, to discover that my review article was rather wide-ranging and essentially complete. Inevitably, it drew heavily upon and reproduced a number of magnetic surveys and profiles acquired by others. I was anxious that I should not only have their permission to do this but also that the work should be published. This was true but for the notable exception of the *Eltanin*-19 profile. Should I withdraw it from the paper or wait until it was published? Jim very generously suggested that we should try to arrange to publish simultaneously in *Science*, and I was happy to agree to this.

The paper on the *Eltanin* profiles by Jim Heirtzler and Walter Pitman was published two weeks ahead of mine, on December 2, 1966; this struck me as entirely reasonable.[33] There were many rumors circulating at the time regarding the lobbying and discussions that were going on in relation to these two papers. The only hard evidence I ever had of this, apart from the delay in publication, was a comment from Harry Hess, who was a good friend of Phil Abelson, the editor of *Science* at the time. Abelson had asked Harry whether he thought that my paper was worth publishing; apparently he felt that *Science* was carrying too many earth science articles at the time.

At the meetings held in New York and San Francisco in November

1966, at which I presented the content of this paper, Lynn Sykes, of Lamont, gave presentations of his recent work on focal mechanism solutions for earthquakes associated with mid-ocean ridge crests.[34] This
confirmed the validity of Tuzo Wilson's transform fault hypothesis, and
provided further, and entirely independent, evidence for the reality of
sea floor spreading. Together with the symmetry of the magnetic anomalies about ridge crests and their correlation with the geomagnetic reversal timescale, this result finally convinced most earth scientists of the
validity of the hypothesis of sea floor spreading, and hence of continental drift.

POSTSCRIPT

In early 1963, when Drum Matthews and I first discussed the possibility
of combining sea floor spreading with reversals of the earth's magnetic
field to explain oceanic magnetic anomalies, we could not have dreamed
that the idea would be spectacularly confirmed within less than three
years. The magnetic data that we were working with showed no symmetry or regularity, except for a large, central anomaly over mid-ocean
ridge crests, and the earth's magnetic field, if it had reversed at all, was
thought to reverse at a fixed interval (possibly between a half to one million years). Taken together with the variable width of the stripes of the
northeast Pacific, reversals with a fixed periodicity implied that the
spreading rate must be very irregular. Ultimately, of course, it transpired
that it is the time between reversals that is very irregular and that the rate
of spreading has been remarkably constant for millions of years. The first
problem, of the complexity of the magnetic profiles, was more serious,
although as geologists we did not find it surprising. It implied that new
crust is formed by a process of intrusion and extrusion of basaltic material, and by faulting, over a zone perhaps a few tens of kilometers in
width. This seemed very likely by analogy with the central zone of active
volcanism and faulting on Iceland, which lies astride the mid-Atlantic
ridge. What we did not realize at the time was that spreading rates vary
greatly, by a factor of ten, around the mid-ocean ridge system, and that
the complexity of the crustal structure decreases with increased spreading rate, presumably as the zone of formation gets progressively nar-
rower. The spreading rates in the North Atlantic and the northwest
Indian Oceans are relatively slow, whereas those on the East Pacific Rise
are three to five times higher, and that for the Juan de Fuca Ridge is twice
that in the North Atlantic. Thus the Pacific ridges behave more like a

tape recorder than we could ever have imagined. Indeed, it has been suggested that fast-spreading crust not only preserves a record of the reversals of the earth's magnetic field but, in addition, information on changes in the intensity of the field with time.

Within ten years of the confirmation of the Vine–Matthews–Morley (VMM) hypothesis, the same sequence of magnetic anomalies, reflecting the history of reversals of the earth's magnetic field during the past 160 million years, had been recognized in all the major ocean basins. By rewinding the tape recorder it was possible to determine the relative positions of the continents, and the sequence of continental drift, throughout this period of time. My one regret, as a geologist, was that this detailed record, written within the 60 percent of the earth's surface covered by oceanic crust, is only available for 4 percent of geologic time. Earlier phases of continental drift would have to be deduced from the more complex and fragmentary geological record within the 40 percent of the earth's surface covered by continental crust.

A surprising aspect of the widespread acceptance of the VMM hypothesis in 1966 was the fact that there was no direct evidence that the magnetic stripes are underlain by bands of normally and reversely magnetized crust. It was only inferred, there being no oriented samples from the volcanic rocks of the ocean floor. It was many years before there was evidence for this: initially from measurements of the magnetic field made very close to the sea floor, and ultimately from the recovery of oriented drill cores. In a similar vein, it was 20 years before it became possible to confirm the rates of spreading deduced in 1966 by an independent technique. By the mid-1980s, it was possible to determine the change in the distance between two points within the interiors of different plates, by making repeat measurements over several years using the satellite laser ranging technique.

Still outstanding is the nature and vertical extent of the magnetic crust that gives rise to the anomalies. The basaltic layer, which is typically less than one kilometer (0.6 mile) thick, is strongly magnetized when first formed, but its magnetization decays with time. It would appear that a lower crustal layer also preserves the magnetic record, and that this magnetization is more stable with time. Its contribution to the magnetic anomalies probably becomes more significant as the crust gets older, but the precise nature, geometry, and thickness of this layer are still not fully understood.

The contribution that I was able to make to this subject, during the 1963–1966 period, was a classic example of being "in the right place at the right time." I was lucky. I think I could claim that, to some extent, I maneuvered myself into an area that struck me as being fertile ground for a possible breakthrough. There is little doubt, however, that the intellectual environment in the Department of Geodesy and Geophysics at Cambridge at the time was an ideal spawning ground for such an idea. In the 1950s, much of the early paleomagnetic work that provided support for the theory of continental drift was carried out there. Teddy Bullard had worked on the origin and nature of the earth's magnetic field. Bullard and Maurice Hill were working in marine geology and geophysics, and Drum Matthews was one of very few people who had measured the magnetic properties of basaltic rocks dredged from the ocean floor. I also had the advantage of having heard talks by Patrick Blackett and Harry Hess on continental drift and sea floor spreading. Basically, there was very little left for me to do.

CHAPTER 4

THE ZEBRA PATTERN

Lawrence W. Morley

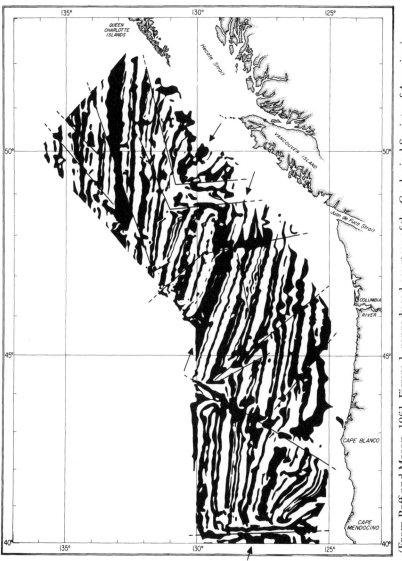

(From Raff and Mason, 1961, Figure 1, reproduced courtesy of the Geological Society of America.)

THIS MAP WAS CHOSEN AS THE CENTERPIECE OF MY ESSAY BE-
cause, in my opinion, it was the trigger that set off the escalation of investigations and ideas that culminated in the theory of plate tectonics. Professor R. G. Mason and his co-author, A. D. Raff, undertook several
voyages, plowing back and forth across the northeastern Pacific in a ship
towing a magnetometer that continually measured the intensity of the
earth's magnetic field.[1] These data were compiled and presented in the
figure above. I shall attempt to trace the train of events and discoveries
over the years from 1946 until 1963 that led to the theory of plate tectonics. I shall also try to show that it was this map and its later interpretation that provided the increased interest and research activity that led
to the legitimization of the theory of continental drift and plate tectonics. Of necessity, it will be limited to my personal experience in recalling
the facts as they were revealed to me through my own work and research
and through the available scientific literature. My direct involvement in
paleomagnetic research cuts off in 1963, when my consuming interest
switched to remote sensing. Anyone prominent in this field after 1963
should not feel slighted by my lack of reference to their work.

EARLY INTERDISCIPLINARY TRAINING

My interest in geophysics goes back to my undergraduate years at the University of Toronto, Canada. In 1938, I had enrolled in mathematics and
physics and was headed toward a career in actuarial science. During the
summer after my first year, I worked with a life insurance company calculating policy dividends on a mechanical calculator. The work was so bor-
ing that the thought of spending my career in that environment caused
me to choose the option in my second year which was honors physics and
geology. I soon found out that this course was intended as an interdisciplinary experiment, that there were only two of us enrolled, and that previously there had only been one graduate in the past six years, J. Tuzo Wilson. He later became my Ph.D. supervisor and still later he became one
of the group that originated the theory and the name *plate tectonics*.

I also soon learned about the huge gulf that existed, at the time,
between the geologist's mind and the physicist's mind. Physicists looked
down upon geology as being a descriptive and qualitative subject that did
not really qualify as a science, and geologists thought that physicists were
"egg heads" who did not live in the real world. Needless to say, the two
groups had difficulty in communicating professionally. The physics and
geology course was set up so that half of my lectures were in the Geology

Lawrence Morley, around 1963. (Photo courtesy of Lawrence Morley.)

Department and the other half in the Physics Department. It would be a few years before I could figure out how to fit the two subjects together, so different were their cultures. Throughout my career, this interdisciplinary training put me in the habit of reading the new literature from both disciplines, which I realized, in retrospect, enabled me to envisage the hypothesis that would explain the zebra pattern.

In my studies, I had become quite interested in geomagnetism, especially in its application to mineral prospecting. In 1940, in the middle of my third year, I left the university to join the British Navy as a radar officer at sea. In later years, this experience in electronics helped me devise instruments for my Ph.D. thesis. It also aroused my interest in oceanography. After World War II, I graduated in 1946 and took a job with a small geophysical prospecting company doing magnetometer surveys using the most sensitive magnetometer known at the time. It was an Askania magnetometer, invented in Germany and designed for prospecting for magnetic ore. It measured the changes in the intensity of the vertical component of the geomagnetic field, which deflected a magnetized needle balanced horizontally on a knife edge. This was high-tech at the time. It would take a whole day to collect a mile-long magnetic profile of data. This tedious work, combined with unbelievable clouds of mosquitoes in the Canadian Shield, led me to think that perhaps I had made a mistake in dropping actuarial science.

THE AIRBORNE MAGNETOMETER

I remembered from lectures that the Gulf Research and Development Corporation, a research subsidiary of the Gulf Oil Corporation, was working on a new invention – an airborne magnetometer that had been used experimentally during World War II to detect submerged submarines.[2] I dreamed of the possibility of being able to collect a line-mile of data with an airborne magnetometer in 20 seconds instead of taking a whole day on the ground. What a powerful prospecting tool this would be! It could be used to cover extremely large areas very quickly and cheaply. With these thoughts in mind in the fall of 1946, I emerged from the bush and headed for the Gulf Research and Development Corporation located in Harmarville, Pennsylvania, just outside Pittsburgh, to see if I could get a job connected with their airborne magnetometer. They were putting the final touches on the latest version that Victor Vacquier (now at the Scripps Institute of Oceanography) had adapted for use as an airborne instrument to be towed on an 80-foot cable behind an aircraft to avoid being magnetically contaminated by stray magnetic fields originating in the aircraft itself. They didn't hire me because "I didn't have a Ph.D." However, they introduced me to an executive of Fairchild Aerial Surveys, Los Angeles, named Max Phillips, with whom Gulf had just contracted to magnetically survey the major part of the Llanos areas east of the Andes in Venezuela and Colombia. To my surprise and delight I was instantly hired as the party chief for the two-year project (1947–1948). As it turned out, that instant determined my career for the next 20 years.

The Llanos area was totally virgin territory as far as knowledge of the geology was concerned. It was mostly covered with thick jungle and there were few open areas or rock outcrops to map the geology. Thus, the dream about the power of this instrument for conducting reconnaissance surveys over very large areas quickly and cheaply was realized a lot faster than I could have imagined. This was the world's first commercial aeromagnetic survey, and it effectively demonstrated the power of the instrument. As a result of this survey, sedimentary basins were discovered that today have become important oil-producing regions. (Most of the potential oil and mineral areas of the world, including the continental shelves, have now been covered by aeromagnetic surveys.) Ten years later, the same magnetometer was adapted by R. G. Mason and A. D. Raff for use as a shipborne instrument that could be towed on a long cable to avoid being affected by the ship's magnetization. It was this instrument that was used to cover the very large area in the northeast Pacific from which the zebra pattern map was produced.

I returned from South America to Toronto in September 1948 to work with Dominion Gulf Company, a Canadian subsidiary of the Gulf Oil Corporation. They wished to exploit the use of their new "toy" in the exploration for minerals as well as for oil. In the first flight over a portion of the Precambrian Shield in northern Ontario, we recorded a magnetic anomaly so intense that it ran the recorder off-scale. For the next ten years, there was a feeding frenzy of prospectors looking for magnetic ore deposits in the Canadian Shield. The end result was that enough iron ore was discovered to meet the demand for the next 50 years.

I joined the Geological Survey of Canada (GSC) in 1952, as their first geophysicist, to take charge of an aeromagnetic mapping project. In the GSC the magnetometer proved useful for assisting geological mapping in areas covered by lakes and swamps. In 1960, we initiated a contract project to cover the whole Canadian Shield, at half-mile spacing at a cost of about $30 million over 17 years. More than 5,000 aeromagnetic maps at a scale of one inch to the mile were published and made available to prospectors for 75 cents each. The benefits of this survey to the mining industry in Canada have never been calculated, but they must be more than several billion dollars and are still going strong.

POSTGRADUATE THESIS WORK

In 1949, I left Dominion Gulf to return to the University of Toronto for postgraduate training because of a remark made by one of Gulf's senior geologists: "geophysicists are great at going out and gathering and compiling data, but when it comes to interpreting the maps, they look like a bunch of monkeys trying to read the *New York Times*." This remark certainly described me, so in an effort to understand and interpret the geological causes of magnetic anomalies that were depicted by contour lines on aeromagnetic maps, I began the study of rock magnetism under Professor J. Tuzo Wilson, who had recently been appointed head of the Geophysics Department. Wilson was mostly interested in macro-geology: the theory of continental growth, mountain-building, and geosynclines. In studying rock magnetism, I was very much on my own at the time: Wilson was not a "drifter," a word used to describe "the lunatic fringe" who believed that continental drift might have been possible.

Understandably, I got no help from either the Physics or Geology Departments because none of the staff in either department had any idea what I was trying to do with such an esoteric subject. I also found the international literature very thin in this field. Previous investigators had begun to use the term *fossil magnetism* because they found that remanent magnetism appeared to be 'frozen' into the rocks in the direction of the earth's magnetic field prevailing at the time the rocks became consolidated.[3] They postulated that the process was analogous to the fossilization of sea life, so the term *fossil magnetism* began to be used. Later, when its potential began to be fully realized, it was given the more erudite Greek description *paleomagnetism*.

French investigators in the 1920s had measured the remanent magnetization of ancient earthenware pots found by archaeologists. The purpose was to determine their ages, or alternatively, if they knew the age of the pot, they could estimate the angle of declination between the geomagnetic and geographic north poles at the time the pot was fired. It had been shown experimentally that if an earthenware pot were fired in a kiln to a temperature above the Curie point for hematite (675°C) and subsequentially cooled to room temperature, it always acquired a remanent magnetization in a direction parallel to the earth's field.[4] As pots are always placed vertically on their bottoms when being fired, it means that it is possible by measuring the direction of the remanent vector, to calculate the magnetic orientation of the earth's field. The French scientists used this technique to try to track the secular movement of the earth's magnetic poles during history, a parameter of great interest in the days before satellite navigation, as marine navigators depended on this knowledge.

Before starting my thesis research, I was mostly interested in measuring the content of the mineral magnetite in rocks of the Canadian Shield, because it has long been known that it is chiefly the varying magnetite content in the rocks that causes the magnetic anomalies. To measure the magnetite content of a rock in the field, I used an adapted wartime mine detector. It measured magnetic susceptibility, which had been shown to be proportional to the magnetite content of a rock. I calibrated it by mixing known amounts of magnetite grains with non-magnetic sand. I was going to devote my thesis entirely to this work. However, shortly after I started my research in 1949, I came across John Graham's paper, which added "another string to my bow," namely, the measuring of magnetic remanence in rock samples.[5] He referred to a paper by R.H. Johnson that described a 'spinner' magnetometer adapted for measuring the direction of remanent magnetization in small rock cubes that had been cut from spatially oriented rock specimens broken away from solid bedrock.[6] Using this technique, Graham demonstrated that sedimentary rocks have the ability to hang onto their primeval remanent magnetizations. He did this by taking a number of closely spaced rock samples from a sedimentary layer that had been folded into an anticline. Even though the rock was folded, it managed to hang onto its original magnetization and was said to be magnetically stable. By studying the magnetism in rocks, it would be possible to unravel geological folds and faults in complicated structures. But what inspired me was this: Graham also wrote that it might be possible to prove whether or not the continents had drifted! Geologists and geophysicists in the 1920s thought they had buried Wegener's idea of continental drift, which he first published in 1912.[7] How can continents drift like ships through solid rock across oceans? they asked. Yet the circumstantial evidence was mounting. Graham was right in the end: *paleomagnetism* did give quantitative proof to continental drift, but not in the way he expected.

Off I went into the Grenville subprovince of the Canadian Shield in the summer of 1951 with my spinner magnetometer and the magnetic susceptibility meter. I labored for three months, spending most of my time sawing about 300 rock cubes with an inadequate diamond-studded circular blade. When I returned to the laboratory in Toronto, I remeasured all my samples, and to my dismay and disgust, most of the measurements had changed radically. It was so frustrating. I found that by merely banging them on the table, I could change the measured direction of magnetization by as much as 90°! I fooled around for a bit, magnetizing and demagnetizing my samples with a permanent magnet, but basically I considered the paleomagnetism part of my thesis a failure. There was, however, an interesting facet. The fine-grained volcanic rocks held their original fossil magnetism, whereas the coarse-grained batholithic rocks did not. Furthermore, the ratio of the magnetic remanent component to the component induced by the earth's field was higher in the volcanic rocks. This fact later became important in the interpretation of the zebra map.

The thing that interested me was that all the mathematical models, developed for interpreting magnetic anomalies, took into account only the induced component.[8] They ignored the remanent component, because there is no way to know the direction of the remanent component without extracting and measuring a rock sample. This is not very practical when you are interpreting an aeromagnetic map hundreds of miles square. The induced component is always in the same direction as the earth's field. I used to dream "wouldn't it be wonderful if we could pull the main switch on the earth's magnetic field?" We could then do our paleomagnetism without having to cut rock samples out of the outcrops. After receiving my Ph.D., I left this research in order to join the Geological Survey of Canada in June 1952 to manage their aeromagnetic program.

POLAR-WANDERING CURVES

In 1953, I was able to attend my first meeting of the American Geophysical Union. The keynote speaker was Dr. Keith Runcorn, a recent Ph.D. graduate from Cambridge, England. To my surprise, he was speaking on paleomagnetism. To my knowledge, he had not published the results of his research at the time, so I knew nothing about him or his work. He was a very dramatic and articulate speaker. He presented the paleomagnetic results from the Torridonian Old Red Sandstones in the United Kingdom. They were so spectacular that, from that time on, earth scientists sat up and took notice of paleomagnetism (the first time I had ever heard the term used). He related how the magnetizations of these sediments were unusually stable, although he did say one had to have a "green thumb" in selecting samples that were stable. Also, an idea new to me was that *he calculated the position of the geomagnetic North Pole* from the direction of the magnetization at the time these sediments were laid down or had become consolidated. This implied that if the same thing were done on other rocks of the same age from all over the world, the pole should be in the same position. He went one step further and stated that if many rocks of many ages from all over the world were measured, we would be able to plot the position of the pole for the whole of geological history and *thereby create a polar-wandering curve*. Because of continental drift, however, a polar-wandering curve for each continent would have to be constructed. Plotting the polar-wandering curves for each continent from that moment on for the next 12 years became the "holy grail" for most paleomagnetists.

A REVERSING EARTH'S FIELD?

Runcorn admitted that there was there was a slight problem with the results he had obtained. A number of his samples had a reverse polarity of exactly 180°. Without skipping a beat, he said that this was not a big problem. He simply counted them as though their polarities were not reversed – "The Earth's field was probably reversed at the time." Wait a minute: was he saying that the earth's magnetic field was actually flipping inside out throughout geological history, that the North Pole periodically became the South Pole and the South Pole became the North Pole almost instantaneously? The American researchers working in paleomagnetism were there: John Graham of the Carnegie Institute and the Jim Balsley–Arthur Buddington team from Princeton University. During the question period, they spoke up, strongly disagreeing that polarity reversals were happening. Their explanation was that certain rocks had a peculiar mixture of ferromagnetic minerals that somehow caused a so-called self-reversal during the time of their emplacement. The debate went on for about 15 years before it was settled. In the end, Runcorn was right. It is now known that the earth's field has been reversing throughout geological history.

My thesis supervisor, Tuzo Wilson, was sitting beside me and he urged me to get up and "say something" about my thesis results. I remarked at the meeting that whereas Runcorn was working on sedimentary rocks, I had been working on igneous rocks in the Precambrian Shield and that my rocks were quite unstable – that I could change their magnetic orientation by banging them on the table. This attracted some attention, because it placed some doubt on the reliability of paleomagnetic data. I was invited the next summer, in 1954, to attend a National Science Foundation conference at Idyllwyld, a mountain retreat outside Los Angeles. There were 25 attendees, about half of whom were young paleomagnetic investigators, the other half were senior scientists, representing a broader area of expertise, including famous physicists with expertise in ferromagnetism, geologists, and geophysicists. We were honored by the presence of the Nobel laureate, Linus Pauling. The conference lasted three days. Its main objective was to try to cast some light on whether or not the earth's field had been reversing during geological history. All the paleomagnetists presented their results and conclusions. Long discussions and disagreements ensued, but nothing was resolved.

In 1955, the year after the AGU meeting in Washington, I had a chance to visit John Graham in his laboratory at the Carnegie Institution, Department of Terrestrial Magnetism. He set me straight on the instability of my igneous rocks by explaining that all rocks had two kinds of magnetization, a soft component and a hard component. The hard component was the one we were after because it had consistent results from within the same geological formation. It is considered to be the true paleomagnetic magnetization. The soft component, which may have been caused by any number of things – from lightning strikes to glacial scraping – had to be eliminated before measuring. He had devised a method of doing this by subjecting the sample to an alternating demagnetizing field at the center of a Helmholtz coil over the period of a few minutes while gradually diminishing the field to zero.[9] After such "magnetic washing," the samples showed a consistency in their results. I don't know whether or not he was the first to do this, but from then on all investigators magnetically "washed" their samples before measuring. This technique meant that almost all rocks could be used paleomagnetically and that one did not have to have a "green thumb."

At the Geological Survey in Ottawa, I decided to start some research in paleomagnetism and devised another spinner magnetometer like the one I had made at the University of Toronto three years earlier. Philip Dubois, whom I met at the geomagnetic retreat in Idyllwyld, California, and who had completed his doctorate at Cambridge University under Keith Runcorn, joined me at Geological Survey of Canada in 1956 for a year or two to do paleomagnetic research. In 1957, he had published the first comparison of the polar-wandering curves for Europe and North America.[10] It showed a separation of the two curves equal to what one would expect from the presumed drifting apart of the two continents. This was the first study I had seen that tried to fulfill John Graham's 1949 dream of using rock magnetism to measure continental drift. Because of the uncertainty about the position of some of the ancient poles, however, a lot of skepticism about the results remained. In any event, loyal followers of the polar wandering school continued to gather such data from all over the world.[11] The more data gathered, the greater the confusion became about polar-wandering curves. The theory seemed to work for the younger rocks, but fell apart with the older rocks, whether sedimentary or igneous. The British directionalists spread their gospel to the United States, where they also seemed to ignore the reversal problem at first. This philosophy dominated the paleomagnetic literature until about the mid-1960s, when plate tectonics was born.

During the 1950s, however, a few paleomagnetists concentrated on the problem of reverse magnetization. They were intrigued by the suggestion that the earth's field may have been reversing periodically throughout geological history. In 1955, Tr. Einarsen and T. Sigurgeirsson examined the polarity of a large number of samples taken from a thick section of lava flows in Iceland, which is part of the mid-Atlantic ridge.[12] They showed that there were as many with reversed polarity as there were showing normal polarity. I had observed several reverse polarities in volcanic rocks in my own thesis work in the Canadian Shield. In 1958, Andre Larochelle joined the Geophysics Division of the GSC. I urged him to return to McGill University for his Ph.D. and to study the reverse polarity of the rocks in Yamaska Mountain, a volcanic plug southeast of Montreal. A negative anomaly had shown up in an aeromagnetic survey we had done over the area. I was pretty sure it was caused by inverse remanence, which Larochelle confirmed by his paleomagnetic sampling. In addition to this, we had observed many other negative anomalies occurring in the vast areas we had surveyed in the Canadian Shield. The criterion I used to identify an aeromagnetic anomaly that was caused by inverse remanence in the underlying rock was as follows: if the negative anomaly occurred to the north of a larger positive anomaly, the effect was not ascribed to inverse magnetism but to the northerly dip of the earth's magnetic field. If, however, there was a negative anomaly without the associated positive anomaly to the south, it was interpreted as negative polarization. Yamaska met this criterion, as did numerous other anomalies we had surveyed. Together with all the other evidence in the literature of reverse polarity, even in sedimentary rocks, this put me firmly on the side of those who advocated a reversing of the earth's magnetic field. The year was 1957.

THE WONDER OF THE MID-OCEAN RIDGES

The most important ingredient in the formulation of the sea floor spreading and the later plate tectonic theories was the knowledge of the existence of the mid-ocean ridge system. The presence of mid-ocean undersea mountains had long been known, but recognition of their full extent into a connected worldwide system had to await the invention and use of the echo sounder, or fathometer. Even when its existence as a unified system became known, few geologists had the opportunity to consider its implications because of its inaccessibility. It remained to oceanographers and geophysicists, with an interest in ocean basin geology, to explore and explain. They discovered that the floor of the ocean was covered largely by relatively recent volcanic rocks, as opposed to the continents, which have a mixture of very young and very old rocks. The seismologists supplied the information that both the mid-ocean ridges and the continental margins were seismologically very active. This information implied that the mid-ocean ridge system was a significant part of the earth's basic structure. It suggested that the earth is analogous to a cracked egg, intermittently leaking hot liquid out along the extent of the ridge system. It must have been these thoughts that led Hess and Dietz to the theory of sea floor spreading – a concept that is fundamental to plate tectonics.[13]

MASON AND RAFF

I do not know what led Ron Mason and Arthur Raff to undertake their magnetic survey of the northeastern Pacific in the 1950s. Because seismologists had plotted a large number of earthquake epicenters in the vicinity of the mid-ocean ridges as well as near the edges of the continental shelves, a lot of curiosity had been aroused about the largely unknown geology in the ocean basins. There had also been a lot of speculation about the origin of the continents: did they grow outward along their margins, as Tuzo Wilson was espousing, or did they split apart from one large supercontinent, by the process of continental drift?[14] Most of the scientific establishment had long since poured cold water on Wegener's theory. Now they began to wonder if perhaps he had been at least partly correct.

Another factor that must have led Mason and Raff to do this survey was that there was available to them an airborne magnetometer which they could adapt to tow behind a ship.[15] The airborne/shipborne magnetometer was perhaps the only instrument capable of revealing some basic reconnaissance information about the geology of the ocean basins over large areas. I was familiar with aeromagnetic data over the continents and, like Mason and Raff, we at the GSC were gathering magnetic data from a shipborne magnetometer over the continental shelves off Newfoundland and Nova Scotia. However, I never thought of operating over the ocean basins, as there was no economic incentive for the discovery of oil or minerals in the deep ocean basins.

Mason and Raff completed their work and published their results in the form of the zebra maps. They admitted they had no explanation for the cause of these linear anomalies. They tried correlating them unsuccessfully with gravity data and with sea-bottom topography. The data remained in the literature for about four years with no plausible explanation as to their cause. Then, along came Hess and Dietz.

OCEAN FLOOR SPREADING

The concept of ocean floor spreading was first envisaged in detail in 1960 by Professor Harry H. Hess, a renowned professor of geology at Princeton University.[16] Unfortunately, his paper was in the form of a report to the Office of Naval Research and, as it was supported by a contract, it did not appear in the open literature. The following year, in 1961, Robert S. Dietz published essentially the same hypothesis in the journal *Nature*.[17] In his paper, he clearly described the "spreading sea floor theory." In essence, he stated that the earth's mantle (the main solid part of the earth outside the liquid core) is comprised of very viscous rock of pitchlike consistency, moving in a number of convection cells that are in a state of constant, slow movement, fueled by heat generated by the decay of radioactive minerals. This movement forces the rock up from the deep part of the mantle and spews it out all along the mid-ocean ridges in the form of lava. At the ocean ridges, the material bifurcates, half moving out from the ridge one way and the other half moving the other way. As more lava arises, the solidified lava that was there before moves out, making room for the new material. This process continues so that the whole ocean floor behaves like a wide conveyor belt until it reaches the edge of the continental shelf. At this point, the heavier ocean floor sinks under the continental shelf, and continues down to the depths, eventually turning back toward the ridge from which it originated, thus completing the convection cell. Dietz suggested that there are a number of convection cells operating independently, all related in position to the mid-ocean ridges and continental shelves.

An ingenious and outlandish hypothesis, it nevertheless provided a mechanism for continental drift. Instead of the continents having to push through solid rock in their migrations, they ride around like so much scum on the top of a boiling porridge pot. The ocean basin rocks pushing up against continents also explained the process of mountain-building, which geologists had been debating for 100 years. And it explained why all the rocks in the ocean basin are comparatively young lavas. The really old rocks are all in the continents. Like continental drift itself, it was a nice dream, but how were we going to prove it? I was not aware of this concept until after I saw Raff and Mason's zebra pattern map. Unbeknown to me, Arthur Holmes had actually proposed such a theory in the 1920s.

THE EUREKA MOMENT

Raff and Mason's zebra map first appeared in August 1961. I literally freaked out when I saw it! I had been studying aeromagnetic maps from all over the world – both on the continents and on the continental shelves – and had never seen such a regular linear pattern of positive and negative anomalies stretching for 600 miles (1,000 kilometers) or more. All the continental maps with which I was familiar had anomalies in a sort of random bird's-eye maple pattern, nothing like these long, linear features, and Mason and Raff were unable to explain them. They might just as well have been maps of features on Mars because the geology was so unknown. Mason and Raff at first thought they might be caused by long ridges and troughs in the ocean bottom, but no correlation was found. They thought that there might be a series of long, dike-like bodies. This was checked by gravity surveys, but none was found. The data remained in the literature for a year and a half with no plausible explanation.

I had these maps on my mind for nearly two years before I spotted the Dietz paper on ocean floor spreading. Eureka! The idea came to me so suddenly that I sat down and submitted the following (unexpurgated) paper to *Nature* in February 1963.

> Several investigators and authors writing on the subject of continental drift and convection currents in the earth's mantle have referred to the puzzling linear magnetic anomalies in the Eastern Pacific Ocean Basin reported by scientists of the Scripps Institution of Oceanography.
>
> If one accepts, in principle, the concept of mantle convection currents rising under the ocean ridges, travelling horizontally under the ocean floor and sinking at ocean troughs, one cannot escape the argument that the upwelling rock under the ocean ridge, as it rises above the Curie Point geotherm, must become magnetized in the direction of the earth's field prevailing at the time. If this portion of the rock moves upward and then horizontally to make room for new upwelling material and if, in the meantime, the earth's field has reversed, and the same process continues, it stands to reason that a linear magnetic anomaly pattern of the type observed would result. This explanation has the advantage, over many others put forward, that it does not require a petrologically, structurally, thermally or strain-banded oceanic crust. It requires a convection cell whose axis of rotation is at least as long as the linear magnetic anomalies and whose horizontal distance-of-travel stretches from ocean rise to ocean trough. In addition to this it requires a large number of reversals of the earth's magnetic field from at least the Cretaceous period to the present (since no older rocks than the Cretaceous have been found in the ocean basins).

R.L. Wilson reported that Mrs. J. Cox, in a recent search of the palaeo-magnetic literature, was able to find 136 normally polarized cases and 141 reversely polarized from the Carboniferous to the present. Since there is no evidence to suggest that the earth's field should have been 'normally' polarized for any more periods or for longer periods than it has been reversely polarized, it is entirely possible that there may have been as many as 180 reversals since the Lower Cretaceous. This would be one reversal about every half million years on the average (a figure which T. Einarson(5) gives from his investigation of Icelandic lavas). He also suggests that the time taken for a reversal of the field is geologically very short – a few centuries to 10,000 years.

From an examination of the Scripps magnetic maps, the width of a complete positive and negative cycle, averaged over the widest part of the available surveyed section, is about 35 kilometers. To travel this distance in 1,000,000 years (time of two reversals), the convection current must have a rate of about 3.5 centimeters per year. This figure is only good to an order of magnitude, because no accurate data are available on the length of the periods of reversals. A better way to arrive at the rate of con-vection travel and the reversal period would be to measure the ages of rocks at widely spaced locations in the Pacific and to count the number of reversals occurring between these points.

Mason and Raff (1) report that some of the many guyots which were detected on the echo sounder produced magnetic anomalies, while oth-ers apparently had little or no effect. It seems unlikely that these guyots would be divided into two classes – those containing magnetite and those containing little or none. A more likely explanation would be that the ones which give little or no effect are negatively polarized to an intensity which nearly equalizes their magnetization induced by the pre-sent earth's field. If the 'non-magnetic guyots' always occur in the nega-tive anomaly bands, and the magnetic ones in the positive bands, this would be evidence that they cooled below the Curie Point at approxi-mately the same time as the rock surrounding them, because they were magnetized in the same direction. Indeed, since at that time they would have been in the shallow water of the ocean ridge, they would have protruded above the surface and have their tops flattened by ero-sion. As they proceeded along with the mantle convection current, they would pass into deeper water. This is an alternative explanation of ori-gin to that suggested by Darwin for the flat-topped guyots in the deep Pacific.

There are a few difficulties. The seismic results postulating 3 layers above the Moho must be incorporated into the theory. Mason and Raff (1) offer three models to marry the seismic and magnetic results:

(1) A 2 km-thick slab of intensely magnetized lava of K = .015 units underlain by a relatively non-magnetic crustal layer.[18]
(2) A topographical plateau 2 km high composed of material K = .015 underlain by a main crustal layer of the same magnetic susceptibility.
(3) A 6 km-thick slab extending from the bottom of the unconsolidated sediment to the Moho composed of 2 seismic layers, but all of the same magnetic susceptibility K = .005.

From measurements of several thousands of basaltic lavas from the Canadian Shield(6), none have been shown to possess a magnetic susceptibility of as great as .015 c.g.s. This would mitigate against accepting models (1) and (2). On the other hand, many lavas have a susceptibility as high as .005. This is not to imply, however, that we are postulating nothing but lavas down to the Moho. The other seismic layer in between must be explained. If this layer were unaltered ultramafic rock, it would not be sufficiently magnetic to cause the observed anomalies, nor would it have a significant seismic velocity contrast with the mantle material. Hess (7) has suggested that the main crustal layer beneath the oceanic basalt could be serpentinized ultramafic rock. This would satisfy both the magnetic and seismic requirements, since the serpentinization process both increases the magnetic susceptibility of the ultramafic rock and lowers the seismic velocity.

Thus Mason and Raff's (1) model number 3 is favored, with the modification that adjacent prisms would be magnetized oppositely. The prism producing the positive anomaly would be normally polarized with a total magnetization (remanent plus induced) equal to > .005 c.g.s. The prism producing the adjacent negative anomaly would be inversely polarized, so that the remanent magnetization would approximately cancel the induced."

The purpose of this letter is to point up the possibility of calibrating the frequency and duration of reversals of the earth's field in geological history from a study of the ocean basins, and the idea presented is considered to support the theory of convection in the earth's mantle.

REFERENCES

1. Mason, R.G. and Raff, A.D. 1961: Magnetic Survey of the West Coast of North America 32 degrees N Latitude to 42 degrees N Latitude; G.S.A. Bull. 72. 1259–70.
2. Vacquier,V. 1962: Magnetic Evidence for Horizontal Displacement in the Floor of the Pacific Ocean, Chapter 5, Continental Drift, Academic Press.
3. Wilson, J. Tuzo, 1963: Continental Drift, Scientific American, v 208, No. 4. pp. 83–100.

4. Wilson, R.L.1962: The Palaeomagnetism of Baked Contact Rocks and Reversal of the Earth's Magnetic Field; *Geophy. Jour. of the R.A.S.,* V.7 No. 2, 194–201.

5. Einarsson, T. 1957: Magneto-geological Mapping in Iceland with the Use of a Compass; *Advances in Physics,* V. 6.

6. MacLaren, A.S., personal communication.

7. Hess, H.H. 1946: *Amer. Jour. Sci.* V. 244, p. 772

THE PALEOMAGNETIC ESTABLISHMENT REJECTED THE IDEA

I never had any doubts about the concept. It locked three disparate and unproven theories together in a mutually supportive way: the theories of continental drift, sea floor spreading, and the periodic reversing of the geomagnetic field. It was like finding the key piece to an enormous jigsaw puzzle that made everything fit together. And it was based on actual quantitative data spread over thousands of square miles. To me, this was such a straightforward idea that I wanted to get it into print in a widely recognized journal as soon as possible, before someone else thought of it. I therefore put it in the form of a short paper that I submitted to *Nature* in February 1963. I received a rejection notice two months later stating that they did not have room to print it. I then immediately submitted it to the *Journal of Geophysical Research* in April but received no answer until late August. During this hiatus, I had assumed that my letter would shortly be published, but as I was impatient to get some feedback, I presented the idea at the June 1963 meeting of the Royal Society of Canada held in Quebec City. During the presentation, about 40 geologists and geophysicists were present. While I had not expected a standing ovation, I was somewhat surprised that not even one question was asked. In retrospect, I realize that there was probably no one there who had read anything about sea floor spreading or paleomagnetism and that I had not taken enough time to explain. I now know that this is a common fault among inexperienced authors.

In August 1963, I attended the San Francisco meeting of the International Union of Geodesy and Geophysics. Again, I still thought I was safe to talk about it before publication, so when I saw Runcorn and Hess engaged in conversation, I went up to them and briefly explained my idea. Runcorn was either bored or distracted, because he was obviously not listening, but Hess, who had been pushing his ocean floor spreading theory, was very interested and expressed the desire to meet with me again. Unfortunately, he died before we had the opportunity to do so.

Toward the end of August 1963, I received a rejection notice from the editor of the *Journal of Geophysical Research,* accompanied by a note from the reviewer with his name cut off from the letter. It stated: "Found your note with Morley's paper on my return from the field. His idea is an interesting one – I suppose – but it seems most appropriate over martinis, say, [rather] than in the *Journal of Geophysical Research.*"

I received this bad news at the end of August and was thinking of publishing elsewhere, but in the September 7, 1963 issue of *Nature,* the now famous article by Vine and Matthews entitled "Magnetic Anomalies over Oceanic Ridges" appeared.[19] It contained essentially the same idea that I had unsuccessfully attempted to publish twice. In the parlance of the time, I was "scooped." Obviously I could not publish elsewhere because I could have been accused of plagiarism.

I felt frustrated with the system. I knew that when a scientific paper is submitted to a journal, the editors choose reviewers who are experts on the topic being discussed. But the very expertise that makes them appropriate reviewers also generates a conflict of interest: they have a vested interest in the outcome of the debate. We could call this the "not invented here syndrome": scientists may be biased against good ideas emerging from someone else's lab. In retrospect, that is exactly what happened.

The hypothesis was not generally accepted until about 1965, the date commonly regarded as the birth of plate tectonics. At the time I took less interest in geophysics because I had received approval to set up a satellite remote sensing branch for the Canadian government, and soon left the Geological Survey of Canada to manage the Canada Centre for Remote Sensing.

In 1982, Professor Fred Vine and I had lunch together in London. He told me then that his paper also was not generally well received at first. However, by chance, both Professors Hess and Tuzo Wilson were taking a sabbatical at the same time at Cambridge University in 1965. The three of them got together. Wilson had envisaged a new kind of 'transform fault' that explained the structure of the mid-ocean ridges. I understood that it was at that time that the fully integrated theory of plate tectonics was put forward.

CONCLUSION

It seems to me that junior scientists are often cowed by the self-assurance of recognized authorities. In retrospect, what stands out for me is how Mason and Raff frankly admitted to having no plausible explanation for the "zebra stripes," which they had spent so much time and effort to acquire. If they had suggested that one of their interpretations was indeed correct, I would probably have accepted it, thinking that they had the situation in hand. Instead, their forthright approach created a space for a young scientist like myself to attempt an explanation of their data.

But if junior scientists were generally afraid to challenge accepted wisdom, what of the senior scientists? For two years, the "zebra stripe" data remained in the open literature, with no plausible explanation. Were senior scientists in turn cowed by what the recognized authority of their day, Sir Harold Jeffreys, had said 40 years earlier: that continental drift was physically impossible?

Plate tectonics ultimately required integration of evidence and insights from various fields. Many people missed the boat because they lacked the breadth to see the larger picture. Most geophysicists at the time, being grounded in physics, were not in the habit of reading geological literature, and probably would have missed the papers on sea floor spreading that countered Jeffreys' dictum about the physical impossibility of continental drift. Likewise, few geologists understood anything about the geomagnetic field, ferromagnetism, or rock magnetism.

Hence a word to humble earth science students: don't be cowed by the experts, and don't be too narrow in your reading habits. Science has only scratched the surface of the natural world. Opportunities for new important discoveries are limitless.

CHAPTER 5

ON BOARD THE *ELTANIN* -19

Walter Pitman

THE SHIP WAS THE *ELTANIN*, A MILITARY SEA TRANSPORTATION service freighter converted to do oceanographic work in Antarctic waters. At the time, early September 1965, we were headed due south through the far southwestern Pacific, toward the ice that fringes the Antarctic. We were there to work eastward along its edge studying the water column, the biota, and the geology beneath it all.

We had an ornithologist on board, who spent his time on the wings of the bridge, notebook in hand, peering through binoculars, scanning the sky for Antarctic birds. Suddenly he informed the captain that we were within a few miles of the ice. He had spotted some sort of bird that lived only along its fringes and in half an hour we were there, at the edge of a vast expanse of pancake-ice, a mosaic of flat slabs, large and small, covering the ocean surface as far as we could see. Most were several feet to tens of feet in diameter, riding several inches above the water and ground to a near roundness as they bumped and rubbed against each other. The spaces between the larger were filled with the smaller pieces of ice debris.

The ship had a reinforced hull and other modifications to make it seaworthy in the pancake ice. After some preparation we headed south again into this great expanse of brilliant white, pushing aside all the icy pieces large and small to see them gather together again astern of us. Then, abruptly, after an hour or two the ship came to a halt, all the deck lights out, the power shut down. We were dead in the water. A chunk of ice had been sucked into the water intake system used to cool the engines. In fact, this particular piece of ice had a fish frozen inside it. So the ship had to be shut down while the water intake was cleared. All rotating machinery was secured except for an emergency generator somewhere in the ship's bowels. We were adrift.

Aboard the *Elantin* research vessel. (Photo courtesy of Lamont-Doherty Earth Observatory)

Walter Pitman. (Photo courtesy of author)

It was near the time of the equinox, so although the sun was up for almost 12 hours each day, it barely skimmed along the horizon. A disk of deep red in a gray sky, it yielded little light and no warmth. But we didn't have even that cold warmth, for it was nighttime, and the seascape (if that is what it can be called), was well lit by reflection. It was a vast whiteness, rising and falling, almost imperceptibly, ever so slowly, ever so gently, with the rhythm of the ocean. There was only a gentle breeze, and because we were not too far from the open water, the air was warm, perhaps right about the freezing mark. The light was as at dusk; it seemed to come from everywhere. The silence was profound, intensified in our minds by the contrast with the weeks we had been on board, living with the constant hum and rumble and banging of machinery and equipment, people yelling and shouting, the sea rushing by the sides of the ship. Even when hove-to, a working ship like this one is a box of noise. Now, there seemed to be no sound at all, except the grinding of slabs of ice against each other and against the ship. The heavy layer of ice damped the waves, but they still rolled slowly, lifting the ice ever so gently a few inches and down again, perpetually groaning, as if the ice itself were alive.

The overall mission of the *Eltanin* was to do oceanographic work in the areas surrounding Antarctica, to gradually obtain data over a number of years to understand the geologic history of the region, the circulation of its waters, and its marine life. This particular leg of the *Eltanin* cruises, *Eltanin*-20, was conducted mostly for the purpose of obtaining physical oceanographic data in the Antarctic waters. Initially we were to make one geophysical traverse of the Eltanin fracture zone system, and later to make two crossings of the Pacific-Antarctic Ridge. I was considering a study of the Eltanin fracture zone as a possible doctoral thesis topic. Although this particular leg of the *Eltanin* was to make only one more or less incidental crossing of the Eltanin fracture zone, I was on board to "pay my dues," to be part of the data-collecting process. The satisfactory completion of this ritual would give me access to other *Eltanin* data.

When the engine intake was cleared, we headed back north a bit, out of the ice, and then eastward along 60 degrees south latitude, then to about 63 degrees south latitude, and then along 105 degrees west longitude, all the while taking a range of measurements of the physical and biological environment. We towed a magnetometer, and a single-channel seismic system that measured the sediment cover on the ocean floor. When we would heave-to, which was frequently, we gathered oceanographic and biological samples, and drove pipes into the bottom to obtain sediment cores.

Part of the cruise plan was to make two long geophysical traverses of the Pacific-Antarctic Ridge, obtaining magnetic, topographic, and seismic reflection data. This was a segment of a continuing study of the mid-ocean ridge system conducted by several scientists at Lamont Doherty Geological Observatory. For years, there had been puzzlement about the strong magnetic anomaly (irregularity among the measurements) that seemed always to be associated with the ridge axes, particularly in the North Atlantic. So wherever possible, the research ships over which Lamont had any control were oriented to make clean passes over the ridges. We headed north, from

60 degrees south and 105 degrees west, then northwest across the ridge axis, then east across it again, and then northeast to Valparaiso, Chile. Little did we know what treasure lay beneath us in these crossings, or the heat and smoke that would arise from the magnetic data we collected.

The crossings were quiet and uneventful, conducted as part of the ship's routine as we rolled and pitched along through perpetually rough seas beneath a gray sky. We crossed the ridge twice, and although we plotted the magnetic data by hand, nothing striking showed, mainly because of the scales we used, and because the ideas of Vine and Matthews had not yet really sunk in.

In 1965 I was a graduate student and only vaguely aware of the hypothesis of "sea floor spreading" that had been proposed by Harry Hess, at Princeton, and by Robert Dietz at the Naval Electronics Laboratory and the Scripps Institution of Oceanography. The suggestion of Vine and Matthews – that the spreading at the ridge axes combined with magnetic reversals would generate a pattern of magnetic stripes parallel and symmetric with respect to the ridge axis – in the end proved to be a brilliant insight. But the data they presented, taken from over the Carlsberg Ridge in the northwest Indian Ocean, were not very impressive with regard to correlatability, linearity, and symmetry. I was skeptical and dismissive, as were most others. I certainly had no expectation that there might be such a very definitive set of magnetic anomalies, all lined up, one after the other, in sequential "stripes" running parallel to the axis of the Pacific-Antarctic Ridge system.

To many scientists, continental drift was still a kind of mythology. During the 1950s, it had received a substantial boost because of the paleomagnetic work initiated by P. M. S. Blackett, and continued by Keith Runcorn and Ted Irving.[1] Although the paleomagnetic data seemed to strongly support the theory of continental drift, there was still much doubt. In the process of doing the paleomagnetic experiments, a number of instances were found in which the magnetic field seemed to have a reversed polarity. This observation contributed greatly to the ongoing skepticism. The idea that earth's magnetic field could reverse polarity seemed so fantastic that the meaning and significance of the paleomagnetic data were disputed by many. Still, the nagging question remained: how could continents move about through the oceanic crust without leaving a trace? "Sea floor spreading" provided a system of crustal motion that solved this problem.[2] In effect, the ocean was regarded as drifting *with* the continents. All the tectonic action was seen to take place at the very narrow zones of separation where new oceanic crust was formed, at narrow convergent zones where mountain-building occurred, and at very narrow boundaries where transcurrent (trans-form) motion took place. But to many, "sea floor spreading" was a fantasy, as was the corollary that there would be a consequent pattern of magnetic anomalies symmetric and parallel to the spreading ridge axis.

After the final ridge crossing, the ship headed for Valparaiso. It had been a long cruise, 60 days in the southerly waters. I would return home to my family and to a child who, when I left, was speaking in sentences, and two and a half months later, was talking in pages. Little did we know that in the belly of our vessel we carried a treasure trove. No one could have possibly guessed the bounty that the *Eltanin*-19 and -20 cruises would bestow.

The processing of these data was carried out as part of a routine that had been developed at Lamont, handling the data from three ships, each logging 30,000 to 40,000 nautical miles per year. In particular, all magnetic, gravity, and bathymetric data were processed and formatted in the same way; thus they could be retrieved and plotted in a number of different configurations that could be accessed over and over again.

I had read Arthur Holmes' proposal of a "purely hypothetical mechanism for 'engineering' continental drift" in his 1945 edition of *Principles of Physical Geology*.[3] The scheme envisioned a convective upwelling at a mid-ocean swell and a convergence beneath and at the edge of continents, transporting both the continents and the oceans. Some of the basic elements of sea floor spreading were already there. As already mentioned, "sea floor spreading" had been proposed in the early 1960s by Hess and Dietz; its corollary of transform faults had been posited by Toronto's J. Tuzo Wilson.[4] Fred Vine and Drummond Matthews, working at Cambridge, published a paper in 1963 which suggested that if crustal separation were taking place at the ridge axis and if the earth's magnetic field were reversing polarity, then there should be a pattern of magnetic anomalies parallel to and symmetric with respect to the ridge axis.[5] It was envisioned that the volcanic material magnetized by the earth's magnetic field as it was intruded at the ridge axis and transported laterally away by the sea floor spreading process would preserve like a tape recorder the polarity history of earth's magnetic field. This would give rise to a pattern of magnetic anomalies, appearing like stripes, which were symmetric to the spreading ridge axis and reflected the history of magnetic reversals.

It took us a month to get the magnetic data digitized. When we ran preliminary plots from *Eltanin*-19 and -20 (starting with *Eltanin*-19), there it was: this marvelous section from over the Pacific-Antarctic Ridge, a pattern so very symmetrical. In its central part, it was very well correlated to another magnetic profile constructed from the Juan de Fuca Ridge (off the Pacific northwest coast of the United States), the results of which had just been published by Vine and Wilson in the December 1965 issue of the journal *Science*.[6]

The development of the potassium/argon (K/Ar) dating technique allowed very precise age determination of volcanic rocks. This technology was seized by Allan Cox and colleagues to determine the age of reversal boundaries of suites of young volcanic rocks from a number of sites around the world. They were able to develop a magnetic reversal timescale, and to prove the occurrence of polarity reversals once and for all.[7] Fred Vine and Tuzo Wilson had tried to match the magnetic anomaly profile from the central axial area of the Juan de Fuca Ridge to the known history of magnetic polarity reversals.[8] But their experiment was not persuasive. One problem was that the Jaramillo event, a normal time interval that occurred just less than one million years ago, had not yet been discovered. It was missing from the known polarity timescale, and it seemed that the models did not fit the data. It was only the most central part of these axial profiles, the centermost 40 miles (100 kilometers), that seemed to correlate. Still, in retrospect, it is difficult to see why Vine and Wilson's paper did not attract more attention. Several anomalies did seem to correlate, especially when taken in conjunction with the (by then) famous Art Raff and Ronald Mason lineations described in their side-by-side scientific papers of 1961.[9]

The attitude toward continental drift seemed to be divided into two camps: believers and non-believers. And it seemed that with each new idea or discovery, and the growing compendium of paleomagnetic data, the sea floor spreading mechanism, together with Vine and Wilson's tentative proof, drew a reluctant few more into the camp of supporters.[10] A particular target of criticism was Vine and Matthews' idea that separation was taking place at the ridge axis along a very narrow line.[11] If the accretion of new material was symmetric about the line of separation, if the separation rate was steady for a long time interval, and if Earth's magnetic field reversed polarity, then there should be a pattern of magnetic anomalies parallel to the ridge axis and symmetric to it. Moreover, these anomalies should reflect the pattern of known reversals as found by Allan Cox and his colleagues.[12] If sea floor spreading occurred, the evidence would exhibit a degree of orderliness not previously known in major geologic processes.

A lot of "ifs." The astonishing fact was that the *Eltanin*-19 profile fulfilled all of the "ifs."[13] The very precision of the symmetry and the precipitous nature of the reversal boundaries, both qualities that were reflected in the anomaly profiles, put off some scientists. In a sense, it was too perfect. The central portion of the profile was compared to the known history of the reversals of the past 3.4 million years (except for the more recent Jaramillo event, which had not yet been found). The profiles actually could be modeled assuming an almost constant spreading rate, and the correspondence between the modeled anomalies and the data showed that the reversals were abrupt. At the fast-spreading ridges, the separation of plates occurred within a very narrow zone, and the accretion of new material was symmetric on either side of that zone with respect to the separating plates.

The presence of symmetric and correlatable anomalies beyond 3.3 million years clearly implied a history of reversals beyond the known paleomagnetic studies made on lava flows.[14] It would be a simple matter to extrapolate to the end of the profile, defining polarity events out to 10 million years ago. Lava flows are notoriously discrete events, and although they may give a continuous appearance for thousands of years, they are much more episodic: occurring here, and then ceasing, and then occurring there. In contrast, the patterns of mid-ocean ridge flows were stacked side by side in an arrangement so that the layers were systematically older the farther they were from the ridge axis. Furthermore, some events apparent in the magnetic anomaly profile had not been detected yet by paleomagnetic techniques in use, simply because the resolution was not precise enough. One of the problems that the lava flow K/Ar dating method would encounter was that the error in the age determinations was often larger than the individual polarity reversal events themselves.

The *Eltanin*-19 profile stirred a profound reaction among the Lamont scientists. Some responded negatively, but most responded positively. Neil Opdyke had come to Lamont the year before, to set up a Paleomagnetics Department, mostly for the study of tectonic movements and continental drift. One of his graduate students, John Foster, had developed a slow spinner magnetometer for the specific purpose of making paleomagnetic measurements on sediment cores.[15] This was the right opportunity at the right time and place. The Lamont core collection in 1967 exceeded 10,000 samples from all over the world. It would be a matter of finding one that contained clean, continuous characteristics back to some 2 million years ago. Cores that were mostly composed of sediments would yield a better record of Earth's polarity than cores composed of lavas. We located several such cores from the Lamont collection, and to augment these samples, the Lamont research ships were deployed to find and retrieve new, longer sediment samples. The cores soon yielded the sequences of known magnetic reversals, together with the very distinct Jaramillo event. In addition, polarity events beyond the range of the 3.3 million year mark could be correlated with the Eltanin-19 profile – nearly, but not quite to its end.[16] (By this time in early 1966, Dick Doell and Brent Dalrymple had found the Jaramillo event.) These data, particularly from the cores with a high sedimentary deposition rate, demonstrated clearly that the reversals were quite precipitous, that changes in polarity took place in less than several thousands of years.[17]

Following Tuzo Wilson's predictions of undersea transform faults off-setting the mid-ocean ridge, Lamont's Lynn Sykes began his first earth-quake studies that would show compelling correlations of earthquake activity along the ridge axes.[18] In fact, it was realized quickly that the plate boundaries were the locus of most of the world's earthquakes: that normal faulting earthquakes took place at separating boundaries along the ridge axes. Strike-slip earthquakes occurred at transform faults, and a more complex combination of thrust mechanisms occurred at sub-duction boundaries. Lamont's Bryan Isacks, Jack Oliver, and Lynn Sykes called this systematic arrangement of plates and their motions "The New Global Tectonics."[19]

The Eltanin-19 profile spanned only the most central portion of the ridge system, out to 10 million years before present. As the magnetic data from the ridge flanks were examined, it became apparent that a pattern of correlated anomalies extended well out over the flanks of the ridge system, right to the edges of the continents. These magnetic anomalies were symmetric, aligned along both sides of the ridge axes, and consti-tuted a unique and readily identifiable pattern.

Fred Vine had used the Raff and Mason lineations to propose an exten-sion of the magnetic polarity events from the 10 million years encom-passed by the Eltanin-19 profile continuously back to the Upper Creta-ceous era, some 80 million years ago.[20] In addition, Vine correlated one of the oldest fragments of the Raff and Mason lineations to patches else-where in the North and South Pacific Oceans. By December 1966, Jim Heirtzler, Geof Dickson, Ellen Herron, Xavier Le Pichon, and myself together were able to correlate the entire set of Raff and Mason lineations from the North Pacific with equivalent lineations from the South Pacific Ocean, the Southeast Indian Ocean, and the South Atlantic Ocean.[21] From this more extensive data set, a timescale was derived, which differed in its Upper Cretaceous to Lower Tertiary section from Vine's. All of this, and Jason Morgan's paper on tectonics on a sphere, essentially set forth a whole earth treatment of plate tectonics, and defined the geometry of the plate motions in a rigorous mathematical framework.[22] Together, these concomitant ideas were presented at the spring American Geophysical Union meeting in 1967. It was a watershed event.

For me it was a time of great excitement but also wonderment. I had started out my thesis studying time variations of the earth's magnetic field, in particular diurnal variations and micropulsations, and switched over to marine geology and geophysics only at the beginning of 1965. So the problems were all new to me. We had been fed some notion of the idea of continental drift sometime back in grammar school geography class ("see how South America and Africa seem to fit together like pieces in a puzzle?"), but it was not until 1965 that I began to be aware of the immense controversy this idea had provoked. In effect I was learning about the problem as I was helping to solve it. It would take some time for the full magnitude of what had happened to sink in.

The *Eltanin*-19 profile was first recognized in December 1965. By December 1966 there was a far more extensive magnetic anomaly pattern reaching from the mid-ocean ridge axes down the flanks, from the present back to 80 million years before present. The pattern had been identified in half the Pacific Ocean, most of the southeast Indian Ocean, and half the South Atlantic Ocean. With this accomplishment came the confidence that the same or even older patterns would be found in the rest of the oceanic areas. Thus within those 12 months ideas that had seemed to many no more than dubious, and at best controversial, proved irrevocably the hypothesis of Alfred Wegener. The insights and speculations provided by Harry Hess, Bob Dietz, Tuzo Wilson, Fred Vine, and Drum Matthews were transformed into marvelous reality. Not only did sea floor spreading explain the origin of the oceans and provide the mechanism for continental drift and mountain-building, but the precision of the process was astonishing. Plate motions could be described with geometric rigor. It had been a very good year.

CHAPTER 6

THE BIRTH OF PLATE TECTONICS

Neil D. Opdyke

THE FACT THAT THE EARTH'S MAGNETIC FIELD HAS CHANGED magnetic polarity repeatedly through geologic time is one of the most important discoveries of paleomagnetism, the study of the history of the earth's magnetic field. Early work by M. Matuyama (1929) and P. L. Mercanton (1936) had pointed to possible polarity changes in the earth's magnetic field.[1] The story that I want to describe is the demonstration that the earth's magnetic field did reverse polarity, a demonstration achieved by paleomagnetic studies in Europe, North America, and Australia during the 1950s and 1960s.

In the 1950s there were small groups of paleomagnetists, scientists who study the earth's magnetic field, working in different countries. In France an active paleomagnetic group led by E. Thellier was working at the University of Paris. Thellier made many important contributions to paleomagnetism, in particular in studies of paleointensity, the change in strength of the magnetic field with time, and paleosecular variation, the change in direction of the earth's magnetic field. A. Roche, one of his students, was actively researching reverse and normal rocks in Central France, at Pay du Dome.[2] Thellier, however, was very skeptical of reversals of the earth's magnetic field. Later, when reversed directions were found in marine sediments, he asked me to provide him with samples. This I did, but he never contacted me concerning his results so I expect he obtained the same results as I did.

In England, several paleomagnetic groups were formed in the early 1950s. Two of these groups originated from research by P. M. S. Blackett on the earth's magnetic field. The first of these groups was established at Imperial College, London, under the direction of Dr. John Clegg. The other was at Cambridge under the direction of S. K. Runcorn. A third

a

b

c

Paleomagnetists in the 1950s. a) Jan Hospers, collecting lava samples for magnetic measurements on Mt. Hekla, Iceland, in the early 1950s. b) Keith Runcorn, collecting red sandstones for magnetic analysis in Bryce Canyon, Utah, summer 1957. c) Ted Irving, in Canberra, Australia, at the Australian National University, 1956. Photos a) and b) courtesy of Neil Opdyke, c) courtesy of Ted Irving.

group studying paleomagnetism was earlier established at London University under the direction of J. Bruckshaw. In Japan, active paleomagnetic studies were under way at Tokyo University under the direction of T. Nagata, who was a student of M. Matuyama. The Japanese group was to provide important information concerning self-reversing minerals in rocks.[3] In the United States, an active group of researchers was studying the paleomagnetism of rocks under the direction of John Graham and Merle Tuve at the Carnegie Institution of Washington. Work at the Carnegie Institution had begun before World War II and was reestablished there in the late 1940s. Another research group was established at UC Berkeley under the direction of John Verhoogan, a widely respected earth scientist. Another paleomagnetic group was established before 1955 at the Bernard Price Institute in the Republic of South Africa under the direction of Anton Hales and his student, Ian Gough.

One of the most important and influential studies to be published in the early 1950s was that of J. Hospers, a Dutch student who was studying at Cambridge University. Hospers initiated paleomagnetic studies on the long sequence of lavas exposed in Iceland, resulting in important inferences relating to polar wandering and field reversal.[4] Hospers interacted with S. K. Runcorn, whose paleomagnetic laboratory was just beginning to be assembled, and Hosper's collection of samples from Iceland were measured not at Cambridge but at London University in Bruckshaw's laboratory. The results of this study were published in 1951–1953/54 with the title "Reversals of the Main Geomagnetic Field."

The studies of these lava sequences showed clearly that changes of polarity were not random in time, but that groups of normal and reverse lavas followed each other in stratigraphic order and appeared to be time-dependent. In all cases recent lavas possessed normal polarity. Another piece of evidence for reversal of the magnetic field was the fact that where lavas flowed over soil, baking it, the direction of magnetization of the baked zone agreed with that of the lava flow. The third piece of evidence that supported field reversal was that 180° reversals of polarity had been observed in sediments, initially in pre-Cambrian Torridonian sediments from Scotland studied by E. Irving.[5] This observation was duplicated quickly in sediments of other ages and areas by John Graham in the United States and Ken Creer in the United Kingdom. The fact that 180° changes in polarity were observed in lavas, where the magnetism is a thermal magnetization, as well as in sediments that acquired their magnetization during sedimentation or diagenesis, would seem to confirm the hypothesis that changes of polarity of the geomagnetic field were taking place.

A complicating piece of evidence, however, became available from Japan in 1951, where Professor T. Nagata found a rock, the "Haruna dacite," which was self-reversing.[6] When a sample of this rock was heated and cooled in a field of known direction, it would become magnetized in a direction opposite that of the applied field. A complete study of self-reversal in this rock was carried out by Seiya Uyeda in his doctorial thesis.[7] L. Neél, a French theoretical solid state physicist, had suggested several ways in which self-reversal could take place in igneous rocks. The Japanese scientists, being intellectual descendants of Matuyama, were inclined to favor reversal of the main field and were embarrassed to find this self-reversing rock.[8] In North America, A. F. Buddington and Jim Balsley argued for self-reversal for the origin of reversed magnetization observed in metamorphic rocks from the Adirondacks. Buddington was a very influential mineralogist from Princeton University and his opinion carried considerable weight in North America. Balsley was in charge of the aeromagnetic survey of the United States – the study of the variations of the intensity of the earth's magnetic field by airborne magnetometers – and wanted to understand negative magnetic anomalies associated with reversely magnetized igneous rocks.[9]

In the mid-1950s the paleomagnetic community divided into two camps. In North America, John Graham was very skeptical of field reversal, as was much of the geophysical community. They did not favor either reversal of the geomagnetic field or continental drift. The two hypotheses were related. After being rejected in the 1920s and 1930s, continental drift was now being revisited on the basis of paleomagnetic evidence of apparent polar wandering. Since both this newfound support for continental drift and the argument for field reversals rested on paleomagnetic observations and were often argued by the same people, they were lumped together as radical and unreliable. In France, Thellier was skeptical of magnetic field reversals and remained so until the mid-1960s. The other point of view was favored by groups in England, especially the Cambridge group led by S. K. (Keith) Runcorn.

In 1955, I became a field assistant for Runcorn working in Arizona, essentially by accident. Keith Runcorn was in New York in June 1955 preparing to go to Arizona to collect orientated rock samples. A student from Princeton was slated to be his assistant, but he withdrew at the last minute. This turn of events set Keith looking for a replacement. Runcorn was a keen swimmer and went to the Columbia University pool. It so happened that a friend of mine, Max Pirner, was in charge of the pool. Runcorn asked him if he knew of any geology students who were big enough to carry rock samples out of the Grand Canyon, preferably a football player. Max and I had played football together and it was certain that I was the only student that met all the qualifications. I had graduated that spring and was home in Frenchtown, New Jersey, working for my uncle as a carpenter. Runcorn called on a Saturday morning, reaching my grandparents by mistake. I happened to be there and spoke with him. He asked if I would be willing to go with him on Monday morning to Arizona to be a field assistant. At first I thought that it was a hoax. After a while it became clear that it was a real offer. I said that I would have to talk it over with my uncle and the rest of the family. We all agreed that I should take the opportunity. I called Keith and told him that I would go with him, and gave him instructions how to get to Frenchtown. I told him to ask anyone where I lived and they would know how to find me. (The population of the town was only 1,150 at the time.) On Monday morning I was sitting in front of my grandparents' home on the only route into town from New York. At 11 o'clock a car pulled up and this red-haired guy in a T-shirt rolled down the window and said, "I say, do you know a chap called Neil Opdyke?" I said, "It is I." We drove to my parents' home and had lunch. Then we started for the west. Keith drove over a stump that was just in front of the car and spent the first 10 miles of the trip telling me all about paleomagnetism. Unfortunately, he had neglected to shift into third gear. I sat there thinking: Shall I tell him or let him find out for himself? Finally I could not stand it; I had to tell him that he was driving in second gear. It was the start of my scientific career.

The fieldwork went well that summer and as a result Runcorn invited me to go to Cambridge as his student. I was to work on paleoclimate as it related to paleomagnetism. Subsequently, I went with Runcorn to the University of Newcastle, where I received my Ph.D. in 1958. It turned out that the first paper I was to author with Runcorn was on the reversal sequence of the lava flows around Flagstaff, Arizona. That article appeared in the journal *Science* in 1956, entitled "New Evidence for Reversal of Geomagnetic Field near the Plio-Pleistocene Boundary."[10] Following this foray into field reversals, I spent the next few years researching crustal mobility, holding postdoctoral fellowships in Australia with Ted Irving, and in Africa (Zimbabwe) with Ian Gough and Mike McElhinny.

During the late 1950s the situation regarding the understanding of reversals of the field remained more or less what it had been in 1955. However, new players were entering the field and new information became available. In North America Allan Cox and Richard Doell had received their doctorates from Berkeley under the supervision of John Verhoogen. They wrote a very influential review of paleomagnetism that appeared in the *Geological Society of America Bulletin* in 1960, in which they concluded correctly that the paleomagnetic data indicated that both reversals of the geomagnetic field and mineralogical reversals occurred. They pointed out that all late Pleistocene (roughly the last 800,000 years) and recent lavas had normal magnetization, whereas the youngest reversed rocks were all early Pleistocene or late Pliocene in age (1 million to 2.5 million years). One year later, in 1961, they met and persuaded Brent Dalrymple to join them in an attempt to use the radiometric decay of potassium-40 to argon-40 (K/Ar) to date the age of this reversal, and demonstrate that reversals were time-dependent.

The other important observation of the late 1950s was the magnetic anomaly pattern observed in marine magnetometer surveys off the coast of California, by Ron Mason and Arthur Raff in 1961.[11] The linearity and continuity of these magnetic anomalies were remarkable, and the origin was unknown, since they did not correlate with topographic features. They were, however, offset by the large fracture zones such as the Mendocino fracture zone, which at that time was assumed to be a transcurrent fault.[12]

In Russia, paleomagnetic research rapidly led to the use of magnetic stratigraphy for correlation of sedimentary sections in Pliocene and Pleistocene rocks of western Turkmenia by A. N. Khramov in 1958.[13] This work was largely unknown in the West until translations of this work began to appear in the 1960s.

In 1960, I traveled to Canberra, Australia, to work with Ted Irving for one year as a Fulbright Scholar. Irving and John Jaeger at the Research School of Earth Sciences at the Australian National University (ANU) were preparing to do K/Ar dating of young lavas to determine if reversal of the geomagnetic field was time-dependent. Ian MacDougall, who was to do the radiometric dating, had just finished his Ph.D. and went off to California to learn K/Ar dating techniques, while Don Tarling, a student of Irving, began to collect samples from Plio-Pleistocene lavas in the South Pacific to study secular variation of the earth's magnetic field. Attempts were made to set up the K/Ar line at the ANU that year, and Jack Evernden came out to Australia to help. These developments set the stage for the future competition between the ANU (Ted Irving, Don Tarling, and Ian MacDougall) and the U.S. Geological Survey (Allan Cox, Dick Doell, and Brent Dalrymple) to produce the first geomagnetic timescale.

I traveled from Australia to Africa and spent three years at the University of Rhodesia and Nyasaland researching continental drift, which was absorbing and challenging. Meanwhile, important developments were happening in North America and Europe. In 1962 the concept of sea floor spreading was launched by Harry Hess, supported by Bob Dietz.[14] In 1963 the corollary to the sea floor spreading hypothesis was published by Fred Vine and Drummond Matthews. This stated that if sea floor spreading was taking place, then one might expect that linear magnetic anomalies would form parallel to the ridge crest and be symmetrical on either side of it.[15] This hypothesis might explain the lineated magnetic anomalies observed in the Pacific as well as other areas of the world ocean. Larry Morley, a Canadian geophysicist, had put this idea forward independently prior to Vine and Matthews, but his paper was rejected by reviewers in North America who did not believe that reversals of the earth's magnetic field had taken place.[16] In fact, I knew nothing about this paper at the time and did not meet Morley until years later. The Vine and Matthews paper, on the other hand, had been published in the British journal *Nature,* where it received a favorable review and had wide distribution.

The race to establish a radiometrically dated magnetic polarity timescale ended in a tie in 1963, when papers were published both by Cox, Doell, and Dalrymple, and by McDougall and Tarling.[17] These important papers and the others that followed throughout the 1960s supported the hypothesis of reversal of the geomagnetic field, because the reversals appears to be time-dependent. Nevertheless, geophysicists in North America remained hostile to the idea. If reversal of the field were accepted, then continental drift, as supported by paleomagnetism, might seem more credible. It was easier to discount the whole discipline than to accept findings that might lead to taking the whole subject of drift seriously.

I returned to the United States from Africa in December 1963 and began working at Lamont Geological Observatory of Columbia University. At the time, the Lamont Observatory and the Department of Geology at Columbia University were citadels of belief in permanent continents and oceans. Since paleomagnetists were the high priests of continental drift, it might be a little difficult to understand why the observatory had hired me. I later found out that the reason was very simple. Maurice (Doc) Ewing, the observatory's founder and one of the most prominent geophysicists of the 20th century, had been the prime mover in the observatory's acquisition of large numbers of piston cores from the world's oceans. Ever alert, he had decided that it might be possible to correlate these cores using reversal of the earth's magnetic field. He was always interested in the paleoclimatology of the Pleistocene, and I believe that he badly wanted to correlate marine microfossils with climatic proxies such as ice rafted detritus, and, if possible, to date them. I was hired to do just that. I had previously become acquainted with paleomagnetic research on marine sediments carried out at Scripps Institution of Oceanography by Chris Harrison under the direction of John Belshé, whom I had known at Cambridge. Belshé had presented a lecture in 1960 at the ANU, which I had attended. In that lecture, he presented data from short gravity cores that showed a reversal of the field.[18] However, the data were not clear-cut and I knew that Chris had encountered many difficulties. I was therefore not terribly enthusiastic about working on cores, but I knew it could be done.

The Magnetic Department in which I was to do my work was under the direction of Jim Heirtzler, who had, in fact, hired me. Jim was heavily involved in research on small changes in the earth's magnetic field called *micropulsations,* and most of the students in the department were researching this field. Jim was also in charge of the acquisition and curating of marine magnetic data. This data set was very extensive and widespread geographically, due to the way the Lamont research ships were routed. However, it was not regarded as a very important data set, since everybody supposed that the observed magnetic anomalies only reflected topography, which at the time was the only known way of producing magnetic anomalies.

Shortly after my arrival at Lamont, a meeting was called to consider if continental drift was possible. This was done at the instigation of Bill Donn, an atmospheric scientist, who with Doc Ewing had put forward a theory to explain the Ice Ages. Donn had attempted to apply his hypothesis to the great Permo-Carboniferous glaciation (300 million years ago) on the Gondwana continent but failed because of their dispersal. I was not scheduled to speak at this meeting, but was there as an observer and resource person. During the course of the discussions, paleomagnetic data were mentioned. Joe Worzel, the associate director of the observatory, made a statement belittling paleomagnetic research and results, implying that paleomagnetists were dishonest and their results could not be trusted. I was incensed and demanded an apology. I believe that he had entirely forgotten that Lamont had employed a paleomagnetist, and that I was actually there. Doc Ewing tactfully calmed everyone down. I left the meeting with the feeling that perhaps I would not be around too long. I was mistaken.

During my first year at Lamont, most of my time was engaged in setting up the paleomagnetic laboratory, which became operational in March 1965. I also did fieldwork during this time, and when the laboratory came online, there was much to do. In my first year at Lamont, Hans

Wensink, a Dutch scientist, was a postdoctoral researcher working with me and he helped set up the laboratory.[19] We subsequently carried out a study of the White Mountains igneous rocks of Vermont and New Hampshire. As part of his Ph.D. thesis in Holland, Hans had done a study of the magnetic stratigraphy of Pliocene and Pleistocene lavas of Iceland, building on the work begun by Hospers. He gave a scientific report of these results at the Lamont seminar series. The talk was quite well received and Manik Talwani and Xavier Le Pichon waylaid Hans on his way back to the office and asked many questions. However, nothing seemed to develop from this conversation, at least in the short term.

I also acquired my first graduate student that year, John Foster, who was an excellent instrument designer and very helpful in setting up the laboratory. He developed a slow-speed spinner magnetometer of high sensitivity that used fluxgates to measure the direction of magnetization in rock specimens. This instrument and its derivatives was widely used in paleomagnetic research until the advent of the cryogenic magnetometer in the late 1970s.[20] John was given the job of researching the paleomagnetism of marine cores; research on cores began slowly in the summer of 1965 on some piston cores that had been taken by Walter Broecker on the 20th cruise of the research vessel *Vema*, to the Central Pacific Ocean.

Another interesting study had been under way in the Magnetic Department. Scientists from the Naval Research Laboratory in Washington, D.C., were collaborating with Heirtzler on the study of magnetic anomalies along the Reykjanes Ridge south of Iceland.[21] An airborne survey had been taken during a time of quite severe magnetic storms, and these effects were being subtracted from the data set. The U.S. Navy was interested in the magnetic field around Iceland because seaborne magnetometers were used in antisubmarine warfare. One day, Jim Heirtzler called me in to show me the results. Lo and behold: a beautiful set of linear magnetic anomalies had emerged from the noise. These magnetic anomalies were parallel to the ridge axis as required by the Vine–Matthews hypothesis. I commented that I guessed that Vine and Matthews were correct. Although he was impressed by the anomalies, Jim was as yet unwilling to accept the reversal hypothesis for their origin.

Another incident took place during these early days at Lamont. I was asked to be a reader on a paper by Xavier Le Pichon on the magnetic anomalies associated with the mid-Atlantic ridge, on which he did his thesis. In the paper, Xavier had maintained that the basalts of the ocean floor did not have a high enough magnetic intensity to give rise to magnetic anomalies caused by field reversals. I was aware of a *Nature* paper by Jim Ade-Hall that indicated otherwise.[22] Xavier did not alter the manuscript, but the discussion caused me to begin research on the magnetic properties of marine igneous rocks in concert with another French scientist, Roger Hekenian. It turned out that they were indeed strongly magnetic and were resistant to change. It is interesting to note that sea floor basalts were an important element in the extensive dredge collection at Lamont, yet had hardly been looked at until this time.

One of the most important events to affect the marine program in 1965 was that Lamont's Walter Pitman changed his research from micropulsation studies (small variations of intensity) of the earth's field to marine magnetics. In the autumn of 1965, he went to sea on the research vessel *Eltanin*-20 across the East Pacific Rise. On his return, he began to process the magnetic data from *Eltanin*-19 and -20 cruises. Much of the computer work that was done at Lamont was often done at night; one morning I came in to work to see the *Eltanun*-19 profile pinned to my office door. This profile is beautifully bilaterally symmetrical and the entire timescale as known at that time was displayed by the magnetic anomalies. The reversal pattern was unknown beyond 3.5 million years, but displayed on this profile were magnetic anomalies well into the Miocene (up to 10 million years old). The profile also displayed a magnetic anomaly that occurred just below the Brunhes–Matuyama boundary, the last known reversal of the earth's magnetic field. This came to be known as the Jaramillo event, a period of normal polarity about 50,000 years in length, and the profile would become known as "Pitman's magic profile" because of its beautiful symmetry.[23]

At about the time Walter was at sea, important developments were taking place in the paleomagnetic study of marine sediments at Lamont. John Foster, who had been given the task of studying reversals in marine sediments, was talking about his research with other students, in particular with Billy Glass, a student of Bruce Heezen. During the course of their discussions, Billy suggested to Foster that perhaps it would be a good idea to study higher-latitude cores, in particular those around the Antarctic that had been studied by James Hays for his Ph.D. thesis. I told them to go ahead and give them a try, which they did. Sure enough, they were quite highly magnetic and changes in polarity were soon observed. Progress was very rapid, and by early 1966 research on reversal stratigraphy of cores spread to sediments of the North Pacific on the suggestion of D. Ninkovitch, a colleague of Heezen, who was interested in correlating volcanic ash layers in cores south of the Aleutian islands. By February 1966, we had magnetic stratigraphy from fossil-bearing marine sediments to the base of the Gauss normal period 2.6 million years ago, and we had identified a new normal event older than the Brunhes-Matuyama boundary, which we called the *Emperor* event, after the Emperor seamount chain. This event was also observed on the profiles from the East Pacific Rise (*Eltanin*-19 profile). By March 1966, we had all of this information at our disposal, but it was known only to the members of the Magnetics Department, and to Bruce Heezen, who was very enthusiastic about the application of magnetic stratigraphy to marine geology.

In March 1966, Fred Vine visited Lamont to see the *Eltanin* data. I showed him the new results from the cores, and the first thing he said was, "I see you have found the Jaramillo event." I replied, "What's that?" It turned out that the event at the end of the Matuyama (about one million years ago), which I had called the Emperor event, had been found by Doell and Dalrymple and was already in press. They had called it the Jaramillo event after a location in New Mexico. There went the Emperor event right out the window. During this visit, Heirtzler arranged to give the *Eltanin* data to Fred Vine, with the stipulation that Walter Pitman and he would publish first. Vine was elated by the *Eltanin* data: it convinced him beyond a shadow of a doubt that the Vine–Matthews hypothesis was correct.

Soon after this, Walter suggested that I show Doc Ewing the results that we were getting from the sedimentary cores. I made the appointment and with trepidation went to see the great man. I laid the stratigraphy out on the long sheets where we had recorded the data. He was enthusiastic and impressed with the results. For some reason he did not make the connection among sea floor spreading, reversals of the earth's magnetic field, and continental drift, which he still adamantly opposed. At the time, it was clear that the same polarity changes were observed in three different recording systems, (1) terrestrial lavas dated by radioactive decay of potassium to argon, (2) sea floor magnetic anomalies, and (3) marine sediments, which were fully time-dependent. It was clear that reversal of the earth's magnetic field was a fact. However, the details were only known by a small group, mainly at Lamont.

The spring meeting of the American Geophysical Union was rapidly approaching. I had put in an abstract to talk about the White Mountains results and Walter and Jim Heirtzler had an abstract in on the *Eltanin*. There was a lot of discussion in the department about how far they should go in endorsing sea floor spreading. I pointed out that if you present the *Eltanin*-19 profile and say nothing about sea floor spreading, you would look very foolish. So Walter was given permission by Jim to take the giant step forward. I decided that I would cancel the paper on the

White Mountain series and instead give the paper on reversals in cores. I remember going around AGU telling everyone that the Frontiers of Geophysics lecture could be missed, but you had to hear Pitman's talk. It caused a stir, as did the data I presented, but the news was really spread by word of mouth.

This meeting saw the last gasp of mineralogical self-reversal as a serious contender to explain reversed directions in rocks. Rod Wilson at Imperial College believed that he could correlate differences in oxidation state of magnetic minerals with magnetic polarity in lavas. The presentation of the reversal data from the cores essentially ended the argument, since it was clear that reversal of the earth's field was fully time-dependent. After this meeting I don't believe the idea that the earth's geomagnetic field reversed polarity was ever seriously threatened again. At the next meeting of the AGU, over 40 papers on crustal mobility and associated topics were presented to standing room-only audiences.

Following the AGU meeting I flew to Tallahassee, Florida, where the *Eltanin* cores were stored. Jim Hays had previously identified several cores that he believed were late Miocene in age (circa 6 million years old) based on the microfauna; I sampled the cores at 4 inch (10 centimeter) intervals. Norman Watkins had just arrived to set up a paleomagnetics laboratory. In a year we would soon become keen competitors but we were always good friends. I returned to the observatory in the summer of 1966. The direction of magnetization in these cores had been measured, so I worked on the core that went back farthest in time, and as I worked the magnetic pattern was revealed to the base of the known stratigraphy, about 3.6 million years. We had the *Eltanin*-19 profile as a template so we knew what pattern should emerge. When I got to this point in the core I informed Pitman that the pattern that he had observed should soon be revealed – that is, of course, if we were correct in all our assumptions. Walter came to the laboratory with me, shaped the samples, and handed them to me to be measured: sure enough the pattern emerged as expected, just like magic. We were both ecstatic. Walter and I were now secure in our predictions that the reversal history of the field was accurately recorded in both the sediments and the sea floor basalts.

The activity in the Magnetics Department was soon brought to the attention of members of the Seismology Department, who began rapidly and aggressively to pursue the implications of crustal mobility for seismology. On the other hand, Doc Ewing and the Marine Geology group held firm against continental drift. This ferment resulted in a meeting held at the Goddard Institute for Space Studies in November 1966. The meeting was organized by Paul Gast and Robert Jastrow and was by invitation only. The power structure of North American earth science was there. I was invited only because Allan Cox could not make it. The data that had been accumulating at Lamont were presented and every speaker was aggressively interrogated. The idea was to hold a debate on whether crustal mobility was a real possibility or not. At the end, Sir Edward Bullard was to sum up in favor of crustal mobility and Gordon J. MacDonald against. The open hostility toward crustal mobility gradually receded as the meeting wore on. At the end, MacDonald failed to give a summary, essentially yielding to the mobilists. The summary in favor of crustal mobility was provided elegantly by Bullard. It is my opinion that this meeting set the stage for the coming plate tectonic revolution.[24]

It was an exciting time for all earth science. The next few years were to see a complete change in the way that we viewed the earth. One might ask, are there any lessons to be learned from my experiences? I would say that there probably are. From my perspective I feel that if you fail to take advantage of opportunities offered, then your chances of success are diminished. I could have turned Runcorn down. The second point that is clear to me is that if a scientist wants to make an impact, he or she must seek to solve important problems and master the necessary tools that provide the opportunity to solve them.

The choices that a scientist makes in research are all important. Some are scientific big game hunters; others would rather not take a chance and find that shooting fish in a barrel is more fun. Persistence pays off; if you think you are right, keep going. It is not bad to be a little aggressive.

PART III
HEAT FLOW AND SEISMOLOGY

The confirmation of sea floor spreading led to a rush – some might say a stampede – to put together the pieces of a global story. There were several important lines of evidence, and at first it was not entirely clear if they would fit together. If sea floor spreading at the mid-ocean ridges was caused by convection currents rising from deep within very hot regions in the earth, then heat flow should be highest over these ridges, but scientists found some heat flow values at the ridges that were extremely low. This didn't seem to fit the big picture. Nor did the fact that heat flow over the continents was the same, on average, as over the ocean floors. For some scientists, these were reasons to remain unconvinced. But while heat flow measurements caused a certain amount of confusion, seismic data proved compelling.

Advances in seismology were crucial to illuminating the big picture. For some time, seismologists had been mapping the distribution of global earthquakes and attempting to determine the nature of the motions associated with them. But their data were often sparse, inaccurate, or confusing. The development of the world wide standard seismograph network (WWSSN) to aid in detecting nuclear weapons tests came at just the right time to solve the problems of plate tectonics: accurate locations of earthquakes displayed a fabulous pattern outlining the earth's crustal blocks, and accurate determination of earthquakes' slip directions proved that these blocks were moving in just the ways that global tectonics required.

CHAPTER 7

HOW MOBILE IS THE EARTH?

Gordon J. MacDonald

IN THE 1950S, POLAR WANDERING AND CONTINENTAL DRIFT WERE controversial subjects, often leading to heated discussions between North American and European geophysicists and geologists. Many Europeans favored mobility while most Americans believed in stable continents and oceans. I started serious work on these topics in 1957, when Walter Munk and I began the research and writing for our book, *The Rotation of the Earth*.[1] At the time, Walter was one of the leading oceanographers of the 20th century and a longtime professor at the Scripps Institution of Oceanography; I was starting my academic career as a professor of geology at the Massachusetts Institute of Technology (MIT).

In planning the book, Walter and I had a mild debate on whether or not to include discussions of continental drift and polar wandering. I argued we should, so as to tweak the geologists into considering limitations (imposed by geophysics) on their wilder speculations. George Darwin's analysis of polar wandering provided a start.[2] Furthermore, Walter was always ready for a good argument. The final chapter of *The Rotation of the Earth* takes up the subject of the earth's mobility, as we understood it in 1960.

When I was an undergraduate at Harvard in the late 1940s, my professors ignored or dismissed (with ridicule) speculation that continents move relative to each other, the poles tip, and convection currents constantly stir the interior of the earth. However, I was very much impressed in 1949 by reading Reginald Daly's book, *Our Mobile Earth*.[3] By the time I reached Harvard, Daly had retired, but from time to time he appeared on campus and occasionally engaged in brief conversations with students. Daly's book provided a comprehensive overview of the evidence favoring drift, and, like Alfred Wegener (a meteorologist and early proponent), Daly advocated drift in the 1920s and early 1930s. By the 1940s,

Gordon MacDonald at the Scripps Institution of Oceanography on the occasion of the 65th birthday of geophysicist Walter Munk, October 1982. (Photo by Roy Porello, courtesy of the Scripps Institution of Oceanography, used by permission.)

Daly continued to advance his view that a plausible mechanism for mountain-building involved blocks of crust moving as a result of gravitational forces, crushing sediments deposited along the coast.[4]

Whatever sympathy I had for Daly's notion of continental drift was overwhelmed by the work of two giants of 20th century geophysics: Cambridge professor Sir Harold Jeffreys and Harvard professor Francis Birch. In fact, Walter Munk and I dedicated our book to Jeffreys. Walter, in his many visits to Cambridge, England, had come to know Sir Harold and Lady Jeffreys quite well; my personal interactions with Sir Harold were very much more limited. In the early 1950s, he presented a course on fluid dynamics at Harvard. Despite the broad title of the course, its subject matter was basically the mechanics of boiling porridge. While Sir Harold's writings were admirably clear and concise, these talents did not carry over to his lectures. Typically, he would write down a number of equations on the blackboard, wander over to an open window, and discuss the peculiarities of the mathematical properties of the equations that describe the motion of fluids. Apparently, Sir Harold felt that a flock of Cambridge pigeons would appreciate his insight far more than his human audience of four or five scruffy Harvard graduate students.

Sir Harold's book, *The Earth*, defined the subject of the physics of the earth's interior.[5] In my view, the early editions, particularly the first (published in 1924), were the most influential. The main objective was to compare theory with observations in a quantitative way. How well theory fit observation was a major theme in much of Sir Harold's research; it led him to write his most important work, *The Theory of Probability*, published in 1939.[6] In recent years, this masterpiece, demonstrating the wide range of applications of the work of 18th century mathematician Thomas Bayes, has been rediscovered and has spawned literally hundreds of books that discuss Bayes' approach to probability.

Not surprisingly, geologists in the pre–World War II period found that *The Earth*, the book, was much more mathematical than the real thing. Jeffreys' answer was simple: the results sought were quantitative and there was no way of obtaining quantitative results without mathematics. The later editions, beginning with the fourth, published in 1959, suffered as the vast amount of new data relevant to the earth's interior rolled in as a result of great advances in instrumentation. One person, even one as brilliant as Jeffreys, could not have easily assimilated such a wealth of new information.

I found Jeffreys' reasoning about the strength of the earth, based on gravity observations and the existence of mountain ranges and ocean trenches, to be convincing. In essence, he argued that the earth was too strong to permit significant crustal mobility. It was rigid, not flexible. I carried this view into the 1960s discussion of the earth's rotation and into my later investigations with respect to continental drift and polar wandering.

I was fortunate enough to know Francis Birch well. In his laboratory, I began to understand the art of research on high pressure. Birch received a bachelor's degree in electrical engineering from Harvard in 1924. He returned to Harvard in 1928 as a graduate student in physics, working chiefly on high-pressure experiments in the laboratory of Percy

W. Bridgman, a 1946 Nobel Laureate for his work on high-pressure physics. In the early 1930s, Bridgman and Reginald Daly, an almost daily visitor to Bridgman's laboratory, combined to set up a committee on experimental geology and geophysics that would be devoted to revitalizing a program of seismology, the study of the earth's interior using observations on the propagation of sound and other waves, and to developing a program for comprehensive high-pressure studies to shed light on conditions deep within the earth. Birch was invited to lead the high-pressure research program, and research on the physical and chemical properties of the earth's interior became his lifelong task.

Probably the most significant among Birch's many contributions appeared in a now classic paper published in 1952, entitled "Elasticity and the Constitution of the Earth's Interior."[7] In this paper, Birch demonstrated that (1) the mantle is predominantly composed of silicate minerals; (2) the upper and lower mantle regions are each essentially homogeneous but with somewhat different compositions, separated by a thin transition zone associated with a silicate phase transition; and (3) the inner and outer cores are composed of crystalline and molten iron respectively. While these ideas had been earlier proposed by others, including Jeffreys and Danish seismologist Inge Lehmann, alternative interpretations were still in play at the time. For example, some geophysicists thought the core might be composed of liquid or solid hydrogen. By 1965, the available evidence confirmed the essentials of the Birch model.

Birch felt that his demonstration of the homogeneity of the mantle in both the upper and lower regions ruled out large-scale convective motions. His position was that, given the homogeneity of the two regions of the mantle, there would be no driving force for large-scale convection (the motion of a fluid driven by density differences resulting from temperature differences; according to this theory, hotter, less dense material would move upward against gravity while cooler materials would sink). Birch argued that convection was unlikely to break through the transition in density between the lower and the upper mantle. I agreed.

Elastic materials have what physicists call a "finite" strength, which means that upon the application of a stress (a force over an area) they will deform a certain amount in proportion to that stress. But no matter how long the stress is applied the deformation is limited. The material does not deform infinitely; it has a finite (non-zero) strength. In contrast, fluids continue to flow and deform as long as a stress is being applied. Therefore, they have no finite strength; their strength is zero.

The viscosity of a fluid is a measure of its rate of deformation: more viscous fluids flow more slowly than less viscous fluids under a given stress. Most of us know this from common experience: honey is more viscous than water, and so takes longer to flow out of a jar. The question with respect to the earth was this: is it a viscous fluid, like honey, with no finite strength, which will continue to flow under stress indefinitely, or is it an elastic material of finite strength, resisting deformation after a certain point? Based on my readings of Jeffreys and my close interaction with Birch, I concluded that the earth indeed possessed a finite strength. This had to be taken into account in any discussion of the earth's interior, and it seemed to argue against crustal mobility – even if geologists considered mobility essential to the explanation of many of their observations.

In the 1960s I found the contradiction between the geophysical evidence for a strong earth and the mobility required by some geologic observations deeply troubling. My work as a young geologist in New England and the Alps convinced me that during periods in which mountains were formed layers of rock behaved plastically, with the layers wrapped up in giant folds, and breaks or faults in the rocks permitting large displacements of masses of rocks – a view shared by most field geologists. I could explain these deformation patterns in terms of the properties of rocks at high temperatures and pressure with the rocks having a high water content: rocks would deform like viscous fluids under such conditions. But these local and regional observations could not be extrapolated to the motion of whole continents, because continents had physiographic features such as the Rio Grande Valley that have persisted untouched by large-scale deformations over hundreds of millions of years.

In the 1950s and 1960s, the structure of the sea floor had not been mapped in the detail that the continents had, but it was nonetheless clear that volcanic activity, with massive flows of liquid lava, had taken place in the ocean basin. Given the abundance of volcanic activity in the basins, I considered the possibility that the sea floor in the interior of the ocean basin was mobile compared to stable continents, which maintained their relative geographical position over time. Perhaps the sea floor moved without displacing the continents?

In the early 1960s, new observations and interpretations of the sea floor data led to the theory of plate tectonics, which permanently altered the discussions of the mobility of the earth. Harry Hess, Bob Dietz, Lawrence Morley, Frederick Vine, and Drummond Matthews contributed to the invention of the theory of plate tectonics in various ways.[8] Plate tectonic theory postulates that the surface of the earth can be described by breaking it up into plates with a rigid upper layer, the lithosphere, comprised of crust and upper mantle. The asthenosphere, supposed to behave as a viscous fluid, underlies the lithosphere. According to this theory, low-intensity, long-term stresses drive the horizontal motion of the plates. Newly defined plates replace continents as moving objects near the surface of the earth. In most versions of the theory, mantle convection provides the motive force for plate displacement.

THE DEEP STRUCTURE OF THE CONTINENTS

In the 1960s, following completion of *The Rotation of the Earth,* I took on the issue of whether geophysical knowledge of the earth's interior was consistent with large-scale displacement of continents. This research also involved the related question of whether the concept of fixed continents was consistent with geophysical evidence. The most pertinent observations at the time were those that dealt with the subcrustal structure of continents and ocean basins. I argued in two papers that the large-scale differences between continents and oceans were not restricted to the upper few tens of miles, as plate theory would have it, but extended to several hundred miles' depth.[9] One cannot, of course, directly observe that continents extend to great depths; it is a conclusion I believed was demanded by the analysis of: (1) heat flow and gravity observations taken over the surface of the earth; (2) seismological data that included newly observed free oscillations; and (3) new and very important information secured from observation of the behavior of earth-orbiting satellites.

Heat flow is a function of both the earth's internal composition – primarily the concentration of heat-generating radioactive elements – and the processes such as convection and conduction that transfer heat from the earth's interior to its exterior. Gravity as measured at or near the surface depends upon the underlying mass of rocks or fluids. If the rocks underlying a particular observing station have a higher density than those under nearby stations, the pull of gravity will be higher there. Free oscillations are the vibrations produced by a disturbance within the earth, such as an earthquake or underground explosion. The time it takes for elastic waves to travel from a disturbance to the recording device (i.e., a seismometer) depends upon the properties of the rocks lying along the path from the disturbance to the instrument. If the disturbance is large enough, then the whole earth rings like a bell; the periods of oscillation are much longer than those the ear can detect (minutes to hours for Earth versus milliseconds for the bell), but they can be detected by seismometers. Heat flow, gravity, and free oscillations can be used as indirect measures of the earth's internal composition and structure.

Today, satellites permit determination of these properties with great accuracy. For example, if the earth were a perfect sphere having uniform density and lacking an atmosphere, and if there were no moon, sun, or planets, an orbiting satellite's path would be a fixed circle about the center of the earth. Obviously, these conditions do not hold. The equatorial radius of the earth is greater than the polar radius, and the equatorial bulge perturbs the orbit of earth satellites. Other perturbations arise from irregularities in the distribution of mass within the earth. Precise measurement of the satellites' path about the earth therefore gives valuable information about the distribution of mass under continents and oceans. Similarly, measurements of the strength of gravity made by instruments at the surface are strongly influenced by local irregularities in the distribution of mass. Observations of satellite orbits provide data on the large-scale irregularities of mass and are much less sensitive to the local variations than observations made on the surface. Both surface and satellite gravity observations indicate that the total mass of material per unit area in the earth's upper layer is equal. This condition is called *isostasy*, a term coined in the late 19th century by American geologist Clarence Dutton meaning "equal standing."[10] Gravity measurements show that the rock mass contained in a high mountain is compensated by a deficiency of mass in rocks at depth under the mountain – and the net effect is almost as if there were no mountain at all.

In the 1960s, these properties could only be measured at the surface, but the basic result was the same as today: gravity measurements pointed to mass differences below the continents and oceans that compensated for the observable topography. If this difference between continents and oceans extends to great depths, then a theory of continental drift must be very different than that first imagined by Alfred Wegener. In Wegener's view, the motion of continents required a weak oceanic crust extending downward into the mantle; continents were stiff plates pushing through that weak crust. In my view, this was inconsistent with the evidence for deep differences between the oceans and the continents.

My arguments can be simply summarized:

1. Isostasy generally prevails, but deviations from isostasy show that the mantle possesses a finite strength, as argued by Harold Jeffreys. That is, the mantle is not completely weak like a fluid.
2. The surface rocks of continents are less dense than those of the ocean floor. The deeper rocks of continents are denser than the near-surface rocks, a necessary condition to obtain observed approximate isostatic balance.
3. The surface heat flows of continents and ocean basins are approximately equal, but the content of radioactive heat-generating elements is higher in continental rocks than in rocks underlying the oceans. The approximate equality of heat flow in continents and ocean basins requires that higher concentration of radioactive elements in near-surface continental rocks be compensated for by lower concentrations at greater depths.[11] Therefore the high concentration of such elements in the continental crust must be compensated by lower radioactive concentration in rocks deep under continents.
4. Earthquakes are clustered along ocean-continent borders and earthquake foci can be as deep as 450 miles (720 kilometers).
5. The agreement between calculated and observed oscillations caused by earthquakes indicates a homogeneous lower mantle.

These arguments are elaborated below.

GRAVITY

Studies of the earth's gravity field have long shown that despite the difference in densities between the uppermost layers of rocks under oceans and under continents, the amount of mass per unit area under continents very nearly equals that under the ocean, a condition know as *isostasy*.[12] In the early 1960s, knowledge of the earth's gravity field increased significantly. Observations of the orbits of satellites coming close to the earth's surface established that regional variations of gravity exist over large horizontal distances. As Walter Munk and I noted at the time, these cannot be accounted for in terms of the near-surface crustal structure.[13]

The analysis of the earth's gravitational field proceeds by expanding the external gravitational potential in terms of a series of *spherical harmonics*, a form of mathematical analysis used to describe a pattern on the surface of a sphere. The analysis works by breaking the large pattern down into a sequence of simple patterns in a hierarchy of increasing complexity. "Low-order" spherical harmonics describe large-scale variations; for example, order one represents a function that is positive in one hemisphere and negative in the other. A "zonal" spherical harmonic does not vary with latitude; a "sectoral" harmonic does not vary with longitude; "tesseral" harmonics have neither of these symmetries.

If the earth were fluid, devoid of strength, certain zonal harmonics would vanish, and the earth's rotation and internal distribution of density would determine the remaining harmonics. The tesseral harmonic and the sectoral harmonic would also vanish for a fluid earth. But this does not happen. A comparison between the observed values and the values appropriate for a fluid earth, rotating at the same velocity as the earth and having the same radial density distribution as estimated by seismologists in 1963, provides a measure of the stresses supporting the deviations from a fluid earth. The harmonics indicate an earth that is not a fluid, but has finite strength.

A comparison between the observed gravity field and the field expected from an isostatically compensated crust shows two striking features. For the low-order harmonics (those representing large-scale variation), the gravitational anomaly expected from the continental crust is small compared with the observed potential and is of opposite sign. The small magnitude of the crustal contribution indicates that the mass anomaly that gives rise to the observed values must come from somewhere else, and the only plausible alternative is the mantle. The failure of the continental crust to account for the observed values of gravity requires density inhomogeneities that extend well below the oceanic and continental upper layers. The correlation of the large-scale anomalies in the external potential with the continent-ocean structure requires a different density distribution for the rocks at depth under oceans and under continents. In short, the mantle under the continents appears to be different from the mantle under the oceans.

HEAT FLOW

In the 1950s, Roger Revelle and Arthur Maxwell at the Scripps Institution of Oceanography, studying the Pacific, and Teddy Bullard at Cambridge University, studying the Atlantic, found the surprising result that the heat flow through the surface of the ocean floor is virtually the same as the heat flow through the surface of the continents.[14] The surprise arose from the known difference in radioactive element composition of the rocks under the ocean and under the continents. The continents are composed of granitic rocks and sediments derived from them, which contain high concentrations of heat-producing radioactive elements. In contrast, the ocean floor is composed of basalt, which contains relatively low concentrations of radioactive elements. Therefore, it had been assumed that heat flow through the continents would be much greater than heat flow through the oceans. By 1963, approximately 1,000 heat flow measurements had been made, a significant majority of them in oceanic areas, and there was no significant differences in the heat flow between the oceans and the continents.

Based on these data, I concluded that major differences in chemical composition must exist between the subcontinental and suboceanic portions of the mantle.[15] Together with the gravity evidence, the heat flow data suggested that the continental structure extends to considerable depths within the mantle, and that the continental crust cannot be imagined as a thin block overlying a homogeneous mantle.

SEISMIC EVIDENCE

In the mid-1960s, studies of the propagation of body and surface seismic waves were insufficient to yield a detailed picture of the differences between mantle materials underlying continents and those underlying oceans. The classic distributions of elastic wave velocity in the mantle as determined by Jeffreys and by Caltech seismologists Beno Gutenberg and Charles Richter deviate principally in the upper 300 miles (500 kilometers) of the mantle; at greater depths, the two distributions are very similar.[16] Evidence that the mantle at greater depths greater is homogeneous on a large scale came from the radial velocity distribution shown by the studies of free oscillations of the earth – the ringing of the earth as a whole. At the time, I was one of several researchers who obtained close agreement between observed and calculated oscillations.[17] This finding clearly indicated that, on average, the deep mantle can be described by velocity distributions that are functions of depth alone. The low-order oscillations are not at all sensitive to variations in mantle properties at shallow depths.

It had long been known that the principal earthquake zones of the earth are at the boundaries between continents and oceans. Moreover, at a depth of about 400 miles (650 kilometers) the frequency of earthquakes decreases rapidly, and no earthquake foci are found below approximately 450 miles (720 kilometers).[18] The association of earthquake foci with continental borders is particularly marked for those earthquakes whose foci are at depths greater than 180 miles (300 kilometers). The interpretation of the association of earthquake zones with continental borders and the limits of earthquakes to the upper reaches of the earth depends on the mechanisms by which earthquakes are generated. Earthquakes release stray energy developed through processes that elastically deform the earth. Differences in heat production and heat loss lead to the build up of thermal stresses. In one paper I argued that the difference of heat source distribution between continents and oceans is primarily responsible for the concentration of earthquake zones along continent-ocean boundaries.[19]

The limitation of earthquake foci to the upper 450 miles of the earth can be interpreted in terms of either a change in mechanical properties of materials at this depth, or to the vanishing of the effects of differential thermal losses. Thus, if the continents were formed by the upward concentration of mantle material and if this process had affected the upper 300 miles (500 kilometers) of the mantle, then the thermal stresses associated with continental ocean boundaries would be limited to the upper few hundred miles of the earth. Below this depth, radioactive sources would be horizontally homogeneous, thermal stresses due to differential thermal losses would not develop, and there would be no earthquakes.

FINITE STRENGTH

The central issue in the dynamics of polar wandering and continental drift is the problem of the earth's strength. As discussed above, if the earth can be treated as a highly viscous fluid, then it has no finite strength and a very small stress can turn it around, on a timescale that depends on the earth's viscosity. As Walter Munk and I discussed in our book, a fluid earth without finite strength is unstable, because an infinitesimal stress could produce large changes in the position of the axis of rotation with respect to surface features.[20] Since this does not happen, the earth must have some measurable (finite) strength.

By the early 1960s there was abundant evidence from laboratory experiments that a wide variety of silicate materials resist both fracture and plastic flow up to some definite stress difference. Once this stress is exceeded, the materials fail either by fracturing or by undergoing a large plastic deformation. But laboratory experiments are of short duration and may not be applicable on geological timescales. As Jeffreys emphasized, evidence for the finite strength of materials of the earth is found in the existence of continents, oceans, mountains, and ocean trenches. Major gravity anomalies are associated with Paleozoic mountain chains, so that the stress differences resulting from these distributions of mass must have persisted for long periods of time. Jeffreys concluded that gravity anomalies indicate a strength on the order of 150 to 300 bars in the upper 375 miles (600 kilometers), the range of values depending on whether the material below this level has any strength.[21] Furthermore, the fact that the present axis of rotation is not moving toward the principal pole of the continent or ocean system can be interpreted in terms of finite strength. The minimum strength required to prevent polar wandering is on the order of 10 bars. Enthusiasts for convection argue that the surface features are maintained by frictional drag resulting from the flow of deep convective currents. They emphasize the kinematics (geometry) of the flow but avoid the dynamics.

In the 1950s, evidence against finite strength came from three major sources:

1. The uplift of areas formerly covered by glaciers in Fennoscandia and North America was thought to be evidence for finite viscosity and zero strength. Theories of mantle convection adopt the viscosity obtained from the timescale of uplift of formerly glaciated regions. Jeffreys criticized this interpretation inasmuch as other regions of the earth with similar or larger negative gravity anomalies were not rising.[22]

2. The figure of the earth closely approximates that of an equivalent rotating fluid. This is not necessarily an argument against finite strength; rather, it imposes an upper limit. Satellite data indicate that the eccentricity is greater than the equilibrium value and that a strength on the order of 100 bars is required to maintain this deviation from a fluid earth. The evidence from the pole position and the eccentricity is not inconsistent. The former indicates a minimum strength on the order of 10 bars and the latter a maximum strength of 100 bars. Moreover, these values are not inconsistent with those obtained from gravity anomalies or from observations of higher-order terms in the earth's external gravitational potential.

3. Geological observations require a mobile earth and thus the earth's interior must be in constant motion to support uplifted mountain chains and deep ocean trenches by frictional drag.

MY CONCLUSIONS ABOUT CONTINENTAL STRUCTURE

Heat flow, gravity, and seismic observations available in the early 1960s strongly suggested that continental structure extends to depths of about 300 miles (500 kilometers). Such a deep structure posed a difficulty for any quantitative theory of continental drift or polar wandering. Convection was the only proposed mechanism quantitatively adequate to move continents. (Convection as discussed here would be required at a large scale, and should not be confused with penetrative convection, where regions undergo melting and buoyant lower-density liquids move upward through fractures in the overlying material. There can be no doubt that this latter type of motion occurs, as is evident in the giant flows of basalt in the Columbia River basin and in India.)

But if the continental structure extends to a depth of hundreds of miles, it is not possible to imagine thin continental blocks carried along by a flowing mantle. The large horizontal motions extending near the surface would tend to homogenize the upper mantle and destroy density differences between the continent and oceanic structures. A further difficulty for mantle-wide convection was posed by observations on the mechanical properties of the mantle. If indeed the earth possesses a finite strength, then the convective forces would have to overcome this finite strength.

DEBATES

The early 1960s saw numerous meetings and conferences at which participants reviewed recent developments in plate tectonic theory and implications of the theory for many outstanding issues in geology. I participated in a number of these meetings, arguing that a deep structure for continents presented severe difficulties, perhaps fatal, to surface displacement resulting from mantle convection. My insistence that geophysical constraints must be discussed led many participants and later commentators to dismiss me as a troglodyte who was slowing the convergence in thought that was later to be labeled either as a revolution in geology or a paradigm shift. For example, Scripps marine geologist Bill

Menard described my role as follows: "In 1963, Gordon MacDonald was like Harold Jeffreys in 1923 . . . he was to use his analytical powers to contest continental drift. This time, however, the contest would not be an indecisive side attraction lingering for decades. It would be the main event, brief, and won by a knockout."[23]

Of these many meetings, I remember two vividly. The Royal Society sponsored the first, held in the lecture theater of the Royal Institution, March 19–20, 1964. My favorite combatant, Teddy Bullard, a relatively late convert to drift, presented what he regarded as proof that there was a precise fit between the two coasts of Africa and South America. I joined Joe Worzel in pointing out that the computer reconstruction made it necessary to forget about Central America, Mexico, the Gulf of Mexico, the Caribbean Sea, and the West Indies, along with their pre-Mesozoic rocks.[24] The last session of the meeting was on the physics of mantle convection and I once again argued for deep roots to continents and the difficulties these imposed on any drift scheme. Teddy Bullard, in a masterful putdown, responded, "Many precedents suggest the un-wisdom of being too sure of conclusions based on supposed properties of imperfectly understood materials in inaccessible regions of the earth."

Although I maintained an interest in the structure of the earth's interior, I had actually begun to disengage from the field of continental drift in 1962, when I was asked to chair a National Academy of Sciences Committee examining weather modification. The issue of whether introduced ice nuclei can enhance precipitation from clouds raised serious statistical questions. Most important for me, this review brought me into contact with David Keeling of Scripps. Keeling's meticulous observation on the atmospheric content of carbon dioxide clearly pointed toward long-term changes in climate resulting from human activities. Since then, I have devoted a good part of my scientific career to understanding climate and its changes.

I also became convinced that societal problems required input from scientists. Arguing the reality of continental drift presented an exhilarating intellectual challenge, but neither continental drift nor plate tectonics has had much influence on the health of society. For example, earthquake prediction was impossible before the acceptance of plate tectonics and has remained so afterward. The time demands of the President's Science Advisory Committee, of which I was then a member, required me to move from California to Washington, and in 1967 I became vice-president of the Institute for Defense Analyses.

Nevertheless I still accepted an invitation to participate in what turned out to be the second memorable meeting: the "History of the Earth's

Crust," convened by the Goddard Institute for Space Studies of the National Aeronautical and Space Administration (NASA) on November 10–12, 1966.[25] Many adherents of plate tectonics later regarded this meeting as a defining moment for the theory. Abundant evidence was presented showing that basalts pouring from the mid-ocean ridge, the process driving sea floor spreading, was a global phenomenon that generated the ocean floor. I never questioned the validity of these observations, but my unfamiliarity with the wealth of evidence convinced me that I had drifted far away from the physics of the earth's interior. Because of my new commitments, I had little time to think about the recent developments. I let the organizers know that I could not be present at the summary session and arranged for Teddy Bullard to present my arguments. Having listened to me make the same case a number of times, he imitated my presentation in a delightful and comical way, according to those who heard him.

AFTERTHOUGHTS

In all science there is a strong "herd instinct." Members of the herd find congeniality in interacting with other members who hold the same view of the world. They may argue vigorously about details, but they maintain solidarity when challenged or criticized by those outside their comfortable herd. If individual scientists stray too far from the accepted dogma of the day, that of the herd, they are gently (or not so gently) ostracized. The herd instinct is strengthened enormously if the paymasters are members of the herd. Strays do not get funded and their work, sometimes highly innovative, is neglected as the herd rumbles along. When leaders of the herd decide to strike out in a new direction, the herd often follows. Before the 1950s, the North American herd of geologists found it comforting and amusing to ridicule those foreign geologists who advocated continental drift. In the early 1960s, Harry Hess, Tuzo Wilson, and Bob Dietz, all respected leaders of the North American geologist herd, decided to shift directions and the herd soon followed.

By contrast, if the innovators are not part of the herd it becomes very much more difficult, if not impossible, to change direction. Over the years, I have seen examples of the herd mentality in many fields of science. Currently two cases spring to mind. Thomas Gold, an astrophysicist with a long history of innovative contributions, has over the past two decades forcefully argued that at least some fraction of the earth's natural gas originated abiogenetically, having been captured deep within the earth at the time of its formation. Gold developed a theory of a deep hot biosphere.[26] The theory proposes that living creatures populate the crust of the earth, down to depths of several miles. The creatures that we see on the surface of the earth are only part of the biosphere; the greater and more ancient part of the biosphere lies at depth, at high temperatures and pressures. Some of the evidence in support of this concept remains controversial, but other parts are not. However, the world's petroleum geologists find it difficult to accept that they may have been wrong, at least in part, in postulating that rotting plants are responsible for deposits of natural gas. The herd of North American petroleum geologists found it incomprehensible that a mere astrophysicist would dare meddle in the herd's business.

A second example, closer to home, involves a modification of the Milankovitch theory of the ice ages. Richard Muller, a high-energy nuclear physicist, and I are not members of the paleoclimate herd, which has fixed on the dogma of Milankovitch's origin of glacial cycles as its worldview. We question one element of the theory – an element that is of key importance: the origin of the 100,000-year cycle found in proxy measures of the volume of glacial ice.[27] We attribute fluctuations in ice volume to the nodding of the earth's orbital plane with respect to the solar system's invariable plane with a 100,000-year cycle. Despite abundant evidence supporting our view, the dominant herd is distinctly nervous about any intrusion into its midst by scientists they do not know and are uncomfortable with.

Are plate tectonics for geologists the equivalent of Bohr's theory of the atom for physicists, as has been claimed?[28] I am certainly no longer a member of the herd of solid earth geophysicists, having switched herds in the mid-1960s. However, I note that a recent study using greatly improved seismic observations shows that subplate structures extend to depths of 150 miles (250 kilometers) or more.[29] Moreover, models of convection have not explicitly taken into account the complicated substructure of both continents and oceans.

The theory of plate tectonics offers many more degrees of freedom for geologists than the concept of deep continental structures. Rather than working with six continents, the geologist now has 11 plates and can suggest more if the geologic evidence points that way. The loosening of geophysical constraints has led to many speculative explanations of geological observations. But with a large number of degrees of freedom, it becomes increasingly difficult to test whether the postulated processes are even self-consistent. The lack of any discipline required by geophysical observation places few, if any, limits on the creativity of geologists in interpreting the past. Long ago, the geologic herd overcame the physicist herd in the battle about the age of the earth. They now have a comfortable confidence that they have found truth in plate tectonics, even if there are a few troublesome details yet to be dealt with.

CHAPTER 8

HEAT FLOW UNDER THE OCEANS

John G. Sclater

THE INTERNAL ENGINE OF THE EARTH IS DRIVEN BY RADIOACTIVELY generated heat and heat left over from the formation of the planet. The ultimate source of energy for the elevated temperature in mines, volcanoes, hot springs, earthquakes, the uplift of mountains, and global plate motions is this heat from the interior. The thick continental crust is known to have many more heat-producing radioactive elements than the much thinner oceanic crust. As a result, it was expected that the heat flowing out through the ocean floor would be significantly less than that flowing through the continents.

The outward flow of heat from the earth is determined as the product of the temperature gradient and the thermal conductivity. In the oceans, the temperature gradient is measured by forcing temperature-sensing elements into the soft sediments of the ocean floor. The thermal conductivity is determined directly by measurement on the sediments in situ or on a sample brought to the surface in a coring tube.

The early measurements of heat flow in the oceans made by Roger Revelle, the director of Scripps Institution of Oceanography, and his student, Art Maxwell, gave values that were very similar to those on the continents.[1] Sir Edward ("Teddy") Bullard, who had designed the instrument they used, proposed that the extra heat was created by slow-moving convection currents in the upper mantle beneath the oceans.[2] Teddy and many others argued that these currents were the same as those responsible for moving the continents. Dick Von Herzen had participated as an undergraduate in some of the original Scripps expeditions on which heat flow measurements had been taken. He returned to Scripps for graduate studies under the direction of Russell Raitt to make the first systematic measurements of the flow of heat through the ocean floor.

After finishing his thesis Dick accepted a job as a scientific attaché at

John Sclater on the occasion of receiving the Rosentiel Award of the University of Miami for outstanding work by a young scientist in marine geology or geophysics, 1977. *From left to right:* Walter Munk, Freddie Sclater, John Sclater, Judith Munk, and Chris Harrington. (Photo courtesy of John Sclater.)

UNESCO in Paris. I first met Dick there, in the summer of 1964, on my way to join the research vessel *Argo* on the (Scripps) DODO expedition to the Central Indian Ocean. At the time I was a graduate student in the Department of Geodesy and Geophysics, Cambridge University, based at Madingley Rise. My thesis involved the use of a new type of heat flow instrument built by Clive Lister, who had preceded me in the marine group at Cambridge. Bob Fisher of Scripps, who was on sabbatical at Madingley Rise, had arranged that I take the Cambridge heat flow equipment to Mauritius for use on *Argo*. Dick and Victor Vacquier, the chief scientists for this leg of the expedition, had agreed to my participation. They planned to determine the relation of the axis of the Central Indian Ridge, as defined by the central magnetic high, to the heat flow field. This was to be my first expedition at sea on my own without the support of the Cambridge seagoing group run by my supervisor Maurice Hill.

The professionalism of the crew, technicians, and scientists on board *Argo* impressed me. Everyone from the captain on down worked to maximize the use of the ship for scientific purposes. Their combined efforts ensured the reliability and efficiency of the shipboard operations. Dick and Vic had a limited but clear set of scientific objectives. These were to maximize the number of crossings of the ridge and to take as many heat flow measurements near the crest as possible. Further, they had thought through carefully just what compromises were needed to attain these objectives. I saw, firsthand, the low-key but highly efficient American way of operating at sea. This contrasted favorably for me with the more structured but much less focused operation I had observed on a three-month expedition to the Indian Ocean on the *Discovery* the previous year. On my way back from sea I had the opportunity to work at Scripps for two months (November and December 1964). Returning to foggy, frozen southern England in December after the balmy Indian Ocean and a November in southern California was a true culture shock!

I next met Dick in November 1965, when he came to Cambridge to be the external examiner for my Ph.D. thesis defense. Teddy Bullard was my internal examiner. When preparing this essay I found a copy of my thesis that included Dick's handwritten comments.[3] It was his copy from my defense. The Cambridge University Librarian had very exacting standards for the format of theses, and in the days before word processors, these standards made rewriting both difficult and time-consuming. I now agree with many of Dick's comments, but I have to admit that I did not include most of them in the final version of my thesis. To do so would have caused too many problems with the university librarian and I wanted to return to Scripps and California as quickly as possible; there I had a job to follow Dick as the person directly responsible for the heat flow program at Scripps. One of the first things I found on arriving at Scripps was a copy of Dick's thesis.[4]

After his stint as scientific attaché with UNESCO in Paris, Dick returned to the United States to a position at the Woods Hole Oceanographic Institution. There he resumed his career as an active seagoing marine scientist. In this long and distinguished career he had many accomplishments. He was the first to document the high heat flow anomaly at the crest of the mid-ocean ridges.[5] He was co-chief scientist with Art Maxwell on Leg III of the Deep Sea Drilling program that showed the linear increase in age of the ocean crust away from the crest of the mid-Atlantic ridge.[6] His student, Dave Williams, together with Dick, Roger Anderson (a student of mine), and Vic Vacquier, and me, discovered the first hydrothermal vent emanating from a mid-ocean ridge.[7] Finally, he and others found higher than expected heat flow over the oceanic swells, which they attributed to thermal anomalies in the upper mantle.[8]

In the first part of this essay, I concentrate on Dick's contributions to the early heat flow measurements at sea and the influence they had upon Harry Hess in his development of the theory of sea floor spreading.[9] In the second part, I attempt to answer two questions of interest to those who study the history of scientific discoveries. Why did the heat flow community not come up with the idea of sea floor spreading before Harry Hess, and why, once he had published his ideas, did this community take so long to apply the concept to interpreting the heat flow data? Later in this volume Dan McKenzie, who overlapped with me at Madingley Rise, presents an approach to interpreting the occurrence of advances in the earth sciences. His approach stimulated me to reexamine these questions from his observation-oriented point of view. At the end of his essay, Dan raises his own question regarding the history of plate tectonics. "The paleomagnetic observations did not have the impact in retrospect that they should have had. . . . Why they did so remains for me a puzzle and also the most interesting historical question to be raised by the discovery of plate tectonics."[10] Using my explanation of the difficulty that the heat flow community had with sea floor spreading as a basis, I attempt to answer this question. I finish by presenting a summary of my own views on how advances occur in the earth sciences.

VON HERZEN AND THE EARLY HEAT FLOW MEASUREMENTS

Roger Revelle enlisted Teddy Bullard in the early 1950s to set up a program at Scripps to measure the heat flow through the floor of the oceans. In 1956, Teddy, Roger, and their student Art Maxwell summarized the results of measurements made at Scripps and those made by Teddy through the National Physical Laboratory in England.[11] Most of the measurements gave values very similar to those on the continents. However, on Scripps' 1952–1953 *Capricorn* expedition, Revelle and Maxwell had observed two much higher values at the crest of the East Pacific Rise. It was these measurements that led Dick to concentrate his systematic investigation of oceanic heat flow values on the relation between high heat flow and the rise.

In his first scientific paper, published in *Nature,* Dick summarized the measurements he took, and those reported by Bullard and others.[12] He plotted all the values on a topographic chart of the East Pacific Rise. Three features stand out: the large percentage of high values near the crest of the East Pacific Rise, the close–to-average values elsewhere in the oceans, and the scattered low values that occur even very close to the crest of the rise. Three years later, Scripps graduate students Bob Nason and Willie Lee used the Scripps heat flow equipment to show similar very high values at the ridge crest on a crossing of the mid-Atlantic ridge.[13]

In 1962, on Scripps' *Risepac* expedition, Dick and Seiya Uyeda, from

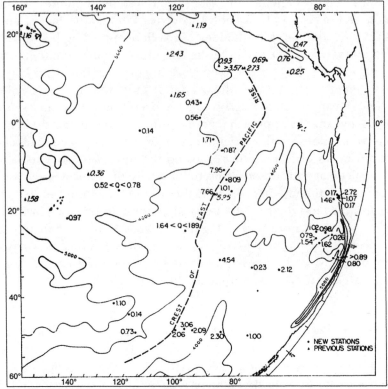

Map of heat flow stations with generalized 4,000 meter and 5,000 meter bottom depth contours. Note the high values (> 2 cal/cm^2 sec) near the crest of the East Pacific Rise, the close–to-average values (1.0–2.0 cal/cm^2 sec) elsewhere in the oceans, and the scattered low values (< 1.0 cal/cm^2 sec) that occur everywhere, even very close to the crest of the rise. (Von Herzen, R., 1959. Heat-flow values from the South-Eastern Pacific. *Nature* 183: 882–883. Reproduced with permission of *Nature*, http://www.nature.com)

the Earthquake Research Institute of Tokyo University, ran a series of heat flow stations at 30 mile (50 kilometer) spacing across the crest of the East Pacific Rise at 14°S.[14] Values up to five times average occurred near the crest of the broad swell of the East Pacific Rise. However, a number of average or below average values also occurred within 30 miles (50 kilometers) of the highest values. In addition, they found low values, some less than one-third of average, at greater distances from the crest of the rise. They devoted considerable space in the resulting manuscript to an unsuccessful attempt to explain these values.

In a study published in the same year in the journal *Science*, Dick compared the heat flow values he had taken in the Gulf of California with

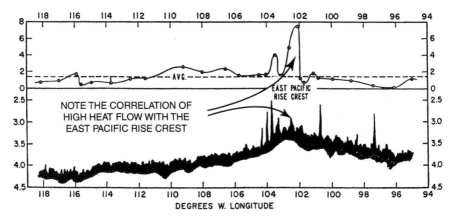

Heat flow and topography across the East Pacific Rise. Note the correlation between the high heat flow values and the crest of the East Pacific Rise. (Von Herzen, R. P., and Uyeda, S., 1963. Heat flow through the eastern Pacific Ocean floor. *Journal of Geophysical Research* 68: 4219–4250. Copyright 1963, American Geophysical Union. Reproduced by permission of the American Geophysical Union.)

those from the Gulf of Aden.[15] In both areas, he observed values significantly higher than the worldwide average. In the abstract, he stated, "[The gulfs] . . . closely coincide with the intersection of oceanic rises with continents and have likely been formed under tensional forces, which suggests an association with mantle convection currents." The figure on the next page, taken directly from his thesis, presents his idea of how these convection currents affect the surface heat flow field.[16] This concept with more geological embellishment appears in Scripps professor Bill Menard's 1964 book, *Marine Geology of the Pacific*.[17] Clearly the idea of an upwelling convection current beneath the rise was generally accepted at Scripps at the time. Both Dick and Bill knew from the absence of a major gravity anomaly that the East Pacific Rise must have a low density root. They inferred that higher temperatures caused this root and postulated that the upwelling limb of a major convection current within the mantle lay beneath the crest of the rise. The upwelling of hot mantle material created both the high heat flow anomaly and the elevation.

The Hypothesis of Sea Floor Spreading

In his now classic paper, Harry Hess used the following line of reasoning to justify sea floor spreading:

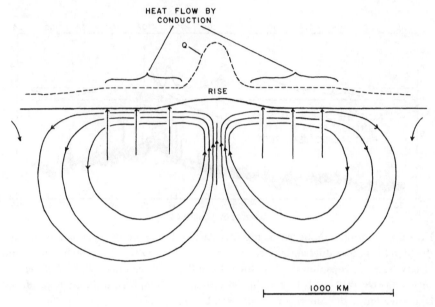

Possible mantle convection pattern beneath the eastern Pacific Ocean. Note the
smooth increase of the predicted heat flow over the upwelling limb of the convection cell. The closed curves represent the motion of the material in the upper mantle. The arrows show the direction of flow. The straight lines with arrows represent
the flow of heat by conduction from the convection current to the stationary crust
above. (Von Herzen, 1960. Pacific Ocean Heat Flow Measurements, Their Interpretation and Geophysical Implications, Ph.D. thesis, UCLA, p. 119.) Used with the
permission of Richard Von Herzen.

The Mid-Ocean Ridges are the largest topographic features on the surface of the Earth. Menard (1958) has shown that their crests closely correspond to median lines in the oceans and suggests that they may be
ephemeral features. Bullard, Maxwell and Revelle (1956) and Von
Herzen (1959) show that they have unusually high heat flow along their
crests. Heezen (1960) has demonstrated that a median graben exists
along the crests of the Atlantic, Arctic, and Indian Ocean ridges and that
shallow-depth earthquake foci are concentrated under the graben. This
leads him to postulate extension of the crust at right angles to the trend
of the ridges. . . . Paleomagnetic data presented by Runcorn (1959),
Irving (1959), and others strongly suggest that the continents have
moved by large amounts in geologically comparatively recent times. One
may quibble over the details, but the general picture on paleomagnetism
is sufficiently compelling that it is much more reasonable to accept it
than to disregard it. . . . Menard's theorem that mid-ocean ridge crests
correspond to median lines now takes on new meaning. The mid-ocean

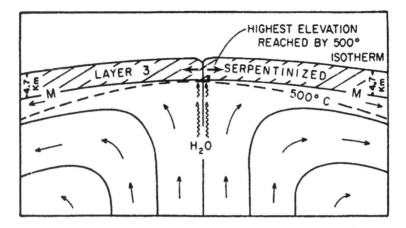

The concept of sea floor spreading as proposed by Harry Hess. Abstracted from a diagram to portray the highest elevation that the 500 degrees C isotherm can reach over the ascending limb of a mantle convection cell, and expulsion of water from mantle which produces hydrothermal alteration, forming the mineral serpentine, above the 500 degrees C isotherm (Hess, 1962, note 9). The shaded area is the oceanic crust that is created by the intrusion of hot molten material. It moves away from the center of spreading with the intrusion of more material from below. Reproduced with permission of the *Geological Society of America*.

ridges could represent the traces of the rising limbs of convection cells, while the circum-Pacific belt of deformation and volcanism represents descending limbs. The mid-Atlantic Ridge is median because the continental areas on each side have moved away from it at the same rate, about 1 cm/yr. This is not exactly the same as continental drift. The continents do not plow through oceanic crust impelled by unknown forces; rather they ride passively on mantle material as it comes to the surface at the crest of the ridge and then moves laterally away from it.[18]

Hess recognized that the high heat flow values could only be explained by massive amounts of intrusion at the crest of the mid-ocean ridges. It is tempting to believe that it was the coupling of these measurements with the insight of Bruce Heezen that led Harry directly to the concept of sea floor spreading.[19] The actual intrusion of mantle material into the crust at the crest of a mid-ocean ridge and the movement of this material away from the crest differentiates sea floor spreading from the mantle convection current hypothesis that preceded it.

The scientific community did not immediately accept the concept of sea floor spreading. For example Dick, Vic Vacquier, and others still found it necessary in 1963 and 1964 to test the relation between high heat flow and the crest of the mid-ocean ridges. In 1963, on the Scripps *Lusiad* expedition, Vic and Dick crossed the mid-Atlantic ridge a number of times in the South Atlantic.[20] They found a striking relation between high heat flow and the crest of the ridge as defined by the central magnetic high. In the following year, they made a series of crossings of the Central Indian Ocean. As expected, they observed the same relation between high heat flow and the crest of the mid-ocean ridge that they had found in the South Atlantic.[21] It was while traveling to join this expedition that I stopped off in Paris to meet Dick. I had just submitted my first scientific paper on heat flow measurements in the Somali Basin and the Gulf of Aden to a Royal Society symposium devoted to geological and geophysical studies in the Indian Ocean.[22] They presented their results in the same symposium.

A question that has puzzled me, in retrospect, is this: why did the three of us not combine at this time to analyze either (a) the two heat flow data sets quantitatively assuming sea floor spreading, or (b) attempt to reinterpret the magnetic data using the hypothesis of Fred Vine and Drummond Matthews?[23] This question becomes even harder for me to understand when I add three other facts. First, I had shared a cabin with Fred Vine for three months on the *Discovery II* the previous year. His hypothesis was not taken seriously by our supervisor and chief scientist on the cruise, Maurice Hill, but I had fought hard in support of Fred so he would get some ship time to test it by surveying a seamount near the crest of the Carlsberg Ridge. Second, I had available to me during the final stages of writing my thesis the manuscript of a paper on the Gulf of Aden by Tony Laughton, from the National Institute of Oceanography in England, to be published in the same symposium as my own paper and that by Dick and Vic.[24] In his paper, Tony, who had been on the *Discovery II* expedition with me, pointed out the necessity for the creation of new ocean floor in the center of the Gulf to fit the two edges back together again. Third, Harry Hess and Tuzo Wilson were both at Cambridge on sabbatical in 1964 and 1965, and I spent many weekends, during one of the coldest winters on record, in Madingley Rise with them in one of the few centrally heated buildings in Cambridge.

It was not until the publication of the interpretation of the magnetic stripes over Iceland's Reykjanes Ridge that the theory of sea floor spreading received overwhelming support, even from among the heat flow community.[25] Why such a significant delay? I believe that the answer to this question lies in the way the heat flow community approached observational data.

AN OBSERVATIONAL APPROACH TO SCIENTIFIC DISCOVERY

Until I started writing this essay I did not realize how much my personal philosophy of how science advances had dominated the way I think about earth science. To understand the answer to the questions I have posed, I need to start from the point at which I developed my own approach to the subject. This happened when I was an undergraduate at Edinburgh University. I planned to work as a geophysicist after I graduated, but such a degree was not available at Edinburgh at the time. As a compromise I proposed to supplement a conventional four-year geology degree with two years of physics. I did much better in my first year in physics than geology. Following the advice of my undergraduate advisor I changed my degree to experimental physics during my second year. I graduated imbued by the hypothesis-testing method of doing science, which I had picked up in the experimental physics courses that I had taken.[26] I started and finished my graduate career at Cambridge with the same basic approach.

At Madingley and Scripps in the 1960s, I became acquainted with most of the major figures in the development of sea floor spreading and plate tectonics. My major effect regarding plate tectonics as a student was negative. In 1964, I successfully convinced Tuzo Wilson that the magnetic anomalies observed on either side of the ridge axis on the recently published *Vema* cruise across the mid-Atlantic ridge in the South Atlantic were not symmetric. Thus I talked him out of developing a magnetic timescale from this profile. Later this profile was to become the basis of the Cenozoic magnetic timescale.[27] When I arrived at Scripps, I worked closely with Bill Menard, Vic Vacquier, Bob Fisher, and Art Raff, and by the end of 1966 my opposition to sea floor spreading had evaporated. In 1967, I shared a cottage with Dan McKenzie in La Jolla when he and Bob Parker, also a student at Madingley, wrote their path-breaking paper on tectonics on a sphere.[28] Dan and I applied the theory to produce the first quantitative tectonic history of the Indian Ocean.[29] Scripps students Jean Francheteau, Roger Anderson, Miller Lee Bell, and I showed that the creation of new plate at a spreading center could account for the heat flow and subsidence of the ocean floor.[30]

As a result of my intimate involvement in the development of such a major advance, I have lost my belief that advances in the earth sciences occur primarily as a result of hypothesis testing. Neither Harry Hess nor Tuzo Wilson was testing a hypothesis.[31] Rather, they were creating new concepts out of the synthesis of poorly-constrained observational information.

They thought their concepts had validity because they explained the patterns they recognized in so many different sets of data. Nor do I believe that advances necessarily result from a paradigm shift during a scientific crisis, as has been advocated by Thomas Kuhn.[32] The advances occurred in the earth sciences before the field even realized that there was such a crisis. It is interesting to note that no advance occurred back in the 1920s when scientists did think there was a crisis. They resolved the crisis by maintaining the status quo and rejected the necessity for a paradigm shift – which is another blow to Kuhn's hypothesis.[33]

THE OBSERVATIONAL APPROACH

Earth science is an observational discipline. However, many processes affect any observation. Earth scientists cannot separate any single process entirely from all the processes that have occurred. Further, except for a limited number of cases, laboratory experiments do not scale to the real world. Thus, unlike physics or chemistry, earth science is not an experimental discipline. Earth scientists, in most cases, observe and describe phenomena rather than conducting experiments to test hypotheses. Synthesizing data and/or recognizing patterns in "noisy" data are in many cases more important than any experiments that they could perform. Major progress occurs by constructing simple physical models that describe the patterns that earth scientists have selected out of the background noise. They have to exercise care with their observations because, occasionally, the background "noise" carries information that is critical to the process or processes under study.

In his observation-oriented analysis of scientific discovery, Dan McKenzie, in a later essay in this volume, separates scientific observations into four categories: (1) observations that are wrong, (2) observations that are correct and can be described by existing theories, (3) observations that are correct but are too complex to be described by any simple model, and (4) observations that are correct but there is no theory that describes them. A scientific advance occurs when a model that accounts for the data in Category 2 also accounts for the observations grouped in Category 4.

It is easiest to see how this approach works by applying it to the evolution of plate tectonics. Tuzo Wilson developed a concept that incorporated sea floor spreading, linear magnetic anomalies, and trenches to explain transform faults, fracture zones, and the fit of the continents.[34] Lynn Sykes, Dan McKenzie and Bob Parker, Jason Morgan, and Bryan

Isacks and colleagues showed that the same concept also explained the type and distribution of earthquakes.[35] In addition, Xavier Le Pichon showed that it provided a self-consistent description of the tectonics of the entire surface of the earth.[36] It was considered a major advance because one simple concept could explain so many different sets of observations, which previously had no theory to explain them.

THE HEAT FLOW COMMUNITY AND SEA FLOOR SPREADING

The early marine heat flow community was made up of seagoing scientists who had the ability to build and run sensitive equipment under adverse marine conditions. All were able scientists who combined physical endurance with a strong physics or geophysics background. They included some of the best marine scientists of their generation, such as Sir Edward Bullard, Art Maxwell, Dick Von Herzen, Marcus Langseth, Victor Vacquier, and Clive Lister.

I believe that, like me, they let their experimental physics background dominate the way they looked at the earth. We knew that the heat flow at the ridge crests was high.[37] In addition, Teddy Bullard and others at Scripps knew that the East Pacific Rise was elevated because it was hot, and that there was a correlation between heat flow and the depth of the ocean floor.[38] However, as experimental physicists, we were stymied by the fact that we could not explain the very low values, especially those found by Dick Von Herzen and Seiya Uyeda near the crest of the East Pacific Rise.[39]

When I was a graduate student at Madingley, Teddy Bullard jokingly complained that he had not accomplished very much in geophysics because his name had not been given to any hypothesis or law. To rectify this omission, the students and junior staff at Madingley, with support from Maurice Hill, created "Bullard's Law." This law asserted, "Never take one marine heat flow measurement within 50 kilometers of another measurement because it is likely that it will differ from the first by at least one order of magnitude." Although humorous, this incident shows just how little respect was paid to the early heat flow measurements by most geophysicists. It illustrates the problem that the community had with the interpretation of their empirical data. Without an explanation of the low values that created huge scatter in the data, no one was willing to attempt to interpret the overall pattern quantitatively. We concentrated instead on trying to understand these low values. As I mentioned previously, Dick Von Herzen and Seiya Uyeda devoted a large part of their text to trying to account for the low values found over the East Pacific Rise.[40] This paper influenced my thesis more than any other. I devoted about half of my thesis attempting to find an explanation for the low values.[41]

WHY THE COMMUNITY MISSED SEA FLOOR SPREADING

We were physicists who wished to test a hypothesis. Led by Teddy Bullard, we wished to discover whether or not the heat flow measurements presented evidence for the upwelling limb of a convection cell beneath the mid-ocean ridge axes. Intuitively, we expected to observe a relatively smooth increase from near normal on the flanks to a factor of four higher than normal over the crest of the ridges. The apparently random occurrence of low values completely confused us, especially those near the crest of ridges, close to values 20 times higher. The scatter in the data was so high that the mean values over the crest were indistinguishable statistically from those on the flanks. The data neither strongly agreed nor disagreed with the hypothesis.

The low values and the scatter in the data became a major concern to both myself and others working in the field. As a consequence we overlooked the pattern in the measurements. We did not see either that the envelope of the high values showed a clear correlation with distance from the crest of all the mid-ocean ridges or that the drop-off rate for these high values varied with the width of the ridge. Thus, we did not realize that differing rates of intrusion at the individual ridge crests could explain the different drop-off rates. We believed our data, but without an explanation of the low values, we were unwilling to interpret the measurements quantitatively. We placed them within Category 3: too complex to be explained by any simple theory. Most geophysicists were less charitable. They could not believe a measurement that could differ by more than an order of magnitude over a distance of only 30 miles (50 kilometers). They placed the measurements in Category 1: observations that were obviously wrong. It took a geologist like Harry Hess to see the pattern in the data.[42] He recognized that what was critical for understanding the earth as a whole was not the isolated lows, but the large number of very high values at the crest of the mid-ocean ridges. It was more important to recognize the consistent envelope of the high values from one ridge to another rather than to be overly concerned with the scatter created by the low values.

THE SUCCESSES

Once Harry Hess and Tuzo Wilson had articulated the key concepts of plate tectonics, the hypothesis-testing approach of the heat flow community moved the field forward very quickly.[43] Even before the concept was applied to earthquakes, Marcus Langseth, Xavier Le Pichon, and Maurice Ewing of Lamont Geological Observatory had introduced the idea of a 60 mile (100 kilometer) thick plate created at a ridge axis to try to explain the heat flow and subsidence data across the mid-Atlantic ridge.[44] From the poor fit of the observed to the predicted subsidence, and the observed to predicted decrease in heat flow with distance from the ridge crest, they argued that the concept did not work. The following year Dan McKenzie recast the problem non-dimensionally and showed that by varying the boundary conditions the same model could be made to match the heat flow data.[45] University of Wisconson professor Ned Ostenso and his graduate student, Peter Vogt, pointed out that Langseth, Le Pichon, and Ewing had omitted the loading effect of the water.[46] Rather than being a poor fit, the model actually gave a reasonable fit to the subsidence of the ridge. (Due to a surprising omission by *Nature,* Peter Vogt and Ned Ostenso have not received the credit they deserve for recognizing the importance of this correction. In the published paper, the entire paragraph that discussed the isostatic correction for the loading effect of the water was omitted. This omission went unnoticed because the paragraph *was* included in the reprints that the journal sent back to the authors!)

The thick plate concept adopted by the heat flow community was the forerunner to the plate models that Dan McKenzie and Bob Parker, Jason Morgan, Bryan Isacks and colleagues, and Xavier Le Pichon developed to establish the quantitative aspects of plate tectonics.[47] It also gave Norman Sleep, Jean Francheteau, my students Roger Anderson and Miller Lee Bell, and me a hypothesis to test.[48] Very quickly we established that the plate model that explained the ridges, trenches, and earthquakes could also account for the subsidence of the ocean floor as the age of the ocean crust increased.

The explanation for the very low heat flow values finally came from the work of Clive Lister.[49] He hypothesized that the low values were due to hydrothermal circulation in the ocean crust. Using Clive's concept as a basis for selecting an area where heat loss by hydrothermal circulation probably did not occur, my students and I showed that these carefully selected heat flow data fit the same plate model that accounted for the subsidence of the ocean floor.[50] In testing Clive's concept, Dave Williams and others found the first hydrothermal vent at a ridge crest.[51] This led to the realization of the importance of hydrothermal circulation on the ocean floor and the discovery of hydrothermal venting at the crest of all of the mid-ocean ridges.

WHY DID PALEOMAGNETISTS MISS SEA FLOOR SPREADING?

Later in this volume, Dan McKenzie raises two questions regarding the history of the development of plate tectonics: (1) Why did the paleo-magnetic community not come up with the idea of sea floor spreading ahead of Hess? (2) Why did the rest of the earth sciences community take so long to accept their conclusions regarding continental drift?[52] I offer here some personal comments based on some obvious parallels between paleomagnetism and heat flow and my experiences answering the same questions for the heat flow community.

Both fields were relatively new at the time of the development of sea floor spreading and plate tectonics. Both fields were based on difficult observations. Like heat flow, paleomagnetism attracted a number of unusually able scientists, for example, P. M. S. Blackett, Keith Runcorn, Ted Irving, Allan Cox, Dick Doell, Victor Vacquier, Neil Opdyke, Christopher Harrison, and Ron Girdler. As a consequence of the difficulties with the technique, in the early stages scientists with a classical experimental physics background dominated the field.

Keith Runcorn and Ted Irving believed that their measurements confirmed the idea that the continents had moved.[53] In addition, they knew of Euler's theorem and called the points about which the continents rotated "pivot points." However, the early paleomagnetists could not explain why they found some rocks magnetized in the opposite direction to that of the present earth's field. This threw some doubt on the reliability of the entire operation. This doubt was exacerbated by the fact that one of the major figures in geophysics at the time in Europe, Sir Harold Jeffreys, did not believe in continental drift and doubted the reliability of paleomagnetic measurements. Indeed, he dismissed them with the following statement: "In studying the magnetism of rocks the specimen has to be broken off with a geological hammer and then carried to the laboratory. It is supposed that in the process its magnetism does not change to any important extent, and though I have often asked how this comes to be the case I have never received any answer."[54] I took the opinions of Jeffreys very seriously, since he was generally credited with having made prewar geophysics into a respectable discipline.

Harold Jeffreys and others who opposed drift used the fact that the reversals were apparently inconsistent and unexplained to disregard the entire category of paleomagnetic measurement. They believed that the paleomagnetic measurements were wrong and placed them in Category 1. Like heat flow scientists, the early paleomagnetists were physicists by training. Without an unambiguous explanation of reversals they could not answer Jeffreys and the opponents of continental drift. Thus they concentrated either on making more measurements of polar wandering to overwhelm the opposition with the weight of the observations, or on trying to understand the reversals. They placed their measurements in Category 3: correct observations that were too complex to be explained by any simple model. Apart from Runcorn they did not attempt to incorporate them into an overall theory of how the earth worked. They were limited further by their lack of understanding of the observations from the oceans that might support drift. Ron Girdler from Newcastle University and George Peter from Lamont showed that positive and negatively magnetized stripes of material on the ocean floor could explain the striped magnetic anomalies in the Red Sea.[55] They came the closest of all to arriving at the hypothesis of sea floor spreading before Harry Hess.[56] However, they did not expand upon this idea of reversals or suggest what process could have created the stripes.

It took a geologist like Harry Hess with training in synthesizing data and pattern recognition to place the results in broad perspective.[57] In a wonderful introduction to the paleomagnetic measurements – "One may quibble over the details, but the general picture on paleomagnetism is sufficiently compelling that it is much more reasonable to accept it than to disregard it" – Harry dismissed the opposition. This permitted him to use the paleomagnetic data to argue that the continents moved significantly over geologic time. As a geologist, he recognized that the important observation for understanding the earth as a whole was that the continents *had* moved. He used these measurements and the dates of extensional processes on the continents bordering the Atlantic Ocean to propose a spreading rate of one centimeter (less than half an inch) per year. That he was within a factor of two of the actual rate shows the power of his reasoning. Hess recognized a pattern that few others were willing to accept. He showed that a model that could explain why the mid-ocean ridges occurred in the center of the ocean basins (Category 2: observations that are correct but can be described by existing theories) could also explain the heat flow measurements and the paleomagnetic results (Category 4: observations that are correct but there is no theory that describes them).

Meanwhile Allan Cox and colleagues in the United States, and Ian McDougall and colleagues in Australia, had been systematically documenting magnetic reversals in rocks.[58] They determined that lava flows of the same age had the same direction of magnetization, but that this direction changed from time to time. They were able to date the time at which the field changed. Fred Vine and Drummond Matthews showed that combining reversals of the field with an intrusion process that added material equally on either side of a spreading center would create positive and negative magnetic stripes of material on the ocean floor.[59] (Unknown to me at the time, Lawrence Morley had developed the same concept but was unable to get his paper published.)[60] These stripes explained the linear magnetic anomalies observed on shipboard profiles and provided strongly positive evidence in favor of the theory of sea floor spreading. In this case, it was the understanding of reversals – the "noise" in the paleomagnetic data – that provided the most powerful confirmation of sea floor spreading: the striped marine magnetic anomalies.

THE PROCESS OF SCIENTIFIC DISCOVERY

The key originators of sea floor spreading and plate tectonics, Harry Hess and Tuzo Wilson, each had a field-oriented and geographically diverse geological background.[61] They demonstrated great ability at both data synthesis and pattern recognition. This gave them the insight to look at the world from a totally new perspective. However, the physics-trained younger generation had skills at developing simple quantitative models from physical concepts. These skills permitted the comparison of the observations with predictions and led to the general acceptance of the theory.

As a result of my involvement in the development of plate tectonics I now believe that advances in the earth sciences occur in three stages. The first involves the origination of the concept; the second, the construction of a model where the predictions can be compared with a set of observations, the third, the application of the model to another set of data. What is common to each stage is the recognition of the importance of the observations. The first stage involves synthesis and pattern recognition; the second and third stages emphasize reliable measurements and the quantitative comparison of these measurements with predicted values.

The first stage can occur – as it did with sea floor spreading – as the result of the synthesis and recognition of patterns in large quantities of data. It can also arrive serendipitously. For example, Marcus Langseth, Xavier Le Pichon, and Maurice Ewing created the first thermal model of the oceanic lithosphere to demonstrate that plate tectonics could not explain the heat flow and subsidence across a mid-ocean ridge.[62] Dan McKenzie generalized the approach, but fit the heat flow observations with a plate that was too thin.[63] However, his generalization showed the power of such an approach and led to a much clearer understanding of the concept.

The second stage involves the testing and refining of the concept. The earth science community accepted the concepts of sea floor spreading and plate tectonics so readily because of the ability of a group of scientists to construct models based on these concepts. The comparison of the predictions of these models with reliable observations permitted a quantitative evaluation of the concepts.

The third stage involves the application of the concept to describe a set of observations that are believed to be correct but are not as yet understood. In plate tectonics this occurred when a concept constructed to explain the features of the ocean floor and the reconstructed position of continents was found suitable to explain the worldwide distribution of earthquakes. For the thermal models, it occurred when it was realized that the concept that accounted for the heat flow data could also explain the subsidence of a mid-ocean ridge.

For the development of a concept, selecting the appropriate and key observations is most important. At this stage geological training involving pattern recognition is at its most valuable. However, in the second and third stages, the ability to construct a physical model and then test it is required. For these stages, the hypothesis-testing methods of experimental physics become more important. The approaches are complementary; for sea floor spreading and plate tectonics both occurred, and this accounts for the speed at which the hypotheses became accepted theories.

CONCLUSION

The problem with the hypothesis-testing approach as applied to the earth sciences is that field-measurement noise often contains crucial information about the process under study. In the case of the interpretation of the early heat flow and paleomagnetic measurements, it was the inability to get past this noise that prevented the observational scientists from moving on to a deeper interpretation of their results.

However, this takes nothing away from the early heat flow community. They made the original measurements under often appalling conditions in rough weather and from very small oceangoing tugs or converted sailing ships. The foremost of this intrepid group of marine scientists was Dick Von Herzen. He made arguably some of the most important measurements ever taken at sea. Harry Hess appears to have developed the concept of sea floor spreading almost immediately after he had read and digested the importance of Dick's discovery of very high heat flow measurements near the crest of the East Pacific Rise.

What appears surprising is that the heat flow community took uncommonly long to try to model the high values at the ridge axis. I believe that they did not press to explain their results quantitatively because of their concern about the scatter created by the low values. Although obviously an oversimplification in light of the efforts of Keith Runcorn and Ted Irving, I argue that the early paleomagnetic community may have behaved similarly because of their lack of a convincing explanation for reversely magnetized rocks. In both cases, the earth science community at large did not believe the early measurements and hence the implications of these measurement went unheeded.

In the 1950s and early 1960s geological-geophysical expeditions at Scripps, and to a lesser extent, the Lamont Geological Observatory concentrated on making theory-relevant observations in a real-world setting. (For example, Bill Menard encouraged Von Herzen to pursue heat flow measurements because he understood their significance for ideas about mantle convection.) The history of the development of sea floor spreading and plate tectonics demonstrates the importance of these thoughtfully taken observations. The concepts were developed as a result of correlating many different types of observations into a coherent pattern. The advances occurred so quickly because of the complementary nature of this basically geological approach, with its emphasis on data collecting, with the hypothesis-testing approach of scientists trained in experimental physics.

As a consequence of my analysis, I do not believe that it was by chance that the Department of Geodesy and Geophysics at Cambridge had such a major effect upon the field. Teddy Bullard, the chair of the department, and Maurice Hill, the leader of the marine group, actively encouraged and hosted the interaction between global-thinking geologists such as Harry Hess and Tuzo Wilson, observational marine geologists such as Bill Menard and Bob Fisher, and their much younger, dominantly physics-educated graduate students. I believe that it was the symbiosis created by this interaction that led so many of the students and younger staff ultimately to contribute so significantly to this major advance in the earth sciences.

ACKNOWLEDGMENTS

An abbreviated version of this essay was presented on December 13,1999 at the annual meeting of the American Geophysical Union in San Francisco at a symposium: "Heat Flow and Oceanic Lithosphere: Honoring R. P. Von Herzen," organized by C. A. Stein and J. Lin. I would like to thank Jian Lin for giving me the opportunity to express my appreciation of the long-term contribution to the heat flow community of Dick Von Herzen. My thanks also go to Dan McKenzie, who asked me to look at an early version of the essay he presents in this volume. It made me think for the first time long and hard about how I actually believe advances in the earth sciences occur. This essay was started while I was on a Guggenheim Fellowship at the Ecole et Observatoire de la Science de la Terre in Strasbourg, France. I thank Marc Munschy for the chance to study in such a work-friendly environment. I thank Jerry Winterer for introducing me to the importance of pattern recognition in geology and Bob Fisher for reviewing this manuscript.

CHAPTER 9

LOCATING EARTHQUAKES
AND PLATE BOUNDARIES

Bruce A. Bolt

THE FOCUS OF THIS ESSAY IS ON THE ROLE OF TWO KEY SEISMO-
logical contributions to the evolution of plate tectonics, based on two
personal experiences. The first, in 1960 at (then) Lamont Geological
Observatory, led to a key advance in reliable locations of the initiation
points of distant earthquakes, what scientists call *teleseism hypocenters*. The
second was after 1962 when I resided at UC Berkeley, the home of the
method of fault-plane solutions for seismic sources, developed there by
Perry Byerly.[1] My discussion aims to demonstrate how the crossing of dis-
ciplinary boundaries was vital to the establishment of plate tectonics and
more generally to geophysics.

As an applied mathematician, I was interested in the construction of
algorithms and computer programs in seismology and, as a student of Sir
Harold Jeffreys at Cambridge, England, and K. E. Bullen at the University of Sydney, Australia, in the incorporation of appropriate statistical
methods in geophysical analysis. My involvement in the plate tectonics
"movement" began with my 1960 publication on the revision of earthquake epicenters, focal depths, and origin times of earthquakes using a
high-speed computer.[2] The main task of seismologists, particularly at
observatories, had long been the determination of the position of the
earthquake center at the earth's surface (the epicenter) and the time of
the seismic wave initiation. It was also important from the viewpoint of
earth structure to know how deep each earthquake source was (the
hypocenter or focus). Until 1960, graphical methods were used to trian-
gulate for the location using travel times (source-to-station) of seismic
waves recorded by seismographs at observatories around the world.
These methods had large uncertainties, resulting in systematic and ran-

Bruce Bolt. (Photo courtesy of Bruce Bolt.)

dom scatter in the maps of the earth's seismicity. In order to improve precision of global maps of earthquake locations and their times of occurrence, I developed a statistical algorithm for use on the newly available fast digital computers. I checked this program for precision and robustness against the known location of some large explosions from atomic weapons tests and, also, against a large, unusual earthquake located deep under the Iberian Peninsula.[3] Successful tests led me to the conclusion that "the program may be useful for research organizations requiring regular or special locations of epicenters." A few other computer programs for earthquake location emerged about the same time, and these programs improved our capacity to locate earthquakes, helping to define the boundaries of the tectonic plates and ultimately leading to visually arresting maps of those boundaries.[4]

While at Lamont my work involved many now-antiquated methods of computer programming. These included the use of data punched onto Hollerith cards, such as the complete 1958 seismological travel time-tables of Jeffreys and Bullen. With the help of James Dorman, then at Lamont, I coded the algebraic scheme in FORTRAN for an IBM digital computer, namely the model 704 at the IBM Research Center in Pough-keepsie, New York. Special aspects of the 1960 algorithm were "bi-weight" mathematical filters to handle scattered observations, smooth-ing devices, and the use of probability distributions estimated from the earthquake travel times reported in the *International Seismological Sum-mary (ISS)*. In general terms, the problem addressed in such computa-tions is an inverse one. In the location problem the inverse can be explained simply as follows: given the earthquake source location, it is a direct application of Pythagoras' theorem to compute the travel times of the seismic waves to a recording station. But the actual problem to be solved is the opposite: given the observed arrival times, what is the best estimate of the location of the source? At the end of 1960, I left the tested program in working order at Lamont, where it was taken up in a most productive way by Lynn Sykes.[5]

By the end of the 1960s, three key seismological underpinnings of plate tectonics were in place. The first one was the world wide standard seismograph network (WWSSN), which was installed in stages from 1961 through 1967 by the U.S. Coast and Geodetic Survey (USCGS) under agreements with station operators in various countries.[6] This network not only greatly reduced the errors in measuring times of arrival of seis-mic phases, but also the new seismograph design and improved clocks aided in picking correctly the various seismic phases on the seismo-grams. Such a system of standard seismographs, distributed worldwide, had grown out of the efforts of the United States to ensure adequate sur-veillance of underground tests of nuclear devices detonated in foreign countries. It also gave a more complete record of natural earthquakes that occur as background seismic sources in Test Ban Treaty monitoring. The data made available by the WWSSN were crucial for the rapid evo-lution and confident adoption of the plate tectonic model.

A second and almost as critical aspect of the relatively quick acceptance of the plate tectonic formulation was the availability of high-speed comput-ers that allowed uniform, rapid, and accurate computations of epicenters.[7]

The third improvement was the ability to make more reliable mea-surements of the motions of the first arriving P (longitudinal) waves on seismograms provided by the WWSSN. These wave polarities were needed to infer mechanisms of the geological fault rupture (or underground explosion) that generated the seismic waves. The basic method had been introduced in 1928 by Perry Byerly and was developed relatively slowly by his students and other seismologists over several decades.[8] For example, A. R. Ritsema in Holland improved the analysis by introducing the artifact of projecting each station measurement on a conceptual focal sphere, and W. Stauder at Berkeley and later St. Louis examined thoroughly the viability of a "double-couple" fault mechanism as the explanation of the observed first motion pattern on such a sphere.[9]

Nevertheless, overall the technique had proved disappointing: solutions from which geological inferences could be made were often questionable because of various uncertainties, particularly of the true polarities of the first P motions, that is, whether the ground went up or down at the beginning of the P wave. Seismograms from the older instruments were so variable in quality that the seismological analyst often found it difficult to pick this direction. Seismographs of the WWSSN, particularly those with the longer period pendulums, permitted Sykes and others to have confidence in the correctness of the choice of first motion directions. In doing so, they found that earlier interpretations of deep-focus earthquakes as strike-slip were wrong; the faults were in fact thrust faults, in which one crustal block overrode the other. For the first time, it was now possible to say with confidence which way the ground was moving.

By the 1950s the compilers of worldwide seismic data – the International Seismological Summary (ISS) – using only hand calculators and some subjective judgment in selection of wave identification and travel times, had fallen seriously behind in cataloguing world earthquakes. Many earthquake locations were determined by graphical methods or by comparison with past earthquake positions. My hypocenter estimation program on a fast digital computer was adopted, after some hesitation, at the *ISS* in 1961, a change made possible by my moving from Lamont for a sabbatical leave to Cambridge University in England during that year. Its use enabled the backlog of the *ISS* earthquake lists to be removed rapidly. Moreover there is little doubt that the subsequent published solutions were more uniformly reliable.

THE EVOLUTION OF THE TWO TECHNIQUES

By 1964, the three seismological tools discussed above were in place and in use by different research groups and at the *ISS*. They provided both improved catalogues of earthquake hypocenters around the world and a growing sample of relatively high-quality polarity data. Standardized programs, using readings from the WWSSN, now provided reliable locations for current earthquakes, and could be employed to revisit earlier earthquake data sets and revise them.[10]

In a parallel development, a new means of graphing the data from the first motions of earthquakes came into use. It now became apparent that the technique that was in general use was misleading. Suddenly, it was clear why seismologists had thought the motions associated with deep-focus earthquakes were strike-slip: the plotting technique had made points from distant sources appear to lie on steeply dipping planes, even when they weren't. Steeply dipping fault planes are usually associated with strike-slip faults, so it was natural to assume that they *were* strike-slip. Now it was evident that the faults were not steeply dipping at all; they were shallow, and consistent with thrust faults and with the newly measured accurate first motions.

Several crucial tests of the plate boundary model could be performed using the new technique for plotting fault planes. In his 1968 editorial review in *The History of the Earth's Crust* (Princeton University Press), Robert Phinney stated that "the convincing confirmation of the predictions [of Tuzo Wilson's transform fault model] in 1966 must be regarded as a major turning point in studies of the Earth." This confirmation was made possible by accurate fault plane mechanism analysis.

The crucial seismological advances were a direct consequence of the improved worldwide earthquake recording network and the availability of high-speed computers for more adequate statistical analyses. The seismicity pattern that helped to define plate boundaries depended upon the use of accurate computer programs for locating earthquake epicenters.[11]

Table 9.1
Milestones in the development of the analytical tools
for locating and understanding earthquakes

World Wide Standard Seismograph Network	Computer Hypocenter Location	Fault Mechanism (First P Motions)
U.S .Coast and Geodetic Survey (1961-1967)	Bolt, Lamont, (1960) U.S. Coast and Geodetic Survey (1960) I.S.S. (1961)	Byerly, Berkeley (1923) Hodgson, Ottawa (1951) Ritsema (1952) Stauder (1962) Sykes (1963, 1967)

The method pioneered by Byerly of finding fault mechanisms by using the polarity of P wave first motions was an early example of remote sensing – detecting earth processes from a distance.[12] It took a long time to be thoroughly accepted in seismology. In his 1958 book, Charles Richter stated, "Byerly pioneered this field as early as 1926 and established principles and methods now generally in use."[13] He remarked further that "In spite of earnest effort, well established results are still too few for worldwide generalization." About the same time, his colleague Hugo Benioff at the California Institute of Technology wrote, "The first motion method . . . for measuring the strike and dip is of questionable reliability . . . evidenced by reports of J. H. Hodgson."[14] Other authorities were similarly unenthusiastic about the Byerly method. For example, in his 1947 textbook, Keith Bullen barely mentioned the fault-plane method; in the 1963 edition he devoted a three-page qualitative discussion to it, but leaned toward "the need for caution in interpreting ostensible patterns of first motion."[15]

From the mid-1960s on, the more reliable and uniform WWSSN observations produced consistent fault-plane solutions. As Sykes concluded in 1967: "Long-period WWSSN seismographs now furnish data of greater sensitivity, greater reliability and broader geographical coverage than were available in previous investigations for mechanisms of earthquakes."[16]

CONCLUSIONS

Diverse interactions of applied mathematics, geology, numerical computer analyses, and statistics contributed to the rapid development of plate tectonics in the 1960s. Perhaps surprisingly, the two key seismological algorithms contained strictly no new theory. Indeed, optimization of the theory involved in computing characteristics of remote earthquake sources and geological structures using seismic waves were not worked through until almost a decade later.[17] From a personal point of view, after 1970 my contributions to these seismological tools continued at UC Berkeley. There I was involved in theoretical improvements in both the algorithms for hypocentral location and for focal mechanism estimation, but these advances were not in time to affect the arguments for the plate tectonic model.[18]

It might be asked: would the convincing arguments for plate tectonics have been significantly delayed if the WWSSN had not been established as a consequence of the U.S. effort to monitor clandestine underground nuclear explosions?[19] Of course, the independent arguments for the plate model from geomagnetism would not have been affected, but the qualitative strength of the seismological ones would have been weakened.[20] More elaborate statistics would have been essential to support the arguments, thus ensuring additional skepticism and controversies. It should be remembered that in the early 1960s there was much scientific opposition to the plate tectonic view. The rapid conversion to the new model was enhanced greatly by the availability and power of the global seismological observations and robust computations. For example, if the non-seismological evidence had been accompanied by global patterns that showed very scattered earthquake epicenters and non-systematic fault mechanisms along the mid-oceanic ridges, the strong arguments so quickly assembled would surely have been much weaker.[21]

Since the 1970s, additional improvements have been made in the speed and convergence of computer-based algorithms for both hypocentral location and estimation of source mechanisms. Both computed locations and fault mechanisms have become much more reliable and complete. We can now depend to a large extent on high resolution of the recorded seismic wave forms and unvarying timing precision at the global digital seismographic stations that replaced the WWSSN. However, it is curious that upgraded computer programs to estimate these important properties, which are embedded in defined probability models, are still seldom used at seismological data centers or in published research. The explanation may be that the plate tectonics model is nowadays so completely accepted in geology that there is no longer a critical demand for global algorithms that embody all the mathematical improvements now available.

CHAPTER 10

EARTHQUAKE SEISMOLOGY IN THE PLATE TECTONICS REVOLUTION

Jack Oliver

Eᴀʀᴛʜ sᴄɪᴇɴᴛɪsᴛs ᴡʜᴏ ᴡᴇʀᴇ ᴀᴄᴛɪᴠᴇ ᴅᴜʀɪɴɢ ᴛʜᴇ 1960s ᴡᴇʀᴇ able to witness, and in some cases to be a part of, the coming of plate tectonics, one of the great happenings in the history of earth science. Each of us, however, saw that unusual event (actually, that series of events) from a different perspective and in a different light. Hence it seems to us that any attempt by someone else to record the history of what went on during that special time is distorted, incomplete, or somehow not quite right. I have no doubt that the story about earthquake seismology in the plate tectonics revolution that I am about to relate will provoke such a reaction in some. All I can say in my defense is that I have tried carefully and painstakingly to make it correct as I saw it, or at least as I remember it more than 30 years later. I also hope that any weaknesses in my story will be compensated by the similar efforts of others to relate that history in this book or elsewhere.[1]

This is not the first time that I have taken on this task. In 1996, my book, *Shocks and Rocks, Seismology in the Plate Tectonics Revolution,* was published by the American Geophysical Union. In 1991, Columbia University Press published *The Incomplete Guide to the Art of Discovery,* in which I explore the discovery of the down-going lithospheric slabs in island arcs that is the key element of subduction as an example of scientific discovery.[2] Subduction is the process by which the earth's crust and uppermost mantle descend into the interior at the so-called arcs, those prominent arcuate structures that are distributed widely over the earth and that incorporate such features as deep sea trenches, explosive volcanism, and shallow and deep earthquakes. I have also given a number of talks on the subject, sometimes followed by published papers. One such paper was

Jack Oliver in his office at the Lamont Geological Observatory, mid to late 1960s. The cartoon on the wall depicts Oliver and Bryan Isacks in Tonga, working on the installation of seismographs to measure slip on down-going crustal slabs in subduction zones. (Photo courtesy of Jack Oliver.)

presented at the AGU symposium in 1992 on the 25th anniversary of the advent of plate tectonics.

As many of the details that I might otherwise feel I had to bore you with are already in print, here I will use thumbnail sketches to describe the essence of what went on in seismology before and during the plate tectonics revolution, and then make some comments about things that I, or we, learned as a result of our experience during those exciting and provocative days of the 1960s.

SEISMOLOGY AT LAMONT

In the mid-1960s I was in my early 40s, a professor at Columbia University and the head of the earthquake seismology group at what was then the Lamont Geological Observatory (now the Lamont-Doherty Earth Observatory). Lynn Sykes and Bryan Isacks, about whom you will read more later, were both 30-ish and former graduate students of mine turned research scientists at Lamont. Jim Dorman was another research scientist, and Muawia Barazangi and Peter Molnar were young graduate students who were part of this story. We had a very active, lively, and innovative group in earthquake seismology at Lamont, and it was rather well funded because of the money that had come into seismological research as a result of interest during the Cold War in a nuclear test ban treaty.[3]

Maurice Ewing, an outstanding seismologist among other things, was the founder and director of Lamont. He had infected us all with the joy of discovery in earth science, and had established certain habits around Lamont, such as extensive archiving of data and working night and day on science, that would serve us well when the plate tectonics excitement arose. Early in the history of Lamont, in the 1950s, Ewing was often active on a day-to-day basis in earthquake seismology. By about 1960, however, his duties as director had grown more burdensome and time-consuming, and the earthquake group had shifted its location to new quarters far removed from its former home near Ewing's office. So he relinquished some of his close ties with that activity, but the overall style of doing science that he had instilled in us carried on.

Ewing was a brilliant scientist and a powerful leader. He was a key factor in, and often the instigator of, the post–World War II exploration of the ocean basins that, in my opinion, triggered what would become the story of plate tectonics. However, he initially did not favor the sea floor spreading and plate tectonics hypotheses. One reason, I think, was that in the early seismic reflection studies of the sea floor, the technique was not well developed and detected mostly the flat and undeformed sea floor sediments, failing to resolve the deformed ones. That sort of evidence seemed to speak against great deformation of the sea floor, and hence against the great movements postulated in plate tectonics theory. Later, of course, the reflection technique at sea was developed further, so that it now detects deformed as well as undeformed sediments.

I have seen and heard it claimed that in one way or another Ewing discouraged his Lamont scientists from publishing pro-plate tectonics studies. I can't speak for all such scientists, but I can state unequivocally that Ewing did not prevent, discourage, or even speak to me in any way against the work of our seismology group. We encountered no opposition whatsoever from him, and we published a good fraction of the most pro-plate tectonics science to come from Lamont. Whatever his own reasons for opposing plate tectonics, he did not impose them on us.

Before turning to a discussion of those Lamont seismological contributions to plate tectonics, let me first mention some relevant contributions by the field of seismology prior to the plate tectonics furor. Seismologists of that earlier era surely deserve some of the credit for what happened later. For example, the creation in the early 1960s of the world wide standard seismograph network (WWSSN), as a result of interest in a nuclear test ban treaty, provided critical and unprecedented data for study of tectonic earthquakes. There were improvements in the techniques for determining precise hypocentral locations, the places where earthquake ruptures begin, by Keith Bullen and Bruce Bolt, and for determining focal mechanisms by H. Honda, Perry Byerly, and John Hodgson. And there were the studies of seismicity by Beno Gutenberg, Charles Richter, and E. Rothé, and the attention to the deep earthquake zones, first found by K. Wadati in Japan and by Hugo Benioff in the United States. Innumerable other seismologists also deserve credit for helping to bring seismology to the point where it was ready to make some major contributions to the plate tectonics revolution. Much of this work was carried out well before the Lamont Geological Observatory was formed in 1949, but, by the early 1960s, when the ideas and concepts that would grow to become the theory of plate tectonics were arising, the Lamont program in earthquake seismology was established and bustling. The atmosphere, the facilities, the data archives, and the colleagues at Lamont made it a very favorable environment for the earthquake-based research that would turn out to be an important part of the plate tectonics revolution.

Let us now focus on Lamont and studies there that affected the revolution. Some were carried out well before the mid-1960s. Studies of earthquake surface waves, those seismic waves that travel through the shallow layers of the earth and hence provide information on the crust and uppermost mantle, by Ewing, Frank Press, and me (among others), helped to show that the crust beneath the deep sea was not subsided continental crust, as some earth scientists had argued. Then Marie Tharp, working with Bruce Heezen on a physiographic map of the sea floor, found that the mid-ocean ridge system that stretches for large distances around the globe had the kind of narrow valley near its crest that suggested that the crust had been rifted apart there, and that many oceanic earthquakes occurred beneath these rifts. Ewing, Heezen, and Tharp described the great extent and continuity of the globe-encircling mid-ocean rift system using seismicity as one key piece of information. That study may have been a factor in Harry Hess's thinking, when he made his great proposal of sea floor spreading: that the sea floor was spreading apart at the mid-ocean rifts where magmas welled up from below to form new oceanic crust.

Hess' sea floor spreading hypothesis was known at Lamont from the time he first proposed it publicly, but it languished for a while. Then the

Lamont geomagnetic group caught fire and brought Lamont's huge supply of magnetic data on the sea floor to bear on the matter, and so strengthened the case for spreading substantially, as has been described elsewhere in papers focusing on the geomagnetic part of the plate tectonics story.[4]

The excitement within the geomagnetic group soon spread to the earthquake seismology group, as Jim Heirtzler and John Foster took pains to pass news of their successes across the Lamont campus to Lynn Sykes and me. At the time, Sykes was working on hypocentral locations and on earthquake focal mechanisms, that is, the orientation of and direction of motion of the earth on opposite sides of the mid-ocean rifts and the transform faults that displaced them. Just the things, it turned out, for the next major step into global tectonics by seismology.

J. Tuzo Wilson had taken up the case of the peculiar steplike offsets of ocean ridges. He called them *transform faults* and proposed a mechanism to account for them; he then published a paper in the journal *Nature* that suggested a seismological test of the hypothesis. Jim Dorman at Lamont called that paper to the attention of Lynn Sykes, who was abroad at the time, and he soon returned to Lamont to take up Wilson's challenge. Lynn was the only seismologist anywhere to do so, at least so far as I know.

Lynn quickly reported positive results based on the patterns of seismicity and of the focal mechanisms at the ridge offsets, and his paper added substance and credibility to the transform fault hypothesis, and hence to the concept of sea floor spreading.[5] That concept, with the addition of the hypothesis by Fred Vine and Drummond Matthews at Cambridge that connected reversals of the earth's magnetic field with magnetic anomalies of the sea floor, was gaining wider attention. Sykes' paper supported the concept and was a clear example of good deductive science: Vine and Matthews had a hypothesis, Wilson deduced a consequence, and Sykes showed it was true.

A major question was then obvious. If new sea floor was created at the ridges, then what? Was Earth expanding to accommodate the new surface area? Or was old surface area being lost as it descended somewhere else? Some said the former, some said the latter, and if crust did descend there was disagreement about where. Some perceptively said at the arcs, the sites of the deep trenches. Some said beneath the continents. Just how and why it descended wherever it did was enigmatic. At Lamont we had the data to provide the answer. With support from the National Science Foundation, Bryan Isacks and I had been operating a seismograph network in the Tonga–Fiji island area of the South Pacific, a region of exceptionally deep earthquake activity. We had begun the project simply to observe and study the poorly known phenomenon of deep earthquakes in inductive style, not to test some particular hypothesis.

We found that the zone of earthquake activity that dipped from just beneath the arc to depths of about 450 miles (720 kilometers) marked a thicker zone of very efficient seismic wave propagation. Seismic waves traveled up this zone with far less attenuation than waves traveling through a comparable range of depths in parts of the mantle far removed from a deep earthquake zone.

When we assumed that the shallow mantle east of Tonga propagated seismic waves similarly, which we now know that it does, and that efficient propagation correlated with strength, we were able to draw a now-famous cross-section showing a layer of strength, the lithosphere, dipping beneath Tonga and hence descending and likely being underthrust there. The basis for the modern subduction model appeared one day on our blackboard, and thus the matter of disappearing surface area was resolved, as were a lot of other things characteristic of island arcs wherever they are found. Those things included the nature and cause of the deep sea trenches or ocean deeps; the explosive volcanism associated with the arcs; the accretionary wedges that were piled up against the arc as the sea floor descended; the grabens, or down-dropped little depressions on the seaward wall of the trenches; the earthquake seismicity and the focal mechanisms of associated shallow and deep shocks; and the gravity anomalies that resulted from the rearrangement of mass as the region of the arc deformed to accommodate the down-going lithosphere.

It was a true Eureka moment for us. Others had suggested before that material descended in the trench areas, and Robert Coats, in a fine paper that attracted little attention until our model appeared, had explained Aleutian volcanism as a consequence of such descent to about 60 miles (100 kilometers).[6] But, so far as I know, no one before us had thought in terms of such a large-scale thrusting phenomenon that moved a 60 mile (100 kilometer) thick slab of lithosphere from near the surface to depths of at least 450 miles (720 kilometers), or had even brought the lithosphere-asthenosphere structure into the picture. The Tonga–Fiji project, as we tended to call it, was a clear and unambiguous example of inductive science and serendipity.[7]

We were more or less onto the concept of the moving plates then, but we called it the *mobile lithosphere model,* using the terms *lithosphere* and *asthenosphere.* These terms had been established much earlier to describe the near-surface layer of strength and the underlying weak zone needed to explain glacial rebound, but had not appeared in studies of large-scale tectonics.[8] The term *plate tectonics,* which is so well known and widely used today, had then not been invented and hence was not in use. We had not worked out the global pattern of the plates and their motions. That was done by Jason Morgan at Princeton.[9] Morgan not only developed the global pattern of the plates, but also gave the basis for a geometrical description of their motion, from which Xavier Le Pichon at Lamont soon provided a global map of the plates with quantitative values for relative plate motion. Morgan's paper on this subject, and our paper on the Tonga–Fiji mobile lithosphere story, were both presented at the outstanding 1967 annual meetings of the American Geophysical Union in Washington, D.C.

As a result of our early successes, Lynn Sykes, Bryan Isacks, and I were flying high, and we set out to relate all relevant information from earthquake seismology to the plate model so as to test and further develop it. We did so rather thoroughly, and our paper, "Seismology and the New Global Tectonics," published in the *Journal of Geophysical Research* in 1968, became something of a classic.[10] It showed that the global pattern of earthquake belts, including the sites of deep earthquakes, the global pattern of earthquake focal mechanisms, and a variety of other evidence from earthquake studies, was in accordance with the plate model (although the term *plate tectonics* was still not in use).

There were many other things special about that paper. For one thing, the order of authors was determined by lot, but not because of any friction among us. We agreed to do so as the work was begun so that the burden would not fall mostly on one individual. All three of us worked hard and made major contributions to the paper, which called for a lot of specialized detail as well as generalizations, as it reported on all evidence from earthquake seismology relevant to global tectonics. We thought, and hoped at the time, that we were writing a basic paper on earth science. The Lamont contribution number assigned in routine fashion to the paper was, coincidentally, "1234"!

The paper included a map of global seismicity prepared by Muawia Barazangi and James Dorman at Lamont, who used computing techniques developed by Bruce Bolt for the IBM704, one of the first digital computers to become widely available in earth science. The paper also included the simple block diagram that illustrates schematically plates and plate motions, and that has been widely reproduced in original or modified form many times. It still appears in most basic geology textbooks. The preliminary sketch on which that figure is based was, incidentally, first doodled on a sketch pad during a rather unexciting UNESCO committee meeting in Paris, a meeting called for another purpose and on a completely different topic. The figure was done privately and not discussed at that meeting.

The "New Global Tectonics" paper was widely read and cited and is probably the principal reason why I was asked to prepare the present essay, although I must say here that, for me, the thrill of discovery was far greater when we came across the down-going slab mechanism in the Tonga–Fiji project. That was the true Eureka moment for me. The "New Global Tectonics" paper was more of a highly successful, broad-ranging synthesis. It produced more or less what we expected and did not have the elements of surprise and revelation that were part of the Tonga–Fiji study.

Many other papers on seismology and tectonics began to appear at about that time or shortly thereafter. William Stauder, at St. Louis University, wrote a fine paper on focal mechanisms along the Aleutian arc.[11] Dan McKenzie and Robert Parker at Cambridge University had written one earlier on focal mechanisms and plate tectonics in one part of the Pacific.[12] Their paper preceded the "New Global Tectonics" paper, but followed the "Tonga–Fiji" paper, which they referenced. Peter Molnar at Lamont and I published a paper showing how seismic waves propagated well within any plate, but poorly across the boundary or gap between plates.[13] Each new seismological paper seemed supportive of the plate model. None challenged it. Now, some 30 years later, although there have been refinements and modifications of that early model, there have not, as far as I know, been any effective challenges to it.

SOME THOUGHTS ON SCIENTIFIC DISCOVERY

What can we learn from our experiences with the coming of the very robust plate tectonics theory that may help guide the direction of earth science, or any science, in the future? Some things have occurred to me, which is not to say that they haven't also occurred to others. I'll try to cast them here in the light of the seismological research on plate tectonics described here.

First, in order to make an important discovery in science, it is essential to be at the right place at the right time. Lamont was clearly a good, perhaps the best, place for a seismologist to be when the plate tectonics story was beginning. Fine colleagues were there, including those working in geomagnetism at just the critical time. There were also archives of what turned out to be key seismological data. We had access to, and training available for, that newfangled device, the computer, a good line of communication to events elsewhere in related fields of science, and an innovative spirit and inner confidence in our capacity to learn more about the earth. Being in the right place at the right time is, of course, often primarily a matter of fate or destiny, but nevertheless a savvy scientist can improve his or her lot by recognizing the features that were favorable to discovery in the past and then seeking optimal surroundings and opportunities accordingly. The Lamont contributions – in fact, almost all contributions anywhere to the development of plate tectonics – were not so much the result of a brilliant idea, that is, a stroke of genius by one human unfamiliar with the observational data of the science. Rather, they were the result of good hard work by very capable people who happened to have, for one reason or another, the right skills, and happened to be in fertile environments with appropriate data and colleagues at the right time.

Second, I think the important role of science in the inductive style must be noted, not only in seismology, but in the development of the early stages of plate tectonics in general. It was surely the exploration of the ocean basins following World War II, almost all done in inductive style, that brought about the events in the earth sciences of the 1960s. The inductive style was important early in seismology through study of seismicity and focal mechanisms and later in discovery of the nature of the subduction process in Tonga–Fiji. I think we must keep a substantial fraction of our research effort in earth science in the inductive style. I really don't have much patience with the peer reviewer who rejects something because the "problem is not clearly stated" without looking at other aspects of what is proposed. For me, a key question is whether an important part of the unknown is being explored.

The other style of science, the deductive style, tends to limit us to what we are capable of imagining. We need not be so constrained. The inductive style of science improves our chances of learning things beyond our imagination. In a sense, the inductive style requires a bit more humility, but it offers the possibility of greater rewards.[14]

In addition to the collection of new observations of the earth in the inductive style, the coming of plate tectonics was also dependent upon the bringing together of many observations, and often very diverse observations. Syntheses, particularly global ones, were an important part of the revolution. I refer not just to "The New Global Tectonics" paper, of course, but also to others such as the maps of sea floor magnetic anomalies, sea floor topography, the global plate configuration, and many others. In my opinion, not enough effort in science is put into synthesis, and not enough effort is put into forms of communication that break the boundaries of modern specialties. In other words, I think too many scientific papers are so loaded with jargon that information is not spread beyond the bounds of that specialty.

It has long seemed to me that a scientist who has what he or she thinks is an important idea or result should write it up in a straightforward style, minimizing jargon, so that the audience will be as large and as broadly based as possible, and the good work will spread beyond just a few specialists. On the other hand, it is easy to suspect that someone who loads a paper with jargon is trying to conceal from many that the work is really not that significant, although, of course, that is not always the case. Perhaps I am overly cynical on this matter.

Icons, simple figures that conveyed immediately the essence of a concept or an idea, were very important in the revolution. I've mentioned the block diagram of the plate model. Others were the global seismicity map, the sketches of the Vine–Matthews model illustrating their hypothesis, the global map of the plates, and Tuzo Wilson's cartoons of transform faulting.

To back up my point on the emphasis on the inductive style of science, I would like to draw further attention to the role of, and the control of science, by observations. After a lifetime of doing science, and of being a part of the world of science, my understanding of the essence of science can be described in a few words. Science is the "organization of observations." Furthermore, that's all it is. Science is basically empirical. Observations are the only truth, or facts, of science. Hypotheses, theories, laws, or whatever, mathematical or not, are merely ways we have devised to organize those facts so that we can comprehend and interrelate them. Any part of the theoretical side of science is subject to revision if observations so dictate. The scope of science is hence determined and delimited by its observations. Therefore one good way, probably the best way, to add to the scope of science is to make reliable new observations of the unknown.

In saying these things, I do not mean to minimize the importance of theory and new ideas for new theories. Our level of comprehension of the world can sometimes make a dramatic advance upon the introduction of a new and cleverly devised theory. Nevertheless, regardless of the beauty, elegance, or sophistication of a theory, the observational database remains in control, particularly over the long term.

To make this point clear, consider the history of the concept of continental drift. It was first proposed in A.D. 1596 by Abraham Ortelius, the famous Flemish cartographer of German ancestry, probably because he was first, or among the first, to see the key observational data, namely reasonably accurate maps of the Atlantic coasts of Africa and South

America.[15] That great idea was quickly forgotten or ignored in the absence of additional observational data to test it. Others later proposed the same idea independently, and were also pretty much ignored, basically for the same reason. It was not until the early 20th century, when Alfred Wegener, a German geophysicist/meteorologist, had the idea of drifting continents, backed it up with sound observation, and communicated it well, that it began to catch hold with some scientists. By then, the geology of the land areas had become reasonably well-observed, and so there was considerable support from geology for Wegener's model. However, Wegener's idea faltered and sputtered for a long while, many decades, until new observations, this time of the geology of the sea floor, saved the day and allowed it to be incorporated into the new concept of geodynamics called plate tectonics.[16]

Some will recognize, correctly, that it was this kind of thinking that led me to attempt to explore and obtain new observations of a poorly known part of the earth, the deep continental crust and uppermost mantle, through deep seismic reflection profiling, that is, echo sounding, of that part of the earth. The continental crust is typically about 25 miles (40 kilometers) thick. The mantle is about 1,800 miles (2,900 kilometers) thick, but I focus here only on the crust and the upper few tens of miles of the mantle. The project, Consortium for Continental Reflection Profiling (COCORP), is designed to probe this part of the earth using the highly sophisticated techniques of the petroleum industry. Industry normally confines its attention to the upper 6 miles (10 kilometers) of the crust that can be drilled for oil, but we are extending our reach to explore the underlying 60 miles (100 kilometers).

I remain convinced of the great potential for major advance in earth science through acquisition of this kind of observational data. For those whose goal is to make a great advance in earth science, perhaps on the scale of plate tectonics, my recommendation is to make a major effort, in the inductive style, to explore and understand that region of earth, the deep crust and uppermost mantle, comprehensively, through application of the seismic reflection profiling technique that we know will work and that we are fully capable of carrying out. The envelope of our world of science that is determined by our observational database will be pushed back. If history is any kind of a guide, new discoveries and a new level of understanding of earth will follow. We are not likely to get to that goal as fast as we might, however, if we confine our efforts to observation only in an area where a "clear geological problem can be solved."

The kind of reasoning that I have just been using for the last few paragraphs is designed to reveal effective strategy and tactics from study of the history of previous successes in science, so that such strategies and tactics can be applied to new topics. Such reasoning, in my opinion, is sound and very likely to produce major discoveries. I think historians should focus on and analyze success stories from the past, and pass those analyses along to active scientists so that they can improve upon their efforts in the future. As a practicing scientist for half a century, I felt little need to know the personal habits of great achievers in science of the past that historians sometimes dwell upon, but I longed to know, and could rarely find, how those achievers thought and planned as they made their great discoveries.

I'm going to leave a few words for young scientists and researchers now that we are in an era well beyond the 25th anniversary of plate tectonics. My message is this. When you hear stories of great advances in science in the past, such as plate tectonics, whatever you do, don't get the feeling that you came along too late and so missed the fun. That's what my fellow students and I thought when we read of some early discovery that we had missed. We were foolish. Plate tectonics was just ahead. Surely, there are other big discoveries in earth science right around the corner and waiting for someone like you. Work hard, develop clever strategies and tactics, try to position yourself appropriately in the world of science (that's not so easy!), and, with luck, you'll be part of a big discovery of the future. There may even come a time in your career when the session on the history of geophysics at an AGU meeting will be about *your* discovery, not about the discovery of plate tectonics!

PART IV
THE PLATE MODEL

By 1967 most geophysicists and oceanographers were either convinced or on the verge of being convinced that the earth's surface was divided into large blocks that were moving en masse: splitting apart at mid-ocean ridges, moving laterally across the ocean basins, and then sinking back into the earth at the boundaries between continents and oceans. Moreover, it was becoming clear that the geological arguments that had been put forward for continental drift more than 40 years earlier were probably largely correct. But those data had been criticized as qualitative and not verifiable – or at least they looked unverifiable in retrospect. It remained to quantify the motions of the crustal blocks, and to show that the motions calculated for any one block were consistent with the motions calculated for their adjacent blocks. The blocks began to be referred to as "plates" – flat, thin, and rigid – and the result was a theory called plate tectonics.

CHAPTER 11

PLATE TECTONICS:
A SURPRISING WAY TO
START A SCIENTIFIC CAREER

Dan McKenzie

Bᴇᴛᴡᴇᴇɴ 1963 ᴀɴᴅ 1968 ᴀ sᴍᴀʟʟ ɢʀᴏᴜᴘ ᴏꜰ ɢᴇᴏᴘʜʏsɪᴄɪsᴛs, ᴡᴏʀᴋɪɴɢ at the Universities of Cambridge, Toronto, and Princeton, the Scripps Institution of Oceanography, and at Lamont Geological Observatory of Columbia University, put together the group of ideas now known as the theory of plate tectonics. Some of these people were already major figures in the field, whereas others, like me, were only just out of graduate school. As the far-reaching success of these ideas became clear, we all rapidly became famous, to a degree that surprised even those of us who were already well known. Nothing any of us did before or after was as scientifically spectacular, although all of us spent our working lives as active scientists publishing papers.

The published papers that led to the theory were collected by Allan Cox of Stanford University, who grouped them by subject area and wrote excellent introductions.[1] The history of the events is well described by Bill Menard of Scripps, who was closely involved in many of them.[2] Perhaps because these two excellent books provide such a good historical record, the editors of this collection asked us to write about the events from our own point of view. Although this is in some ways easy to do, it is not a normal activity for scientists, nor one at which they have much practice. Furthermore, the events took place more than 30 years ago. But this sociological side of scientific discovery has (rightly) become recognized as of great importance by those, such as Thomas Kuhn, who write about the history of science, even though the formalism that they have generated seems to me at least strange and somewhat artificial.[3] But it is

Dan McKenzie, astride the San Andreas Fault in 1967. (Photo courtesy of
Breck Betts, Scripps Institution of Oceanography.)

no good complaining when so few of those involved have written about
their experiences, since such accounts must form part of the raw material used by historians of science.

I have tried to organize this essay around the events of the mid-1960s,
with an account of how I came to be at the Department of Geodesy and
Geophysics at Cambridge, or Madingley as it is often called, after Madingley Rise, which houses it. Little has yet been written about the history
of this remarkable place, where many of the discoveries that led to plate
tectonics were made, and so I thought a brief history might be of interest. I myself only became involved in plate tectonics after I finished my
Ph.D. in 1966. The final version of the theory was put together very
rapidly during 1967–1968. I thought I should say something about what
major scientific advances look like to those involved at the time, rather

than to those who write about them afterward. The most interesting aspect of the period since 1968 is how the ideas that were so new and revolutionary a few years before were so quickly and thoroughly integrated into our understanding of how the earth has evolved, and how quickly those scientists who were involved in these discoveries became recognized internationally.

HOW I CAME TO BE WHERE I WAS IN 1967

I was born in February 1942, after my parents had decided that Hitler was not going to win the war. My father was a surgeon who qualified just before the war started. His father was also a surgeon, with a practice in London on Harley Street, a large house in Highgate in North London, and a chauffeur. My father, like me, went to Cambridge as an undergraduate. My mother's background was quite different. Her father was a laborer who shoveled coal into the furnace in a power station in Leeds, which was then a grim industrial town in northern England. She was intelligent, and benefited from one of the earliest socialist measures in the United Kingdom, which opened up free university education to about 200 people a year who won scholarships. She won a scholarship and applied to Cambridge, which invited her to an interview where she was made to read aloud. They rejected her because of her thick northern accent, and she never forgave them. She was never entirely happy that I became a Cambridge academic. Until I was 7 we lived in the country but we moved to London when my father, like his father, became a Harley Street doctor. My grandfather was grand enough to have a chauffeur who drove him to work every day from Highgate, but we could only afford to rent a flat above my father's consulting rooms in Harley Street.

My mother, whose maiden name was Fairbrother, later wrote a book about this period, when she was bringing up two children largely by herself in the country, and a second book about our life once we moved to London.[4] My brother and I figure prominently, but she used our middle names, Peter and John, to try to save us from problems at school. I am still embarrassed by her vivid picture of me as an introverted, hesitant child, in most ways less successful than my younger brother. I went to three excellent private schools in London, although I was not at all successful academically until I was about 14. Then I started to learn proper mathematics (I have never been able to add or spell), physics, and chemistry, and to win prizes. Although my father made some unconvincing attempts to deflect me into medicine, I was clear that I wanted to go to Cambridge and be a physical scientist. This I duly did in 1960, after an interview in which I talked to a distinguished classicist about Dostoevsky and British wild orchids, on the strength of which he gave me a place at Cambridge to read physics. Cambridge interviews are now much more serious and professional affairs, and I am sure that my mother would now be accepted.

The natural sciences course at Cambridge differs from those at most English universities because it requires students to take lectures and exams in three sciences (mathematics does not count as a science, although it is required for physics and chemistry). I chose to take geology as my third science, largely influenced by the 19th-century books on the subject by Charles Darwin, Charles Lyell, and Archibald Geikie in the school library. But the geology course at Cambridge was awful: I learned the hundreds of fossils necessary to identify stratigraphic zones, and to draw and name the parts of echinoids, ammonites, and so on. Although I liked the people, and especially the field trips, I thought the course was stupid, and gave up geology after a year to take a degree in physics. It was partly the people, especially Maurice Hill and Drummond Matthews (who was always known as "Drum"), and partly the idea of using physics to understand the processes operating within the earth, that attracted me back to the earth sciences. I joined the Department of Geodesy and Geophysics at Madingley Rise as Sir Edward Bullard's graduate student in 1963.

MADINGLEY RISE

This is not the place to give a detailed account of the history of how there came to be a Department of Geodesy and Geophysics at Cambridge (now the Bullard Laboratories of the Department of Earth Sciences). But this place played such an important role in the development of our present ideas that I think I should give some account of its origins. Fragments of the story are described in the biographies of some of those involved: Sir Gerald Lenox-Conyngham, Maurice Hill, and Sir Edward Bullard, which have been written for the Royal Society.[5] The department was started by two people, Sir Gerald Lenox-Conyngham, a retired colonel from the Indian Army, and Professor Hugh Newall, who was professor of astronomy at Cambridge. The correspondence between them is now in the Bullard Archives in Churchill College, Cambridge. Their idea was to found an Imperial Geodetic Institute at Cambridge, like the Prussian Geodetic Institute in Potsdam, and on a similar scale. But there was no money. The department therefore initially consisted of Lenox-Conyngham alone. He obtained a post for an assistant in 1931 and hired Edward Bullard, who was always known as "Teddy," and who had just obtained his Ph.D. (in quantum scattering of electrons) in the Cavendish Laboratory under Rutherford.

Teddy was a first-class experimental physicist. He started by using gravity to study earth processes, but was soon exploiting seismic methods on land and at sea, as well as the measurement of heat flow. His interest in the earth's magnetic field developed during the Second World War, when he was concerned with magnetic mines. He became frustrated when he returned to Cambridge after the war, where Lenox-Conyngham, now in his late 70s, was still head of the department (there was no retirement age), and he left for the University of Toronto. He did not return to Cambridge until 1956, after the group who worked there with Keith Runcorn had broken up when Keith left to become professor of physics at Newcastle. The paleomagnetic work on continental drift, which was carried out by Keith Runcorn, Ted Irving, Ken Creer, Jan Hospers, and others with such energy and success at Cambridge in the early 1950s, stopped with Keith's departure, and had surprisingly little influence on later developments–even at Cambridge, where it had been so prominent. Later discoveries have clearly shown that the conclusions based on palaeomagnetic measurements were correct, and their failure to convince the wider community of geologists and geophysicists remains to me the most interesting part of the history of plate tectonics.[6]

Teddy and Maurice Hill together developed a very active group concerned with marine geology and geophysics. Although they were in competition with similar larger groups at Lamont and Scripps, they retained excellent relations with the Americans. They attracted undergraduates whose backgrounds were in physics and who wanted to build instruments or do theory, and numerate geologists. I arrived as Teddy's graduate student in October 1963, after Fred Vine and Drum Matthews had published their explanation of how the magnetic stripes in the oceans are formed, and while Teddy Bullard, Jim Everett, and Alan Smith were using a well-known theorem discovered in the 18th century by the German mathematician Euler to reconstruct the position of the continents around the Atlantic before they were rifted apart.[7]

People often say to me how exciting it must have been to be at Cambridge at this critical time. But this is a retrospective view. At the time it looked quite different. Fred and Drum were searching for magnetic anomalies in the North Atlantic magnetic records from the many cruise records available at Cambridge, but found little convincing evidence

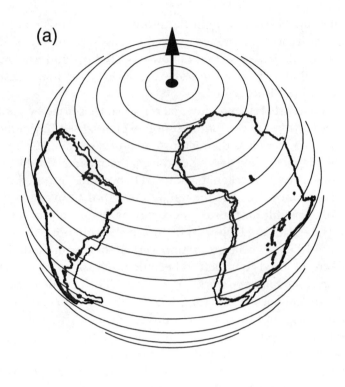

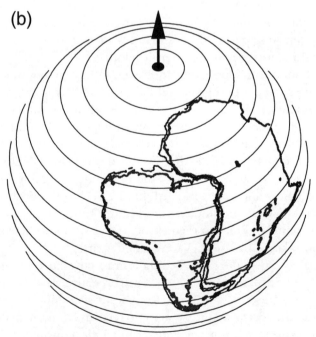

that their suggestion was correct. Why they were having such problems only became clear when Carol Williams, a graduate student at Cambridge, and I reanalyzed the same data in 1971, using what were by then standard methods.[8] The navigation was carried out with a sextant, by taking sights on the sun and stars (satellite navigation only came into general use in 1968). Because the North Atlantic is so often covered with clouds, the navigation errors in the data were sometimes as large as 30 miles (50 kilometers). So it was difficult to match anomalies between different ship tracks.

Another problem is that magnetic anomaly patterns from slowly spreading ridges like the North Atlantic are often variable and hard to recognize without a theoretical model. Teddy tried to find evidence that the boundaries between two geological provinces of different ages in Africa matched similar boundaries on the other side of the South Atlantic in South America. But the evidence was not very convincing, because rifting generally exploits such geological boundaries rather than cutting across them. I wrote a Ph.D. thesis on mantle convection, which taught me fluid mechanics and enough materials science to know that all materials creep at high temperatures and low stresses. Although it now seems strange, neither I nor the other graduate students changed thesis topics to work with Fred and Drum. The only person to join them was Tuzo Wilson from the University of Toronto, when he was on sabbatical at Madingley. He wrote his well-known paper on transform faults

---

The fit between Africa and South America obtained by Teddy Bullard and his colleagues using Euler's Theorem (Bullard et al., 1965). The theorem states that any motion of a rigid plate on the surface of a rigid sphere corresponds to a rotation of the plate about some axis that passes through the center of the sphere. The problem on the Earth is that every point on its surface is on a moving plate, and no rigid sphere exists. So one plate must be chosen and taken to be fixed. Then the motion of any other plate with respect to this fixed plate corresponds to a rotation about an axis. In this figure Africa has been taken to be fixed, so South America moves. (a) shows the location of this axis, marked with an arrow, that Teddy and his colleagues found for the motion between Africa and South America. The circles are lines of latitude about this axis, just like the usual lines of latitude about the Earth's rotational axis. (b) shows the original position of the two continents before the South Atlantic opened, obtained by fitting the edges of the continents together. These edges are under the sea, and are not the present coast lines. As the continents move, every point on the South American plate moves in a direction that is parallel to the latitude lines. This behavior is easily seen by comparing the positions of the latitude lines in the two pictures before and after opening. Their position on South America does not change.

while he shared an office with Harry Hess from Princeton University, and a lesser known paper with Fred Vine on magnetic anomalies while Fred was still a graduate student at Cambridge.[9] As graduate students we were not especially stupid, and three of us (Bob Parker, John Sclater, and myself, in addition to Fred Vine) have since been elected to the Royal Society. At the time it simply was not obvious to us that what Fred and Drum were doing was so important.

OCTOBER 1966–OCTOBER 1968

Teddy persuaded the organizers of a NASA conference to invite me to talk about high temperature creep.[10] It was held in New York, close to Columbia, and was attended by many people whose names I knew from their publications, and to whom I was introduced by Teddy. Two of the papers made a deep impression on me: one by Fred Vine, who had moved to Princeton, showing that the Pacific magnetic anomalies beautifully confirmed his and Drum's suggestion, and the other by Lynn Sykes from Lamont, whom I met for the first time, showing that the sense of motion on transform faults agreed with Tuzo's proposal and could only be explained by sea floor spreading.[11] At last there was good data to confirm the earlier ideas. I returned to Cambridge, completely convinced I should work on sea floor spreading. I stayed for a month, during which I was examined for my Ph.D. I also carried out all the calculations for my first paper on plate tectonics, showing how a spreading ridge can account for the elevated heat flow that Teddy had found to be associated with what we now knew to be spreading ridges.[12] The idea for this paper came from work by Xavier Le Pichon and his colleagues at Lamont, although they had argued that the heat flow observations were not compatible with sea floor spreading. I carried out the necessary numerical calculations at the Seismological Laboratory at Caltech, where I went for six months as a postdoctoral fellow in January 1967, and was pleased with the paper.

Unlike Xavier, I used an analytical solution to the equations, which I obtained by first converting them to dimensionless form. I had learned how to do this from the fluid dynamicists, particularly Adrian Gill, who was then at Scripps, and my approach is still widely used to construct thermal models of spreading ridges.[14] However, what most pleased me was to be able to demonstrate that the heat flow (and, later, the topography and melt generation) of spreading ridges does not require a hot upwelling region of convection beneath the ridge, but can simply be explained by plate separation and the upwelling of hot mantle into the

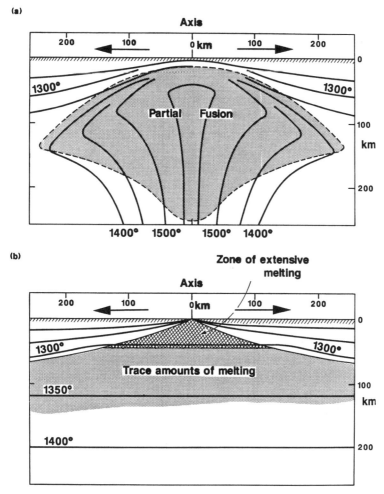

When I first became interested in plate tectonics it was generally accepted that ridges were everywhere underlain by plumes of solid mantle material that was hotter than the surrounding rock, and for this reason moved upwards. This proposal is illustrated in (a), but is very difficult to reconcile with how ridges behave. New ones form suddenly when continents split, their axes are offset sharply in many places by enormous faults, and they sometimes suddenly jump into older parts of plates. In the first paper I wrote on plate tectonics, I suggested instead that ridges are entirely passive, and are formed by passive upwelling of the same hot mantle that everywhere underlies the plates (D. P. McKenzie, 1967. Some remarks on heat flow and gravity anomalies, *Journal of Geophysical Research*, 72: 6261–6273). This model is illustrated in (b), and satisfies all the observations and avoids the problems faced by (a). (Modified from D. McKenzie and M. J. Bickle, 1988. The volume and composition of melt generated by extension of the lithosphere, *Journal of Petrology* 29: 625–679, and R. L. Oxburgh, 1980. Heat flow and magma genesis. In *Physics of the Magmatic Processes*, R. B. Hargraves, ed. Princeton: Princeton University Press, pp. 161–199.)

space in between the two plates. This idea resolved a long-standing puzzle, which I discussed with Tuzo when he was at Madingley in 1965. Africa is surrounded by spreading ridges, all of which have moved away from the continent as they spread. If they were convection cells, how could these cells know how to move at exactly the right velocity to remain beneath the spreading ridges on either side of Africa, thousands of kilometers apart? There is obviously no problem if ridges have no deep structure, but are simply formed wherever plates separate.

Most of my time at the Seismological Laboratory at Caltech I spent learning earthquake seismology. At the time, my geophysical background knowledge was rather poor. I knew a little old-fashioned geology, and some bits and pieces of geophysics I had learned during my three years as a graduate student. But Cambridge had (and has) no courses for graduate students in earth sciences, and no one worked on earthquake seismology at Madingley. So I took Caltech courses, some in earth sciences and some in physics, especially from the physicist Richard Feynman, whose showmanship annoyed me, but whose mastery of analytical methods I found dazzling.

I also went to my first American Geophysical Union (AGU) meeting, in the spring of 1967 in Washington, D.C. I have always found such meetings hard to exploit properly. I try to do so by going through the abstracts of the talks carefully, making a list of where I need to be when. Jason Morgan, from Princeton University, included an abstract of a paper he had recently published, which reinterpreted some Lamont data from the Puerto Rico Trench. I knew several of the Lamont people were cross with what Jason had done, and that they intended to give him a hard time, principally, I suspected, because Jason had reinterpreted their data. I was not involved and not interested, so I left the session just before his talk. As is now well known, he did not talk about his published paper, but about plate tectonics. His talk generated little interest, and I did not become aware of what happened until November of that year.

But two aspects of the meeting did make a deep impression on me. By April 1967, most marine geophysicists had seen the Pacific magnetic anomalies that I first saw in New York, and were equally impressed. They had looked at their own records, found similarly beautiful profiles, and submitted abstracts to the AGU meeting.[15] The result was a sea floor spreading bandwagon – what Drum Matthews always disparagingly referred to as 'me-too' science, and what I think Kuhn means when he talks about scientific revolutions as changes in worldview.[16] There were no new ideas, nor any proper modeling, just an overwhelming sense that the whole ocean floor was covered with stripes. The other was a paper

(a)

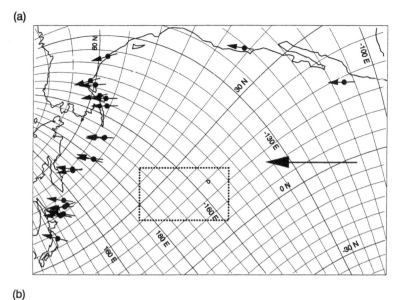

(b)

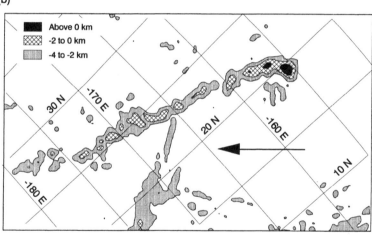

This figure shows the motion of the Pacific plate, obtained from the motion on faults during earthquakes, when the North American plate is fixed. The solid dots show the locations of a number of large earthquakes, produced by the motion between these two plates. The arrows show the direction of motion between the two sides of the faults on which these earthquakes occur when the side of the fault that is part of the North American plate is taken to be fixed and the Pacific side is moving. The map is in a special projection, chosen so that the motion of the Pacific plate is everywhere parallel to the big arrow if the plate moves rigidly. (b) Contours of the depth of the ocean round Hawaii, using the same projection as (a). If the volcano that forms the Hawaiian Ridge is fixed to North America, the ridge should be parallel to the large arrow, which is approximately true.

by Father William Stauder, from Saint Louis University, concerned with the mechanisms of earthquakes along the Aleutian Trench, which I misunderstood. Earthquakes are generated by movement on faults, when one side of the fault slides past the other. The direction of this motion is known as the *slip vector*. I thought Stauder had plotted these slip vectors, whereas in fact he plotted what he believed to be the directions of greatest and least stress. This rather technical issue sounds trivial but is not: plate tectonics is concerned with slip vectors, not with stresses, and the relative motion between plates produces earthquakes whose slip vectors show the direction of their relative motion.

I returned to Caltech, and moved to a postdoctoral research position at Scripps at the end of June. I had sent off the paper on ridges, and was looking around for other good problems. I read all of Harry Hess' papers and had returned to Teddy's paper on the geometric fits because I remembered I had disliked the method he used to fit the continents together.[17] I was reading his section on Euler's theorem when it occurred to me that this was what was needed to describe plate motions. Once I had the idea of doing so, everything else followed, and within a day or two I had thought through the consequences. I saw that I could divide the surface of the earth into plates using the earthquake zones, and use a National Geographic globe with a transparent plastic cap to work out where the relative poles of rotation were from the transform faults and other features on the plate boundaries. My problem was that I had no data, such as spreading rates from magnetic anomalies, because the Scripps data had not been digitized or interpreted. Then I remembered Stauder's AGU talk, and realized I could use the slip vectors from earthquakes that he and his students and colleagues had determined to obtain the direction of relative motion between the two sides of the faults which moved in earthquakes (which I found he had not done), and a mapping program that Bob Parker, who had also moved to Scripps, had written with a projection that made the whole idea simple. I explained everything to Bill Menard at Scripps, who could see nothing wrong and soon became equally enthusiastic.

Shortly thereafter, John Mudie, who was also doing postdoctoral reseach at Scripps, returned from a conference at Woods Hole and gave an account of Jason Morgan's talk there. Although John's account was somewhat unclear, it was immediately obvious to me that Jason had had the same idea as I had, and I asked Bill what I should do. Without hesitation he said, "Publish." I thought I could write a nice compact paper on the idea and used this as a carrot to write up the last part of my Ph.D., a long, dull, worthy paper on the effect of rotation on convection in a rotating sphere.[18] As soon as I finished, I wrote the first published paper on plate tectonics with Bob Parker.[19]

It took only two or three days. Bob and I went down to the main post office in La Jolla in late October to send it off, but were nearly defeated. It was Saturday, so the office was closed. But we got quarters and fed the stamp machine until we had enough stamps to send it to the journal *Nature* in London by air mail. There was just enough space for the address. We never saw the paper again. *Nature* received it on November 14, but we never got referees' comments or page proofs. The paper was published on December 30, 1967.

Soon after Bill Menard told me to publish he received Jason's paper to review.[20] In his book Menard says that he received a pre-print of this paper in the spring of 1967, but he never mentioned this to me, either at the time or later.[21] I don't remember reading Jason's paper at the time, although I may have done so. Jason and I took different approaches to explaining the same ideas, and used different data, from seismology and from marine magnetic anomalies, to show that the ideas worked in practice.

When I wrote the *Nature* paper I was inexperienced, both in writing scientific papers and in presenting data. It was only the fourth extensive paper I had written. Unlike Jason, I made little attempt to explain how the ideas connected with earlier work, and tried to be as economical with words as possible. I did not publish the figure I drew showing the observations themselves. I have always regretted not doing so, and therefore show it here as the figure on page 179. My other regret concerns Hawaii. The projection Bob and I used suggested that the Hawaiian Ridge was formed by a volcano that was fixed to the North American plate, because it was parallel to the direction of movement between the two plates. This result was incomprehensible at the time, and furthermore confused the whole theory, which was constructed in terms of relative motions between plates, and explicitly rejected the whole concept of absolute motions. So I decided to say nothing.

Jason noticed the same thing, and proposed an explanation which has been widely accepted.[22] He argued that Hawaii was on top of a hot part of the mantle that is rising toward the surface, because its density is lower than that of the surrounding colder mantle. Such structures are called *convective plumes*. He proposed that there are a number of such plumes beneath volcanic islands such as Iceland, Reunion, and the Society islands, and that these plumes were all fixed to each other, and did not move around like the plates. The volcanic ridge that extends northwest from Hawaii is then produced as the Pacific plate moves in this direction over the plume. I have never liked this idea, because I do not understand how the plumes can be fixed to each other if the whole mantle is convecting. If the whole mantle consists of moving fluid, how can anything be fixed rigidly to anything else?

I have sometimes been asked whether the approach I used to describe plate tectonics was modeled on the theory of the dynamics of rigid bodies that led Euler to his theorem. It was not. It was, however, strongly influenced by the theory of dislocations, and especially by W. T. Read's book on the subject, which showed how useful the idea of a dislocation was, even though its behavior could not be calculated from the equations that control the electrons and nucleii of which a solid consists.[23]

I left Scripps at the end of October, and drove my old car with all my possessions across the country to Lamont, where I stayed for three months. I found everyone there working on plate tectonics. I discovered that Jason had talked about his ideas at the AGU meeting in the spring, and thought I should try to delay the *Nature* paper so that he would have priority in publication as well as in his discovery of the theory. I was not successful because *Nature* had already set the type. I am now glad that I failed, because our discoveries were independent. I was somewhat dismayed by the number of people working on plate tectonics at Lamont, because I have never liked working quickly and in competition with others. Xavier Le Pichon had finished a global reconstruction of plate motions before he returned to France; Lynn Sykes, Bryan Isacks, and Jack Oliver were writing their monumental review of the implications of plate tectonics for seismology; and Jim Heirtzler and Walter Pitman were systematically going through all the marine magnetic data to work out the history of the ocean basins.[24]

I found Lamont quite different from Madingley: the people were much more competitive and uninterested in theoretical ideas and calculations. People were also secretive about what they were doing. What was clear, however, was that the basic ideas of plate tectonics worked very well in the oceans, but not on the continents, where the deformation is widely distributed. So I decided to ask Lynn Sykes and Peter Molnar, who was then a graduate student at Lamont, to teach me how to construct fault-plane solutions for earthquakes, and see if I could modify plate tectonics to describe continental as well as oceanic tectonics. My colleagues, students, and I have since spent many years working on this problem, which is much harder than plate tectonics.

In the spring of 1968 I moved again, to Princeton, where I at last came to know Jason well. Although I knew that simultaneous discoveries had occurred commonly in many areas of science, I was astonished to discover that he and I had been working on the same problems in the same way for much of the previous year. As Menard points out, most of the discoveries that led to plate tectonics were made independently by two people.[25] But I was still amazed when the same thing happened to me. And it was not just that we had both thought of plate tectonics. Jason had also used the plate model to calculate the heat flow and topography of ridges, and Stauder's slip vectors of earthquakes to determine the poles of relative motion between the Pacific and North American plates.[26] We had both spent the previous year solving the same set of problems by the same methods. This discovery surprised me profoundly: I had not understood that scientific enterprises reach a particular point where the next step is obvious to those who use a particular approach, and that it is a matter of chance who will receive the credit.

The last major theoretical problem in plate tectonics concerned what happens where three plates meet. Jason and I discussed this problem at Princeton without getting anywhere. Tanya Atwater, who was still a graduate student at Scripps, came on a visit and joined in, also without much progress. Even though the problem is purely geometric, we found it surprisingly hard to solve. The solution came to me long after these conversations, in the autumn of 1968 when I had returned to Cambridge. Jason and I published a theoretical paper, which both posed the problem (which did not seem to have worried other people) and solved it, and Tanya worked out in detail the important geological consequences that the evolution of the triple junctions have for the tectonics of western North America.[27]

WHAT IS PLATE TECTONICS?

For me the central idea is the rigidity of plate interiors. It is this property that allows the surface motions of the earth to be described by so few parameters. Early versions of the theory – continental drift, paleomagnetic reconstructions, and sea floor spreading – implicitly or explicitly used this property, but did not recognize its importance.[28] I myself do not describe continental tectonics as plate tectonics, because continental deformation occurs in wide zones where the idea of rigidity is of limited use. Geologists such as John Dewey and Jack Bird recognized that the continental geological record contains structures and stratigraphy produced by plate boundaries, and have sketched plate geometries that could generate the features concerned. But they are unable to show that the motions involved were those of rigid plates, and in many cases I suspect, but cannot yet prove, that they were not. I believe that the type of distributed deformation that is now occurring in the eastern Mediterranean is a more accurate model for most continental deformation than is the tectonics of rigid plates.

A number of later developments are often loosely described as plate tectonics, such as studies of forces that maintain plate motions. Loose use of carefully defined terms is common among geologists, and leads to endless terminological confusion and controversy of the most sterile kind. Plate tectonics was clearly defined as a kinematic theory: one that is concerned with geometry. It is not a dynamic theory: one that is concerned with the driving forces.

In my view, plate tectonics was discovered by the group of paleomagnetists working at Madingley in the mid-1950s, in a specially built nonmagnetic hut that still exists, when they showed that the relative motion of continents could be described by continuous rigid rotations. What Jason and I did was to reduce the ideas to a set of physically reasonable propositions that are sufficient to describe all the earlier proposals that had been shown to work. Who invented the term *plate tectonics* itself is unclear. Several people tried to coin a term, partly (it must be said) with the aim of being able to say that they discovered the theory. None of these proposals stuck. One of the earliest uses of the term that I know of in print was by Jason and myself in our paper on the evolution of triple junctions in 1969. But I certainly would claim only to have written down a term by which the theory was by then widely known.

It seems to me unlikely that plate tectonics will require changes. It is a precisely formulated theory that provides an accurate description of the large-scale tectonics of the earth. Physical theories of this type are with time incorporated into larger and more complete descriptions of processes involved. This process will require studies of other tectonically active rocky planets. Sadly, our solar system does not at present seem likely to be of much use.

REFLECTIONS ON THE PROCESS OF SCIENTIFIC DISCOVERY

Few active scientists take much interest in general models of scientific discoveries, and I am no exception. However, like many physical scientists, I know a bit about the ideas of Francis Bacon, Karl Popper, and Thomas Kuhn, and it is perhaps of interest to those who do study the history and philosophy of science to see whether those who have been involved in scientific advances find such models useful in describing their activities.

Many historians of science have been concerned with theories in physics and chemistry. These are fundamentally different from those in areas like astronomy and earth science, because experiments are not only possible, but are an essential part of the development of any new theory. In many parts of the earth sciences, including tectonics, experiments are impossible: nothing human beings can do can affect the processes involved, and these cannot be scaled properly to conduct laboratory experiments. This difference affects every aspect of the subject: you watch what is happening, but cannot isolate the process that interests you. There is always "noise": processes that are going on which don't interest you. This inability to isolate a particular process also has advantages, because your observations contain information about processes other than the one that you intended to study. So you must always keep an eye open for the unexpected.

Because controlled experiments are impossible in so many areas, hypothesis testing in its strict form is not an activity familiar to most earth scientists. I spend my time trying to construct models that can describe what I and others have observed, using ideas from mathematics and other physical sciences. I judge I have succeeded when I find that my model can account for some well-known observation that was not understood, and which it had not occurred to me to connect with the observations that I was trying to model. For this test to be reliable, it is essential that I separate all observations that I come across into four groups: those that are simply wrong, those whose origin I understand, those that are so complicated that I am never going to know whether or not I understand them, and those that I am sure that I cannot understand using our existing theories. At any time there is usually general agreement that a number of existing observations fall in my last group. For instance, at the present time, I think most earth scientists would agree that our theories of convection in the mantle are in this state, because the seismological observations are best explained by whole mantle convection, whereas the geochemical observations suggest that the upper and lower mantle convect separately.

Another problem in this group is the formation of the solar system. Furthermore, I am sure that major new ideas will come from research in these areas. The last group is therefore the important one, and it is not large; at any time few problems in it are tractable. In the earth sciences, and I think in astronomy also, advances occur when someone realizes that a major problem in the last group has become tractable, generally because of an advance in technology that can be used to make a new observation. There are many examples in both fields of such events. In

the case of plate tectonics, it was our ability to measure the magnetization of rocks, first in the laboratory and then at sea with a towed magnetometer, and to determine the mechanisms of earthquakes, that
rapidly led to the final theory. The paleomagnetic observations did not
have the impact that in retrospect they should have had, because almost
all earth scientists thought they were incorrect, and so put them into the
first, rather than the last, of my groups. Why they did so remains for me
a puzzle, and also the most interesting historical question to be raised by
the discovery of plate tectonics.[29]

I certainly would not describe Jason's and my activities in 1967 as
hypothesis testing: as soon as I realized that earthquakes and their mechanisms were the direct expression of plate tectonics, I knew I was right!
The great and immediate success of the theory was the result of everyone else reacting in the same way.[30] Everyone knew about earthquakes,
and no one knew how they were produced (this is not quite true: most
believed earthquakes resulted from slip on faults, but no one understood how the same fault could slip in the same direction time after
time). So in 1967 most people put earthquakes in my last group. What
made plate tectonics so immediately convincing was that it was principally designed to account for sea floor spreading, continental drift, and
magnetic anomalies. With no further input, it also accounted for the distribution of earthquakes, which in the oceans lie in narrow bands on
plate boundaries.

PLATE TECTONICS AFTER 1968

The final version of the theory was rapidly understood and accepted by
earth scientists everywhere. In the Soviet Union, however, the new ideas
were resisted by a few elderly geologists who were heads of their insti-
tutes, principally I suspect because these people found it hard to think
in the new way. In the west most observations from terrestrial and marine
geology were quickly reinterpreted in terms of the new theory. But at
first most of this work was not sufficiently carefully carried out for me to
be able to put the observations into one of my four categories. The problem was that plate tectonics is a much more precise and rigid theory than
most of those with which geologists were familiar. To test whether it
works requires computers and a variety of programs: to determine poles
of rotation, to reduce magnetic observations to projected profiles of
magnetic anomalies, to calculate theoretical anomaly profiles from the
reversal timescale, to plot fault-plane solutions of earthquakes, and to

digitize and plot maps of reconstructions by carrying out rigid body rotations. Of the main laboratories involved in marine geology and geophysics in early 1968, only Lamont was in a position to test the new theory properly. Several of the important programs had been written by Xavier Le Pichon, and ceased to be used at Lamont when he returned to France. During 1968, these problems slowly became clear, so John Sclater, who was at the time at Scripps, and I decided to do a thorough job on the Indian Ocean, the one ocean that was largely being neglected by those working at Lamont.

My proper education as an observational scientist now began. John and I went to sea to collect our own data. We negotiated with everyone else who had data from cruises. We discovered that Project Magnet, a U.S. military program that had flown magnetometers to make maps of the magnetic field, had flown many profiles across the Indian Ocean, navigated by someone making star sights from the cockpit with a sextant, as mariners do. We digitized all these data, and reduced them to the format used by Lamont, using a small computer on which we could get free time at night (no one would give us a decent grant to do things properly, because neither of us had faculty positions). But the programs ran very slowly. It took several hours to produce a reconstruction of the plates at an earlier geological period on the small computer that we used, which was the size and weight of a large American refrigerator. Each plot required us to pass a few thousand cards through the card reader, containing the locations of the coastlines. Even the main computer, which served the whole campus of the University of California at San Diego, to which Scripps belongs, would have taken an hour or so for each plot, and cost $300 an hour, which we could not afford. I now carry out such calculations in a few seconds on a notebook computer the size of a small book, which is more than 100 times faster than the main campus machine was in 1967. However, the original programs still work fine on the notebook, and I used them to make one of the figures for this essay. We discovered a program for disadvantaged undergraduates that would pay them the minimum wage if we could produce 25 cents an hour, and trained them to reduce and plot the observations. I wrote the necessary programs, some while I was at sea. I also wrote programs that I thought would be able to recognize magnetic anomalies automatically (they could not), and spent months deciding whether I could recognize individual anomalies on the profiles.

The end result of all of this work was a monumental paper that filled one entire issue of the *Geophysical Journal of the Royal Astronomical Society,* and outlined the whole of the plate tectonic history of the Indian Ocean, and the new methods that were required to carry out such an analysis.[31] The main result was that the evolution could be accurately described in terms of plate tectonics. I was very disappointed. As far as I could see, we had done a great deal of work and had discovered nothing new. The entire enterprise therefore belonged in my second category. For this reason, I left mainstream marine geology and geophysics to work on the forces that maintain plate motions, and on continental deformation. Later events, especially from ocean drilling, have confirmed my initial reaction. Five years later in 1976, when I was elected to the Royal Society, I was both surprised and cross to discover that I was elected because of my work on the Indian Ocean, and not because of plate tectonics. The committee involved consisted of field geologists, and they were impressed by someone who could successfully work out the geological history of an entire ocean basin.

By 1969 the present theory was essentially complete. The only important new idea that has been required is that of propagating ridges, which Dick Hey of the University of Hawaii suggested to explain the shapes of magnetic anomalies in the eastern Pacific.[32] The development of the theory stopped very suddenly: in the 1960s continental drift became sea floor spreading, then plate tectonics, as the theory became more precise and as its scope increased. Then, equally quickly, the changes stopped: the theory was complete and rapidly became accepted. Many of the participants, including me, found it hard to adjust to this sudden change. The ideas, methods, and programs that they had developed with considerable intellectual effort a few years earlier became part of the standard undergraduate and graduate courses. One of the first courses on plate tectonics, which I taught at Cambridge in 1970, differed little from what is now taught everywhere. What took me by surprise was how quickly the ideas became detached from their originators as they became accepted. Except for those who were involved, and for historians of science, no one now knows or cares who was responsible for a particular part of the theory. It is even hard for modern undergraduates to understand that the whole theory is so new and caused so much exitement. They, quite reasonably, ask, "So, what did people believe before plate tectonics was discovered?" This is a question that I find unexpectedly difficult to answer, because I cannot remove the understanding that people now have to reconstruct our state of ignorance in the early 1960s. One young faculty member in China knew Dan McKenzie had been one of the people involved in the discovery of plate tectonics, but was astonished to meet me. He thought it happened so long ago that all of those involved were dead.

The effect of these discoveries on the careers of those involved was dramatic, especially on those of the younger people. Several of us had only just finished our Ph.D.s when we were invited to give the major talks at national and international conferences, elected to the Royal Society and the National Academies of the United States and France, and given major prizes by geological and geophysical societies. I found this all very flattering, although I have always regarded my success as a piece of luck. If my parents had not decided so early that Hitler was not going to win the war, I would have been too young to have been involved. This worldly success removed any concern I had about whether I would be able to obtain an academic job when I ceased to be a post-doc. What concerned me much more was whether I was going to be someone who had only one good idea, or whether I would be able to make progress with some of the harder problems in the earth sciences that were still not understood, especially the tectonics of continents. My uneasiness about this question did not evaporate until 1978, when I discovered how sedimentary basins formed, by stretching wide regions of the continents. This idea is the antithesis of plate tectonics, because the deformation is distributed over a wide region, rather than occurring on a single plate boundary. This difference is probably why both I and everyone else were so slow in understanding what was going on. It was also one of our first successes in our efforts to understand continental deformation, which is a harder problem than plate tectonics, and is a less dramatic story that still continues. Perhaps one day I will write an account of this effort, which is so different from the discovery of plate tectonics. But not here!

ACKNOWLEDGMENTS

In the four years after I obtained my Ph.D. in November 1966, I worked at five different universities (Cambridge, Caltech, Scripps, Lamont, and Princeton) for periods that never exceeded six months. Throughout this time I held a research fellowship at King's College Cambridge, and will always be grateful for its support. King's paid, housed, and fed me when I was in Cambridge, and provided a stable point in my peripatetic life. At the beginning of the period, I was generously supported by grants from the U.S. Air Force at Caltech and Scripps; and lived in comfort. But toward the end there was less and less money. At one stage, John Sclater signed a purchase order for 20 days of my services, in the same way as he purchased laboratory supplies. The money enabled me to live at Scripps for four months (extremely frugally and in fear of a knock on the door from the Immigration Service!). I will always be grateful to the people, principally Don Anderson at Caltech, Walter Munk at Scripps, and Harry Hess at Princeton, who gave me the opportunity to be involved in this work at the beginning of my scientific career. The enlightened generosity of these people and of the organizations from which their support came, especially the U.S. Air Force, the Office of Naval Research, and the National Science Foundation, made a deep impression on me, and left me with a lasting admiration for the United States and its way of doing science. This is Earth Sciences Contribution number 6240.

CHAPTER 12

WHEN PLATES WERE PAVING STONES

Robert L. Parker

F OR ONE TIME ONLY, AT ITS INCEPTION, PLATE TECTONICS WAS called the *paving-stone theory of tectonics,* the name Dan McKenzie and I gave to the organization of the earth's surface into a small number of internally rigid bodies in relative motion. When Dan and I wrote the first paper on plate tectonics, "The North Pacific: An Example of Tectonics on a Sphere," we were two unknown new Ph.D.s, fresh out of graduate school.[1] My thesis had been on the mathematical modeling of electrical currents in geophysical systems, and Dan's was on the shape of the earth, so we were both beginners in the science of geology. But we were fortunate to be working as students in a department at the center of a whirlpool of intellectual activity that was bringing about the first true understanding of marine geology and its importance for global tectonics. This essay is a brief review of the scientific and personal events leading up to that first paper.

THE VIEW FROM CAMBRIDGE UNIVERSITY, FALL 1967

In the fall of 1967 I was a postdoctoral fellow at the Institute of Geophysics and Planetary Physics (IGPP), which is part of the Scripps Institution of Oceanography in the University of California at San Diego (UCSD). I had graduated the year before from the Department of Geodesy and Geophysics at Madingley Rise in the University of Cambridge in England. Although I was working in the United States, which was to become my home, my perspective at the time was that of a Cambridge graduate. During my three years as a research student I had not worked on marine geology or indeed anything remotely geological. But it was impossible not to be aware of the great events going on in the department, which seemed

Bob Parker with bicycle, at the Scripps Institution of Oceanography. (Photo courtesy of Bob Parker.)

to be a focus of tremendous creative energy. Let me first survey what I see as the primary influences that went into that paper.

By the middle of the 1960s everyone in the department at Cambridge had been converted to a firm belief in continental drift and the large-scale horizontal mobility of the crust. The head of the department, who was also my research supervisor, Sir Edward ("Teddy") Bullard, had long been what he termed "agnostic" on the subject, but he now became an enthusiastic proponent and supporter of research into these ideas. To him they were a confirmation of his conviction that the most urgent priority of the time in the earth sciences was to correct our almost total ignorance of marine geology. In contrast, we had frequently heard the story that the director of the Lamont Geological Observatory (a department of Columbia University), Maurice Ewing, had decreed to his people that no effort should be spared in an effort to prove continental drift wrong once and for all. Therefore, it comes as something of a surprise for me to learn from recent public reminiscences of the Lamont team that at the time almost everyone there was actually engaged in discovering plate tectonics, with the blessing of boss Ewing himself.

Four years before our paving-stone paper, Fred Vine and Drum Matthews had shown that the magnetic stripes which had been measured in only a few places around the ocean ridges were beautifully explained in terms of an idea proposed by Princeton's Harry Hess that new crust was being created at the ridges, combined with the discovery that the geomagnetic field has been constantly reversing polarity, that is, exchanging the positions of the north and south magnetic poles.[2] The confirmation of the global reversal of the magnetic field by radiometric dating was also new at the time, although reversely magnetized rocks had been known for over 60 years. Fred Vine was a fellow research student and Drum Matthews was a lecturer (I think) on the staff at the Department of Geodesy and Geophysics when they had the extraordinary insight to put these two phenomena together.[3]

While I was a student there, Tuzo Wilson, visiting Cambridge on a sabbatical leave from Toronto during 1964, made his seminal discovery: the nature of the great offsets in the magnetic record on the sea floor. In the eastern Pacific Ocean off the states of Washington and Oregon, great linear gashes called *fracture zones* had been found in the sea floor, several thousands of miles in length. Magnetic patterns on one side of a fracture zone also appeared on the other side almost identical in form, but displaced by vast distances. The traditional geological explanation for such an offset was that the patterns had originally been formed together, but had subsequently been separated by sliding one piece of the crust relative to the other along the line marked by the fracture zone. There were many difficulties with this explanation. For example, nowhere else on the earth had such huge horizontal offsets ever been seen. Moreover, after a shearing motion of this kind, one would expect to see evidence of folding or compression at the ends of the fault zone, yet there was none. To solve this puzzle Wilson invented the idea of *transform faults*.[4] I clearly recall coming out of the first seminar he gave to us on the subject and thinking I had witnessed something profoundly important. In 1965 Wilson published his paper in the journal *Nature* explaining transform faults and their relationship to the fracture zones. The remarkable resolution lay in the totally unexpected finding that the two sets of magnetic patterns had never been aligned, but had actually been created separated by the large offset as we see them today. Transform faults were another key ingredient in the paving-stone theory.

Two years after this, in 1967, Lamont's Lynn Sykes published an article in the *Journal of Geophysical Research* concerning earthquake mechanisms on ocean ridge systems: the mechanism gives the direction of motion of the crust in the immediate vicinity of the earthquake.[5] Sykes discovered that the motion on the offsetting ridges was compatible only with Wilson's model. That model was further confirmed by the absence of earthquakes on the fracture zone traces, which were predicted to be inactive by the transform fault model, but which would be lines of slip in the interpretation of classical geology. Sykes' work was based largely on a newly established seismic network, the worldwide standard seismograph network (WWSSN) of the U.S. Coast and Geodetic Survey – a set of calibrated and nominally identical seismometers globally distributed for the purpose of detecting nuclear bomb tests. In retrospect it was a piece of remarkable good luck that the seismic records from this network were not immediately classified, just as, in later years, enormous amounts of marine magnetic and gravity data collected by the U.S. Navy were kept secret.

In 1965 at Cambridge University, Teddy Bullard, with Jim Everett and Alan Smith (both research students at the time), had already fit the continental shelves around the Atlantic, using internally rigid bodies, moving the continents about their Euler poles of rotation, in a purely geometrical manner.[6] As Teddy was fond of illustrating with everyday objects like books, Leonhard Euler (the 18th-century Swiss mathematician) had proved that it is always possible to move a rigid body from one position to any other by means of a single rotation about an appropriately chosen axis; when one imagines moving a continental plate from one place to another on the earth, the positions where the axis intersect the earth's surface are called *Euler poles*. The continental reassembly project was not an attempt to run time backward and produce a continuous history of the relative positions (something that can be done today). It was simply to fit the edges of the continental shelves together and therefore reconstruct the ancient protocontinent of Gondwana. Although Warren Carey in Tasmania and others had performed similar reconstructions with model globes fitted out with plastic caps, skepticism remained deep about the accuracy of the fits.[7]

Bullard carried out a quantitative fit using bathymetric charts to locate the continental shelves, which he regarded as the proper edge of the continental crust. The astonishing match of the shelves from opposing sides of the Atlantic Ocean convinced most of the unbelievers of the fidelity of the reconstruction. Logically, the validity of the reconstruction required only that the edges of the continental regions should remain rigid throughout geological time and so maintain their present-day shapes. In principle, the interiors would be free to deform, although it is unlikely anyone actually thought of such a physically implausible state of affairs. So interpretation of the very precise fits that Bullard and his co-workers demonstrated regarding the land masses in the geological past was a tacit acceptance of the existence of internally rigid plates moving around on the surface of a globe.

Given all these developments, it is hard to believe in hindsight that the final piece of the puzzle wasn't obvious to everyone in the field. Of course, the full geometrical consequences were eventually realized by two people quite independently, Dan McKenzie (also a student at Cambridge, and with whom I shared an office) and Jason Morgan at Princeton.[8] In the minds of most people, I believe, the clear concentration of activity on the boundaries did not rule out internal deformations of the large, seismically inactive regions. They thought that the boundary regions were places where crust was created at ocean ridges or slid past other crust at the transform faults (the nature of the trenches was controversial), but these were not the only sites of major crustal deformation. I suppose the mental picture was of a generally plastic region, where compression or shear could occur over large areas. But as I remarked earlier, unless the internal deformations preserved the boundary shapes in a most unnatural way, this mental picture is incompatible with the very precise observed match of the continental shelves surrounding the Atlantic. Even a 5 percent deformation of the plate boundary shape would have been very noticeable.

AUTOBIOGRAPHICAL MATTERS

I have already mentioned that I had graduated with my Ph.D. in geophysics from the Department of Geodesy and Geophysics in 1966 and in early 1967 I took up a postdoctoral fellowship at IGPP. One of my research projects was the calculation of electric currents flowing in the oceans due to electromagnetic induction caused by the daily variation of the earth's magnetic field. Surprisingly large amplitude signals had been observed in the time series recorded at magnetic observatories situated on the coasts, and I believed this effect was due to electric currents circulating in the water. As part of my Ph.D. thesis I had solved the associated differential equations in an extremely simple geometry – a thin strip of conductor – and Teddy Bullard had solved problem for a conducting disk.[9] Now we worked together on the calculation in a more realistic model ocean, with the known variations of depth and conductivity and the proper shapes for the coasts to confine the electric current.[10] UCSD's computer center provided a very powerful computer at the time, a Control Data Corporation (CDC) 3600, but the memory of 36,000 words was too small to hold simultaneously the data required to define all the world's oceans at the resolution I needed. So I had to break up the oceans into three major basins: the Pacific, Indian, and Atlantic Oceans. To minimize scale distortions, I decided to represent each region on its own in a map centered in the middle of the ocean basin. For display purposes, and for creating a finite-difference grid (the lattice of points at which numerical values were to be computed), I wrote a program for drawing maps to run on the CDC 3600.

The program was originally named SUPERMAP, after Harold MacMillan, prime minister of Great Britain (1957–1964), known in the press as SuperMac. SUPERMAP was written in Fortran-63 and recorded on punched cards; the database comprised a primitive coastline of about 5,000 points, digitized by hand by an undergraduate student as part of a summer job from a large Mercator projection map provided to me by Bill Menard. Even today, most scientists write computer programs for themselves in the quickest way, with little thought for maintenance or general use. I have always believed it is more efficient in the long run to build programs that can be used repeatedly and that are easily used and upgraded. These days reusable code is heralded as some kind of new discovery, but it was obvious even then what advantages a more forward-looking approach would bring. So, even though I needed only one kind of map projection for my electromagnetic induction problem, I made SUPERMAP a general-purpose program, running under an easily used command language. When I was writing it I had no idea, of course, that it would be soon pressed into service by earth scientists everywhere to perform plate tectonic reconstructions. In those early days its only rival was a program written by Xavier Le Pichon while he was at Lamont; written in the traditional quick-and-dirty mode, it was reputed to be the source of much frustration. In fact, the program SUPERMAP was much easier to use in 1967 than the present-day standard in the earth sciences, the Generic Mapping Tool program (GMT).[11] Thus SUPERMAP was ready for application when plate tectonics came along; all it needed was three or four lines to implement the oblique Mercator map projection, to generate one of the figures in our *Nature* paper. Strangely, I have been unable to locate a single listing or card deck of SUPERMAP. I understand that parts of the code are still active and running in a few computers around the world, but not at IGPP.

Without false modesty, I must make it clear that Dan McKenzie was the creative force behind the 1967 *Nature* paper. Dan spent the year or so after getting his Ph.D. at Cambridge visiting various places in the United States. He was at Scripps for the summer and fall of 1967; he went on to Lamont and Princeton early in 1968. Dan had been thinking about the new, high-quality WWSSN seismic data and what it could say about tectonics. At the time most seismologists were working on getting the directions of the principal axes of stress from the seismic signals, but these proved to be puzzlingly inconsistent and seemed to vary quite unsystematically, even in a small region. The use of earthquake first motions pioneered by Sykes as a diagnostic for tectonics was a break-through. An earthquake breaks the ground along a plane (the fault plane), usually a zone of preexisting weakness. The initial ground motion can be traced back to the site of the earthquake from signals picked up by on seismometers around the world, and the orientation of the fault plane (called the fault-plane solution) can then be inferred. As I mentioned earlier, data had just become available from a well-calibrated global seismic network of long-period instruments, the WWSSN. Suddenly fault-plane solutions were reliable and could be found for many more earthquakes than before, as small as magnitude 6 in size.

Sykes' work at Lamont (where Dan had been and was going back to in 1968) convinced him that the first motions contained important information for tectonics – he started to look at the earthquake mecha-nisms in a systematic way, not just at the ocean ridges. Dan quickly real-ized that it might be possible to treat the interior aseismic (i.e., seismi-cally inactive) regions of the earth as rigid bodies. This meant that the regions under consideration became so large that pictures based on a flat-earth model were no longer adequate. Dan wasn't sure how the geometry of the plane velocity vectors that he was used to would trans-late into a spherical setting. This is where I came in: during his visit to Scripps he told me about the problem, and I worked out for the spher-ical system how to represent the instantaneous velocities through angu-lar velocity vectors and how those vectors were combined at the points where three plates meet. Furthermore, as I have already described, I had on hand my computer mapping program SUPERMAP, which we imme-diately put to work displaying the amazingly compelling results. It was my idea to use an oblique Mercator map projection, which made such a dramatic graphical demonstration in the 1967 *Nature* paper.

COMPUTERS AND THE BIRTH OF PLATE TECTONICS

Nowadays most people will find it hard to appreciate how limited the available computers were at the time. At IGPP we had access to the Uni-versity CDC 3600 in the computer center; the computer had a large mag-netic core memory (36,000 48-bit words), 12 tape drives, a fast card reader and line printer, and, most important, a CalComp plotter for graphical output. I believe we enjoyed one of the best computer facilities at any institute doing research in earth sciences. The CalComp plotter was a simple robust device that moved a ball-point pen across the paper in the *y* direction, while the orthogonal *x* motion was provided by the rotation of an eight-inch diameter aluminum drum; graphs were drawn on long rolls of paper, ten inches wide. I have letters from 1968 in which Dan complained to me that the Lamont computer had too little memory to run SUPERMAP; later at Princeton he could successfully run the program but the computer had no plotter attached to it, so he could not draw the results. Therefore Dan sent the specifications for numerous tectonic cases he wanted to study to me at IGPP through the mail; I ran them on the CDC 3600 and sent the maps back.

It may be interesting to look at how computers had been used in other parts of the early development of plate tectonics. There were of course no general-purpose mapping programs; that is why I wrote SUPERMAP. The base maps in the publications were usually traced laboriously from atlases by staff illustrators or graduate students. Or they were simply haphazardly sketched – this is obviously how Tuzo Wilson's maps were made for the most part. Wilson made no use of the new technology at all; in fact his first demonstration of transform faults was with models made of paper built by Sue Vine, Fred Vine's wife.

The maps published by Bullard, Everett, and Smith were made by first computing the intersection points of selected parallels and meridians on the plate after rotation, and printing a list of these numbers; they were then plotted by hand onto large sheets of paper and joined with curved lines to form an image of the distorted latitude-longitude grid.[12] Then someone transferred the present-day coastline and continental shelf edge from a conventional map in a more traditional projection onto the curvilinear grid. No one thought for a moment that a computer plotter could do a fine enough job to reach the standards of scientific illustration in *Philosophical Transactions,* the journal where the continental fitting work was published. The Royal Society was (and still is) very fussy about diagrams, and would insist on having their own illustrators draw all the lettering and numerals on the diagrams. In contrast, the normal CalComp plotter product was drawn with a ball-point pen, and the one-hundredth-of-an-inch resolution of the stepper motor that drove the drum left easily visible staircases in lines drawn diagonally. But the continental fit of Bullard's paper depended on some heavy computing to minimize the misfit between the segments of the boundaries. In fact, it was the very "arithmetical" nature of the fit that Teddy thought might convince doubters, who saw, probably correctly, too much exercise of artistic license in the sketches and model continents that had been offered earlier as proof.

Somewhat surprisingly perhaps, the earliest analysis of the magnetic stripes due to sea floor spreading depended solidly on the computer models of crustal magnetization, generated by a program in autocode written by Fred Vine; like the continental-shelf fitting programs, it was run on the Cambridge EDSAC II computer in the Mathematics Laboratory. Computer code based on the very same equations is still in use today in the analysis of marine magnetic signals, which have been recorded in every ocean.

On the other hand, I don't know how much, if at all, of the seismic analysis of earthquake mechanisms was computer-aided. I suspect none of it: the seismic traces were recorded on photographic film chips and times of arrival of the seismic waves were estimated by eye. The direction of the earthquake's first motion at the source was plotted on a stereographic or equal-area grid mapping the focal hemisphere, an imaginary hemisphere centered on the earthquake; a master net was drawn accurately once and then copied endlessly. However, I was not closely involved with the process so I cannot be sure when these tedious procedures and the finding of the fault planes were automated.

CONCLUDING REMARKS

Dan McKenzie and I were both undergraduates in physics at Cambridge University. We were the graduate students of another physicist, Teddy Bullard, who was himself a student of Ernest Rutherford. Among physicists at least, the prevailing philosophy at the time was that the touchstone of a good theory was that it should make testable predictions. Sciences that merely made observations and organized them were, in Rutherford's unkind words, "stamp collecting." To us the ability to make quantitative predictions capable of verification, or of falsification, was what made plate tectonics and the paving-stone model so appealing: one could use the information about the direction and magnitudes of the relative motions obtained on the boundaries between plates A and B and between plates B and C to predict what would be happening along the A–C boundary, both qualitatively in terms of tectonic processes and quantitatively in terms of rates and directions of motion. We realized right from the start that the predictive power was restricted to present-day motions, and that once one attempted to extend instantaneous velocities of today back into the geological past, other factors controlled the evolution of the boundary shapes. Initially we hoped some general principle might be discovered governing these factors, but that goal has proved elusive; there appears to be no alternative to a painstaking empirical analysis of the geological record. Nonetheless, plate tectonics succeeded in providing the framework for making sense of the large-scale processes governing the development of the earth's crust.

Subsequently, I did not make the further development of paving-stone theory a major part of my scientific career. I did some minor work on kinematics of plates; I helped my Scripps colleagues delineate the fine details of the marine magnetic anomalies using their near-bottom magnetometer system.[13] Perhaps my only other important contribution was as co-author on the first paper explicitly stating the square-root age rule for sea floor depths.[14] I was caught up in another revolution in the earth sciences going on at about the same time: the creation of geophysical inverse theory. In addition to the pioneers of plate tectonics, the two founders of modern inverse theory, George Backus and Freeman Gilbert, were also sabbatical visitors in Cambridge during the 1960s. I was extremely fortunate to get to know them personally, and to become their colleague in due course. Inverse theory is the set of mathematical methods that allows one to draw sound conclusions from a physical model in the face of severely incomplete and inaccurate measurements – a common situation in earth sciences.[15] Here was a subject in which I could indulge my personal fascination with abstract mathematics to a much greater extent than in plate tectonics. It was clear in 1967 that an enormous amount of work lay ahead to confirm the model of plate tectonics, work that would involve the synthesis of great quantities of geological and geophysical information. I knew my talents lay in another direction.

The term *paving-stone theory* appears six times in the first *Nature* paper, once in the abstract (which the *Nature* editors wrote). Everywhere else (21 times) we refer to the inactive interior regions as "plates" in a completely modern and familiar way. I am not sure who first used the term *plate*, but in any case our name for the new tectonic system, *paving-stone theory*, did not receive popular favor. But even if the metaphor of the paving stone failed to catch on, the concept it described has proved to be much more durable.

CHAPTER 13

MY CONVERSION TO PLATE TECTONICS

Xavier Le Pichon

I HAPPENED TO BE WORKING AT THE LAMONT GEOLOGICAL LABORA-tory (now Lamont-Doherty Earth Observatory) of Columbia University in New York while the plate tectonic model was elaborated, first from September 1959 to September 1960, before my military service, and then from February 1963 to February 1968. Here, I present my views on the plate tectonic conception from the perspective of someone who was at the key acting laboratories. This testimony does not pretend to be an exhaustive and impartial history of the elaboration of the plate tectonic concept. I make extensive use of earlier papers that I have published on the subject.[1] Finally, I briefly place these views within the context of the evolution of plate tectonics from an ocean-based model in the 1970s to a space-based one today.

LAMONT: FIXISTS VERSUS MOBILISTS

The revolution of ideas that led to plate tectonics, from 1955 to 1968, was greatly influenced by the continuous interaction among scientists of three laboratories, Lamont and Princeton University in the United States, and Cambridge University in England. Each of these laboratories was dominated by a strong personality: Maurice Ewing at Lamont, Harry Hess at Princeton, and Edward ("Teddy") Bullard at Cambridge. Although of quite different origins and intellectual capacities, they had in common a deep interest in the geology of the oceans. It was Richard Field, a professor at Princeton, who generated this interest in Ewing, Hess, and Bullard during the 1930s.

Maurice Ewing inherited from Field a burning zeal for the explo-ration of the oceans. With him, marine geology entered a new era. From

Xavier Le Pichon, on the *Vema*, in 1966 or 1967. (Photo courtesy of Xavier
Le Pichon.)

the scattered approach based on discontinuous point measurements,
Ewing moved to a global approach based on continuous measurements.
He was the first to deliberately install himself within the oceanic world,
inventing ad hoc the tools he needed to obtain the maximum amount
of new data on every kind of subject. Although he was a theoretician by
training, he was not comfortable with speculation. He made a religion
of data.

When I arrived at Lamont in 1959, with a Fulbright Fellowship to study oceanography, "Doc," as Ewing was known by his students, sent me around the world on his three-masted schooner, the research vessel *VEMA*. "Oceanography has to be learned at sea," he told me. His deep interest in the exploration of virgin territories probably came from his northern Texas origins. He loved to be where nobody else was. When I told him in 1968 that I had decided to go back to France, he asked me how I could return to such an old country. "If I had to start a new life today, I would go to Australia." Actually, when he did move, he went back home to Texas. But, if Texas was always close to his heart, the ocean remained to the end his real Wild West. I believe that right up to the time of his death in 1974, he still did not accept that plate tectonics had succeeded in revealing the secrets of "his" ocean. In 1970, he confided to me that each time his ship came back, he was waiting for the new evidence that would show that the whole plate tectonic model was wrong: the ocean could not be that simple.

The big thing at Lamont in 1959 was the discovery of the rift valley that runs along the crests of mid-ocean ridges. Earthquakes and volcanic eruptions characterize the whole length of the rift. *VEMA* cruise 16, in which I was going to participate, was supposed to test the continuity of the rift valley from the Atlantic Ocean to the Indian Ocean. Ewing and Bruce Heezen (one of his collaborators who later acquired worldwide fame through the physiographic diagrams of the oceans he made with his associate Marie Tharp) had predicted in 1956 the continuity of the rift valley through the oceans along the mid-ocean seismic belt.[2] This seismic belt had been described in 1954 by a Frenchman from Strasbourg, Jean-Pierre Rothé.[3] We were going to zigzag for nine months above this famous seismic line to test the prediction. As it was estimated to be 37,500 miles (60,000 kilometers) long, the almost unknown rift valley suddenly became the most important structure on Earth. It became clear then that no model of the evolution of the earth that ignored the rift could be considered valid.

This fundamental discovery made by the Ewing team had followed another, of similar magnitude, made in 1955 by Ewing and Frank Press at Lamont.[4] Seismological observations had actually established what had been inferred from gravity measurements by geophysicists: the uplift of the crust-mantle interface (called the Moho after Yugoslav seismologist, Mohorovičić) to a depth of about 3 to 6 miles (5 to 10 kilometers) under the ocean floor. Under the continents, the Moho lies at a depth of about 20 miles (30 kilometers). The oceanic crust is on average four times less thick than continental crust, and the great difference between the continents and oceans pleaded for separate origins and distinct evolutions. Up to that time, most seismologists argued on the basis of the difference of propagation of surface seismic waves across the Pacific and the Atlantic that the Pacific Ocean was the only true ocean, whereas the Atlantic Ocean had an intermediate-type crust. Ewing and Press' discovery of a shallow Moho under the Atlantic eliminated the concept of an intermediate-type crust there. It also laid to rest the possibility that the mid-Atlantic ridge consisted of remnants of continental crust. From then on, the whole debate about the dynamics of the earth would be concerned, first, with the significance of this radical difference of structure between oceans and continents and, second, with the significance, within the oceans, of the rift valley.

At Lamont, two schools of thought prevailed. To Heezen, then a young geologist who had just finished his thesis under Ewing's direction on the morphology of the northern Atlantic Ocean, everything could be simply explained if one accepted the idea of rapid Earth expansion. Warren Carey of the University of Tasmania, at a symposium there in 1956, argued that oceans were geologically recent structures formed by expansion from the rift.[5] This was sea floor spreading without subduction. Heezen was consequently considered a mobilist. Mobilists were either expansionists, followers of Carey, or drifters, followers of Alfred Wegener. But Maurice Ewing thought that the idea of such a fast expansion (a 75 percent increase in the earth's radius in 100 million years) was physically absurd. He remained a fixist; he preferred to explain the tectonic activity of the rift by deep convection currents that did not reach the surface but were the cause of extension and volcanism, without wholesale movement of the crust.

For Ewing, such speculations were premature. What did they bring to science? New facts were within reach of our dredges, corers, cameras, and magnetometers. With his younger brother, John, he was inventing marine seismic reflection, a technique that would continuously record the thickness of the sedimentary layers. This technique was soon to reveal the very thin ocean sediment cover, and its total absence near the rift. But Ewing could not stop Heezen from developing his ideas, and ultimately, conflict between the two men could not be avoided. Ewing could not accept that one of his scientists would act in a completely independent way, without any control of the director of the laboratory. In 1967, it would lead to an open and painful split.

During this whole time, the Lamont team was far from monolithic, contrary to what has often been stated since. There were two schools. One, which was more geologically inclined and included the students of

Heezen, was mobilist and expansionist. The other, which was more geo-physically inclined, and to which I belonged, was fixist and believed in long-standing ocean-continent distribution. Although Lamont faction leaders could hardly work with each other, the younger scientists had many vivid exchanges, especially when they were at sea. The debate was open and always stayed open.

PRINCETON AND THE SEA FLOOR SPREADING HYPOTHESIS

Walter Sullivan, the science chronicler of the *New York Times,* wrote that on March 26, 1957, Heezen presented at Princeton University a seminar about the newly discovered rift.[6] Harry Hess, chairman of the university's Geology Department, stood up to comment. He said in essence: "You have shaken the foundations of geology."[7] It is clear that this curious and open-minded man, who was always ready to reconsider his own hypothe-ses (which he apparently did not want to take too seriously), absorbed a great deal in this seminar. Hess, through a rather complex chain of rea-soning, had become convinced that the oceanic crust is not chemically differentiated from the mantle, but consists of serpentinite, formed from partially hydrated peridotite. In this way, the mantle would crop out on the ocean floor. Hess had also become convinced, following the Dutch geophysicist Vening Meinesz, with whom he had worked on grav-ity measurements in the 1930s, that ocean trenches were convergence zones where the floor of the oceans buckles under the adjacent conti-nents. Combining these hypotheses with the rift expansion concept, Hess revived the idea of convection currents as the driving force of con-tinental drift, as proposed by British geologist Arthur Holmes in the 1920s.[8] Hess' conveyor belt was moving right up to the sea floor: the upper mantle rose along the rift where it became hydrated and moved undeformed from the rift to the trenches, only to plunge back into the deep earth.

This hypothesis takes into account the radical difference in structure between oceanic and continental crusts. It attributes the small thickness of oceanic sediments to the youth of the oceanic basins. The volcanic and extensional tectonic activity at the rift is explained by the divergence of the two conveyor belts – one moving east, one moving west. Hess' model was basically correct, yet it was originally based on one false hypothesis: we now know that the oceanic crust consists principally of basalt and not serpentine. The mantle is no more exposed on the ocean floor than it is anywhere else on Earth.

Hess' model was widely circulated as a contract report in 1960 and 1961, in many places, including at Lamont, although it was not published until 1962. In between, Robert Dietz proposed its now famous trade name of *sea floor spreading*.[9] Hess, with his usual open-mindedness, presented his ideas as a working hypothesis that should not be taken too seriously, as "an essay in geopoetry."[10] His caution may also have been at least partly due to the aggressiveness of the fixist school in the United States. Most of the senior geophysicists would then shoot at sight the few mobilists trying to present their ideas at the American Geophysical Union (AGU) meetings. Six years and one detour through Cambridge would be necessary to establish sea floor spreading as the prevailing model at Lamont as well as in the rest of the United States.

CAMBRIDGE AND THE "VINE AND MATTHEWS" TEST

Paleomagnetism provided the decisive test. This test was proposed independently in 1963, in Canada by Lawrence Morley, and at Cambridge by Fred Vine. Both had good knowledge of the magnetization of rocks. Morley had studied paleomagnetism, and Vine had worked with paleomagneticists. At Cambridge, Edward ("Teddy") Bullard was well known for his interest in the earth's magnetic field and for paleomagnetic investigations. The point was crucial because, in contrast, there was no such tradition and little interest in this domain at either Princeton or Lamont. The only person interested in paleomagnetics in Lamont was Neil Opdyke, a young post-doc who had been hired by Ewing in 1964 to work on the magnetostratigraphy of the core samples. Opdyke has often stated how isolated he felt during his first years at Lamont.[11]

Yet by this time, paleomagneticists no longer doubted the existence of reversals of polarity of the earth's magnetic field.[12] For Morley and Vine, if there was sea floor spreading, the lavas that flow on the floor of the rift valley must be magnetized in the contemporaneous magnetic field, which is alternately "positive" and "negative." Lavas erupting on the sea floor today would acquire a magnetic polarity consistent with the earth's present magnetic field; at other times, in the past, when the field was reversed, the polarity of the lava flow would go the other way. The floor of the oceans must then consist of magnetized stripes parallel to the rift and having alternate polarities. Morley went further than Vine, as he rightly concluded that the resulting magnetic anomalies should be symmetric with respect to the rift. One should therefore be able to use them to measure the rate of sea floor spreading.

There was at the time no existing survey of the magnetic patterns surrounding a properly identified ocean ridge crest. In the North Atlantic Ocean, where the Lamont teams had mostly worked, and in the northern Indian Ocean, where Cambridge's Fred Vine and his instructor, Drummond Matthews, worked, the magnetic patterns were highly irregular. Nice linear patterns had been mapped by two Scripps Institution of Oceanography geophysicists, Arthur Raff and Ronald Mason, during a magnetic survey that was started in 1956 off the U.S. west coast, but no rift was known there.[13] Vine's paper (published with Matthews as a co-author in *Nature* in 1963), as well as Morley's paper (rejected by *Nature* and the *Journal of Geophysical Research* in 1963, probably in good part because it contained no new data) were completely ignored.[14] I remember that Charles Drake at Lamont called my attention to Vine and Matthews' paper but, to my knowledge, the paper was not seriously debated among us. As stated by Vine, when the sea floor spreading magnetic anomaly concept was proposed, few actual observations supported it and, in a way, this concept created more problems than it solved. The absence of supporting observations at the time is demonstrated by the fact that neither Vine, nor Matthews, nor Morley, nor anybody else considered any follow-up to these two papers during the following two years.

Once more, it was Harry Hess who was to open a new pass. I shall not describe in any detail this detour through Cambridge, on which I have no direct information and which is described elsewhere in this book.[15] But one should note that Hess had already played a major role in the elaboration of Vine's ideas when he presented his own ideas at a most remarkable British institutional meeting, the annual interuniversity Geological Congress organized by the undergraduate students, in January 1962. In January 1965, Hess came back to Cambridge for a sabbatical with Tuzo Wilson, a Canadian geophysicist gifted with a stunning vitality and extraordinary intuition. Wilson had been converted in 1963 to sea floor spreading, to which he had immediately proposed a corollary, the "hot spots" hypothesis. The idea came to him while flying over the Hawaiian islands: the Hawaiian islands would have been formed by a deep mantle hot spot acting as a torch on the overlying drifting ocean floor. As the ocean floor drifted over the hot spot, a chain of volcanic islands would be created.

The association of Hess, Wilson, and Vine was a prodigious one, and when the Tuzo Wilson "hurricane" had dissipated, the essential notions of plates, plate boundaries, and transform faults (when two adjunct plates slip as they move against each other at a ridge crest) were established.[16] Starting from Hess' ideas and an intuition of Vine on the

Atlantic equatorial faults, Wilson established the rules of plane plate tectonics.[17] Then, on theoretical bases and following again a suggestion by
Hess, Wilson identified the Juan de Fuca Rise, in the Pacific Ocean west
of Canada, and had Vine identify the magnetic anomalies and the sea
floor spreading rate. The symmetry of the anomalies was rather good,
despite the modest rate of sea floor spreading, but the modeled relationship of the anomalies to the chronology of the earth's magnetic field
reversals was rather poor. This was not surprising, for it was later recognized that the chronology available at the time was incorrect. Yet, the
time had come to test the predictions.

It is somewhat surprising that Bullard seems to have made no contribution to this episode, for in the preceding year he had presented a
paper at the Continental Drift Symposium in London in which he
applied for the first time the rules of motion for rigid spherical caps on
a sphere to the reconstruction of continents before the opening of the
Atlantic Ocean.[18] Before the Second World War, the French scientist
Boris Choubert had made the first fit of the continents precisely along
their continental margins. Later, Carey, whose work influenced Bruce
Heezen, had tried to demonstrate that such a fit required using a globe
with a smaller radius.[19] This symposium had clearly revealed the difference between the British scientists, who were now almost all mobilists
(essentially because of the recent paleomagnetic results), and the American scientists, who were still mainly on the fixist side.[20]

RETURN TO LAMONT: TESTING TIME

What were we doing at Lamont during this time? We were exploring the
world ocean, from rift to trench, from the Atlantic to the Pacific through
the Indian Ocean. Ewing kept two ships at sea permanently. He believed
that the earth could not be understood unless it was studied globally,
using every scientific discipline. He was constantly looking for new tech-
nologies which, more often than not, were introduced for the first time
as a standard tool in the ocean by Lamont teams: underwater photogra-
phy, seismic refraction, continuous magnetic and gravity recording, continuous seismic reflection, heat flow apparatus on piston corers, neph-
elometer (an instrument for measuring particles suspended in seawater),
satellite navigation, and more. Manik Talwani, a Lamont gravity specialist, had organized an entirely computerized data reduction and storing
system, and Lamont was the only laboratory with a complete set of data
on the world ocean that could be rapidly and easily retrieved. Further-

more, the Seismology Department, under the leadership of Jack Oliver, used the global seismographic network installed in 1962 by the United States to initiate a systematic study of global seismology.[21] No other laboratory had similar potential to test Hess' hypothesis, and the autocratic direction of Maurice Ewing imposed a multidisciplinary approach to the study of the oceans that was probably unequaled elsewhere.

When I came back from military service in 1963, John Ewing, a marine seismologist and the younger brother of Maurice, asked me to interpret seismic refraction data obtained prior to 1959. These data had not yet been published because no one knew how to interpret them. It was known that the mid-Atlantic ridge was in isostatic equilibrium: its elevation had to be compensated by a thickening of light material beneath it. One could expect that the crust, which is lighter, would thicken over the heavier mantle. But this is not so. The base of the crust, the Moho, rises as much as the sea floor. Was Archimedes' principle being violated? This was the beginning of a study of the structure of the mid-Atlantic ridge that I did for my Ph.D. thesis in collaboration with several other scientists over the next three years. Once we recognized that the seismic refraction data were correct, we concluded with Manik Talwani in 1964 that the isostatic compensation resulted from a modification of the mantle immediately below the crust, which had to be much lighter than usual, which meant that it was probably significantly hotter.[22]

We began a study of the magnetic anomalies being analyzed by Jim Heirtzler, who was in charge of the Magnetic Department at Lamont.[23] At the ridge axis, and the zone immediately surrounding it, we found large linear magnetic anomalies: zones of intense magnetization caused by volcanic intrusion of rocks that were either highly magnetic or highly susceptible to induced magnetization by the prevailing Earth field. On the other hand, the anomalies on the flanks of the ridge were quite different. They were larger but less intense. This proved that the flank anomalies could not be displaced axial anomalies moved laterally by the sea floor spreading. Our computations showed that the deepening of the sea floor was not sufficient to explain this difference. Only much later was it demonstrated that these differences in wavelength and amplitude resulted from large variations in the earth's magnetic field and from changes in the magnetization of the rocks.

The study of the distribution of the sediments on the ridge, which I made with the two Ewing brothers, revealed large latitudinal variations and a remarkable contrast between the ridge, which had no significant cover, and the basins, which were filled with a thick and undisturbed sediment cover.[24] This was a phenomenon quite different from the regular thickening proportional to the age of the sea floor, as predicted by Hess. We thought that these anomalies were not compatible with the idea of steady sea floor spreading.[25]

But the major stumbling block for us was the presence of undeformed sedimentary filling in some oceanic trenches where Hess had proposed that oceanic crust was being underthrust. The sinking of the conveyor belt below the continent should have accumulated deformed water-saturated oozes and muds within the trenches. Yet the only tectonic evidence was the presence on the oceanic side of the trenches of faults that were obviously due to distension and not to compression.[26]

In late 1964 or early 1965 I noticed, at about the same time as, but independently from Tuzo Wilson and Fred Vine, the remarkable similarity of the magnetic anomalies above what is now known as the Juan de Fuca Ridge, off western North America, with the anomalies above the Reykjanes Ridge, south of Iceland. I pointed it out to Manik Talwani and Jim Heirtzler. It was obvious that a portion of active mid-ocean ridge crest was present to the north of Mendocino off western North America. But how could one explain the remarkable similarity in the magnetic anomaly pattern over any portion of mid-ocean ridge crest, in both the Atlantic and the Pacific? After considerable debate, we considered that too many observations remained unexplained by the sea floor spreading model, and our conclusions were published in *Science* with a fixist interpretation.[27] This appeared in 1966. However, one of us in the Lamont marine geophysics group, Walter Pitman, clearly expressed his dissent with our conclusions at the time. He told me then: "You are going too far; I would be afraid to publish such conclusions." Walter was the first of us to have entered this gray domain where we knew our previous fixist ideas were not right, but were not yet sure that sea floor spreading could work.

The most illuminating example of our dilemmas at the time is the interpretation of the oceanic pattern of distribution of heat flow. With Lamont's heat flow specialist, Marcus Langseth, in 1965, we were trying to analyze and interpret the numerous heat flow measurements he had made in the Atlantic. In particular, I made the first numerical computations of the heat flow pattern that should be produced by Hess' sea floor spreading model. While the overall pattern of heat flow distribution was consistent with Hess' model, quantitatively, the disagreement was obvious. The computed heat flow was three times larger than the measured one. In contrast, deeper convection currents, those that could not reach the sea floor (as proposed by Ewing), would produce a heat flow pattern in good agreement with the measurements. I was convinced then that we had obtained the quantitative demonstration that the Hess model did not work. This is the conclusion we published in 1966.[28] Sea floor spreading should leave a clear heat flow signature – but it was not present. Our computations were correct, our measurements were correct, but our conclusion was wrong.[29]

It is interesting to note that Dan McKenzie, a young scientist from Cambridge who was then working at Scripps and who would soon play a significant role in the elaboration of the plate tectonic model, made the same computations one year later, arguing that the heat flow was compatible with sea floor spreading. To obtain the correct results, he chose a temperature inside the mantle three times smaller – 550°C instead of the 1500°C that we had chosen. This latter temperature was then and is still considered to be much closer to the actual mantle temperature. But McKenzie was already convinced of the validity of the sea floor spreading model, and he preferred to adjust the parameters rather than arrive at an obvious discrepancy.[30] At the time, whether the fixist or the mobilist model was adopted, a certain number of observations did not agree with the predictions. The choice made was heavily influenced by the environment, the working philosophy, and the discipline in which one worked.

THE MAGIC PROFILE: CONVERSION TO SEA FLOOR SPREADING

As far as I was concerned in late 1965, the difficulties that resulted from applying the sea floor spreading model – to the interpretation of the magnetic anomalies and the distribution of sediments, the apparent impossibility of reconciling subduction with the quiet sediment fill in the trenches, and the three-times-too-small heat flow through the mid-ocean ridges – led me to adopt a convection model without sea floor spreading, inspired by the ideas that had just been published by Felix Vening Meinesz.[31] Convection would be confined to the ductile part of the mantle, below about 30 miles (50 kilometers). It would induce fusion of basalt that would in part come to the sea floor and in part create the shallow compensating mass of the ridge. This was the conclusion of my thesis, written in late 1965 and early 1966 and defended at the University of Strasbourg on April 21, 1966.

On February 13, 1966, I left Lamont for Recife, Brazil, to participate as chief scientist in a South Atlantic cruise that would lead me to Buenos Aires and then to Cape Town. I then joined the faculty at Strasbourg, where I defended my thesis. It was April 26 when I returned to Lamont, where many of my colleagues were now "converted" to sea floor spreading. Walter Pitman showed me the "magic" magnetic anomaly profile obtained over the South Pacific ridge crest, the *Eltanin*-19 profile that had been presented by Jim Heirtzler at the American Geophysical Union (AGU) meeting in Washington, D.C., on April 27.[32] My wife still remembers that on my way back from the laboratory, I asked her to get me a drink and told her: "The conclusions of my thesis are wrong: Hess is right."

This extremely painful "conversion" experience has been crucial in shaping my own vision of what science is about. During a period of 24 hours, I had the impression that my whole world was crumbling. I tried desperately to reject this new evidence, but it had an extraordinary predictive power! Why then was the heat flow three times smaller than expected for sea floor spreading? Why were the magnetic anomalies so different over the flanks of the ridge? Why was the sediment fill in the trenches undisturbed? I did not know, but I was progressively forced by the convincing power of the magnetic anomaly profiles to assume that in all these unexplained observations, there must have been hidden parameters that had not yet been taken into account. Since that time, I know that good data and correct models do not guarantee that your conclusions are definitive: the possibility of hidden parameters is always present.

The presentation of the magic profile at the AGU stunned everybody. The 600-mile (1,000 kilometers)-long profile revealed a perfect symmetry with respect to the axis of the mid-ocean ridge crest. Furthermore, it could be interpreted simply and perfectly with the sea floor spreading model, using the Earth magnetic field reversals chronology obtained by the young Lamont paleomagnetic group (led by Neil Opdyke) by measuring the magnetic polarity of oceanic sediment cores. In particular, the magnetic anomaly profile as well as the sediment cores revealed the presence of a new magnetic event that Richard Doell and Brent Dalrymple, at the U.S. Geological Survey (USGS), had just independently identified. They called it the *Jaramillo event*, a short duration of normal magnetic field. With this new event, the correlations from one ridge crest to the other became evident. Suddenly, the balance of phenomena explained or left unexplained by the sea floor spreading hypothesis appeared positive, and acceptable without serious reservation to any scientist familiar with the whole picture. The massive move toward mobilism was then inevitable.

I still cannot understand how I missed seeing this magic profile before my departure from Lamont. I have absolutely no recollection of seeing it, so much so that in earlier retrospective papers I wrote that I had left Lamont in January.[33] But a check through letters sent by my wife showed that I only left on February 13 as just indicated. I guess that I was so buried in the writing of my thesis and the preparation of the cruise that I ignored everything else. Alternatively, I may have subconsciously ignored evidence that clashed so much with the conclusion of my thesis – which I had to defend two months later and which I could not possibly change without delaying the defense. The profile was in any case widely available by mid-February, as Jim Heirtzler showed it to Fred Vine when Vine visited Lamont shortly after he joined Harry Hess at Princeton. Vine included it in his magisterial synthesis published in the December 1966 issue of *Science*.[34] In that paper, he compared the magnetic sections over two portions of mid-ocean ridge in the Pacific with the Reykjanes profile in the Atlantic.

Now we had the key and the data were at our disposal. Immediately, under the leadership of Jim Heirtzler, we started working, one scientist to each ocean. I got the Indian Ocean. Lynn Sykes, in Lamont's Seismology Department, had tested Wilson's transform fault model using earthquake fault plane mechanisms and showed that it worked.[35] Jack Oliver, with his student Bryan Isacks, had demonstrated that the oceanic lithosphere did indeed dive into the mantle along the trenches.[36] In spite of the skepticism of its director, Lamont had moved massively into the mobilist party. Ewing knew that he could no longer contain this rising tide of sea floor spreading: from then on, he would wait silently for the evidence that would demonstrate the falseness of this model.

It was during a conference organized by NASA in New York on November 11 and 12, 1966 that the victory of mobilism was clearly established. Teddy Bullard, who presided, could not find a single scientist to defend fixism.[37] Vine, the Heirtzler group, and Sykes presented their latest work. But it was during the April 1967 AGU meeting that, to use Bob Dietz's phrase, "the total and instantaneous conversion of the American community to continental drift" occurred. It is there, too, that a young Princeton scientist, Jason Morgan, took the critical step toward the present mobilist theory by establishing the bases of plate tectonics on a spherical earth, and not on a flat-plane earth as had been done previously by Tuzo Wilson. Morgan, soon to be joined by Dan McKenzie and Robert Parker, showed the predictive power of kinematic computations. The mobilist model had become quantitative.[38]

ALL TOGETHER FOR THE FINAL CHORUS

I have retained a precise memory of that morning of April 19, 1967, at the spring AGU meeting during which Fred Vine and H. W. Menard, from the Scripps Institution of Oceanography, presided over a "sea floor spreading" special symposium. The large amphitheater was full and expectations were very high. "Sea floor spreading" was the subject of most discussions: 70 abstracts on the topic had been submitted to this AGU meeting. At the end of the session, the program announced that Jason Morgan would present a paper which, according to its title, concerned the formation of oceanic trenches by viscous convection. Manik Talwani and I were preparing to listen very attentively because we had had a vigorous argument with Morgan on this subject. Morgan assumed for his model the absence of any long-term rigidity even at the surface, and we considered this assumption incompatible with the gravity data. But to our great surprise, Morgan announced that he would present a different paper, entitled "Rises, Trenches, Great Faults and Crustal Blocks." Thus the talk he made did not correspond to the abstract he had sent. He was going to discuss the geometric problems concerned with the relative motions of plates (he called them "blocks"), which he assumed to be rigid away from the Atlantic rift. What Tuzo Wilson had done qualitatively on a plane, Morgan was now doing quantitatively on a sphere, establishing the principles of plate kinematics. Morgan has a special gift for disorienting his listeners. This gift was especially well displayed on that occasion, and very few people, if any, actually paid attention to what he said. As for Manik Talwani and me, our dispute with Morgan appeared to be closed, since he now assumed rigid blocks at the surface. We could not understand this "about-face."

Morgan had written an 11-page extended outline of his presentation, including the nine figures illustrating his talk. This short paper was sent to about ten people immediately after the meeting. I was among them. Morgan does not remember all the addressees of his paper.[39] He writes: "I am quite sure I gave copies to Bill Menard (at Scripps) and Tuzo Wilson (at Toronto) and, I am fairly sure, to Lynn Sykes (at Lamont), Carl Bowin (and/or Joe Philipps, at Woods Hole) and Fred Vine (at Princeton). I might have sent one to Jerry Van Andel (Scripps) as I used magnetic profiles from the Circe cruise." Morgan lost his own copy and none of those who received it appears to have made it available to the scientific community. I thought I had lost my copy too. Yet, during an office move, I found it and arranged for it to be published in *Tectonophysics* in 1991 with Morgan's permission. The exact substance of his presentation at the AGU was finally in print, nearly 25 years later.[40]

According to Morgan, "this short description of the main ideas in plate tectonics was written the week before the AGU.[41] The last two pages were written and reproduced the night before the meeting." The extended outline has the same title as the paper later published in the *Journal of Geophysical Research (JGR)*.[42] Eight of the nine figures would later be included in the *JGR* paper, as well as most of the text. In particular, the first two paragraphs make up most of the substance of the abstract of the March 1968 paper. The published version was accepted in revised form on November 30, 1967, seven months after the extended outline had been circulated. Although it is more elaborate, it adds nothing to the spherical plate tectonic model, as defined in the earlier April 1967 version.[43]

On the basis of this document, it seems extraordinary that, in the hall packed with the best geophysicists and geologists in the United States, nobody got excited by or even interested in the implications of Morgan's ideas. They were too new, too different from anything that had been done. Even among those who received the extended outline and had time to digest the new concepts, I apparently was the only one to have considered it sufficiently important to drop everything else and start working along these new lines. As I have written elsewhere, the source of my June 1968 paper was Morgan's 1967 extended outline.[44] I decided immediately to test this kinematic approach, in spite of the skepticism of my colleagues, who considered it more important to continue to decipher the magnetic anomalies. I had to elaborate a rather complex methodology and a system of computer programs, which kept me busy until July. I could then verify that each of the different rift openings behaved according to spherical geometry: plates (as they were later to be called) were indeed rigid, and Morgan was right. Part of my work got incorporated in the 1968 paper by Heirtzler and colleagues on magnetic anomalies and crustal motion.[45] I first extended Morgan's kinematic analysis of the Africa/America accreting boundary to the Antarctica/Pacific, the Eurasia/America, and the Africa/India (actually the Africa/Arabia) accreting boundaries to test his concept. On Heirtzler's suggestion, I used an oblique Mercator projection to test the geometry of opening of these accreting plate boundaries. I also devised numerical search methods to define the magnitude and direction of the plate motion as "Eulerian vectors" – that is, as motions around a hypothetical pole of rotation. By the end of August 1967, this first part of my work was completed. At the time, neither Morgan nor I knew that Dan McKenzie and Robert Parker were working at Scripps on their "paving-stone theory," and Morgan had no knowledge either that I was exploiting his model.

Morgan had spent the months of July and August at Woods Hole, where he finished the version of his paper that was submitted to *JGR* on August 30. This version is close to the revised published paper although its discussion of the Africa-America-Pacific-Antarctica-Africa circuit was not correct. It contained an error in the determination of the Pacific-Antarctica rotation vector that I later pointed out to him. Morgan presented his paper at a seminar in Woods Hole in August, and returned there during September 7–8 to attend a two-day conference. According to Morgan, those attending the conference included Ken Deffeyes from Princeton, John Mudie from Scripps, myself, Walter Pitman, and possibly Heirtzler and quite a few others. In a personal letter that I have kept, I mentioned that people invited from Lamont were Maurice and John Ewing, Joe Worzel, Manik Talwani, Marcus Langseth, and me. Morgan presented his paper. I also presented my kinematic analysis, including the oblique Mercator plots. It was the first time that Morgan heard about my work. From then on, we freely exchanged data and documents. This helped Morgan to rework his Pacific-Antarctica rotation vector and the corresponding Africa-America-Pacific-Antarctica-Africa circuit. This was also a great help for me because at the time I was attempting to obtain a world kinematic model.

Once I verified the rigidity of plates, as Morgan and McKenzie had done for the Atlantic and Pacific, I moved to the next stage, which was to combine the motions of plates to obtain the first predictive global quantitative model. I found that a unique solution could only be obtained by using six plates instead of Morgan's 12. I used Morgan's America/Pacific Eulerian vector to ensure the closure of the model. This six-plate model accounted for most of the world seismicity, as Bryan Isacks and his colleagues would later show.[46] Even now, it is difficult for me to forget my extraordinary excitement the day I realized that my six-plate model worked, and that it could indeed account as a first approximation for the broad geodynamic pattern. I remember coming home early in the morning for breakfast after a night at the computer and telling my wife: "I have made the discovery of the century." Well, I was young and my enthusiasm carried me too far. But this statement is a good indication of how we felt during those days of frantic discoveries.

Finally, I made the first kinematic reconstruction of the evolution of the surface of the earth based on magnetic anomalies. To do this, I had to fit the magnetic anomalies that had identical ages on both sides of the rift in the same way as Bullard and his co-authors had done to fit the continental margins on both sides of the Atlantic.[47] This was the beginning of a paleogeographic method, which has proven to be especially powerful. The fit of the anomalies was done on the computer and involved combining rotations that were no longer small but could reach several tens of degrees. Small rotations can be treated as vectors, whereas this is not true of large ones, which must be treated as matrices. Not knowing that, it took me some time to discover the origin of large discrepancies in my early computations. The rules of spherical geometry were poorly known at the time among geophysicists. Neither Morgan nor McKenzie, according to what they both told me in the fall of 1967, believed that such an approach was possible. They apparently did not have enough confidence in plate rigidity. McKenzie told me then: "John Sclater wanted me to do it – but I did not want to."

On December 13, 1967, I wrote to Morgan: "Thank you for the McKenzie–Parker paper. I am a little bit surprised that they do not seem to know about your paper." Thus it must have been in early December that Morgan sent me a pre-print of the McKenzie and Parker paper that had been received at *Nature*, November 14, 1967, and would be published in December.[48] This is how I discovered that McKenzie had been working on the same subject. The relationship between my paper and Morgan's is quite obvious, but both papers were written completely independently of McKenzie.

McKenzie had arrived at Scripps in June 1967.[49] Allan Cox wrote that "in June, 1967" Dan got the idea of using rigid-body rotations to describe plate motions while rereading the paper by Bullard and co-workers on fitting the continents together. Robert Parker had just completed a general computer program called SUPERMAP for plotting worldwide geophysical data using any conceivable projection.[50] Parker introduced the idea of using a Mercator projection in plate tectonics."[51] As noted above, I independently started using oblique Mercator projection in late May/ early June 1967, and presented the first oblique Mercator maps with the Eulerian pole of rotation as pole of projection at the early September Woods Hole meeting.

In a letter written to me on October 11, 1983, McKenzie explained the relationship between his paper and Morgan's paper: "I was at the 1967 AGU meeting and attended the session in which Morgan spoke, up until the time he did so. But I had read the abstract . . . and thought I would gain nothing from sitting through the talk and arguments and left to go elsewhere. The paper generated little general interest, and I did not hear about it until after Bob (Parker) and I had sent off our paper to *Nature*. When I did, I tried to delay publication, but the editor refused, saying that the issue had been made up. . . . I did not know until I read your paper [a pre-print of Le Pichon, 1984] that Jason [Morgan] had sent you a preprint so early. The first I knew of what he had done was a brief account from John Mudie when he returned from Woods Hole in August. By this time, Bob and I had already produced the Mercator maps of the slip vectors, and John's report acted as an incentive to get something written. I had talked a great deal to Bill Menard about plate tectonics and had convinced him that it worked for the Pacific. *JGR* [*Journal of Geophysical Research*] sent him Jason's paper to referee and, I suspect because of our conversations, he was very critical of it when he showed it to me. I asked him what I should do and he said to go ahead and publish, which we did as everyone knows. When I came to Lamont and Princeton in the autumn of 1967 and discovered what had happened I felt very embarrassed and it was then that I tried to hold the *Nature* paper."

Thus, McKenzie heard about Morgan's work from a brief report of the early September Woods Hole meeting and then, presumably immediately after, from Menard, who received Morgan's paper to review, also in early September. It is then that McKenzie decided to immediately write his short *Nature* paper, probably feeling that his approach (using the horizontal projections of the slip lines of earthquake fault-plane solutions to determine graphically the position of the pole of rotation with oblique Mercator plots), which he had by then been working on for several months, was sufficiently different to justify doing so.[52]

In his book *The Ocean of Truth*, Menard confirmed that he received the early extended outline.[53] He wrote, "Jason Morgan sent me a preprint of his manuscript in its early draft, probably in the late spring of 1967." Menard must have had this extended outline available to him when he wrote his 1966 book, as he quoted the first sentences of this early preprint. Menard added: "The manuscript certainly circulated among my students, and we discussed it. The original draft, however, was difficult to fathom and it did not have the impact of the final publication." Yet, as discussed earlier, the plate tectonic concepts were clearly presented in this early draft, now published, which was not significantly different from the later 1968 version, in spite of what Menard wrote. Actually, it is clear that the concepts were too new and appeared irrelevant to both Menard and his students. Menard, who had co-chaired the AGU session in which Jason Morgan presented his paper, wrote in his 1986 book: "I not only did not remember hearing Jason's famous talk, I didn't remember presiding over the session." Finally, Menard stated: "I believe I also reviewed the paper for an editor."[54] This can only refer to the August 30 version submitted to the *Journal of Geological Research*, which he presumably found upon his return from the *Nova* expedition sometime after September 12, according to the information he gave in his book. At the time, as mentioned by McKenzie in his letter to me, he was "very critical of it."

It is astonishing that McKenzie twice so nearly missed the opportunity to learn about Morgan's model. The first occasion was when he left the room just before Morgan's talk on April 17. The second occasion was when Menard, who had received the extended outline of the April 17 communication in late April, failed to mention it to McKenzie, although they "talked a great deal" together "about plate tectonics" and although Morgan's "manuscript had circulated among Menard's 'students'" and had been "discussed" by them (quote from the book of Menard). But the approach followed by McKenzie was sufficiently different from the one followed by Morgan that it lent credibility to his story.[55]

To me, the most surprising part of it is that McKenzie confined himself to discussing the plate kinematics of the Pacific-America plate boundary based on earthquake fault-plane solutions, and did not consider the kinematics of the Atlantic ridge. In the equatorial Atlantic, good data on transform and earthquake fault-plane solutions were available and the opening of the Atlantic Ocean is the subject of the fit of Bullard's paper, which gave the initial intuition to McKenzie.[56] But he wrote to me in 1988 "that the Atlantic data did not cover a sufficient range of azimuths to determine an accurate pole. So I thought (and still think) that the North Pacific is the best example to use, and it has the great advantage that the same pole produces both spreading and consumption." I did not agree at the time. I wrote in the letter of December 13, 1967, to Morgan: "The main objection I have to the work of McKenzie and Parker is that it may be a dangerous assumption that the continental system of eastern Eurasia, Aleutians, and North America is perfectly rigid. If there is slow deformation of this system, then you might expect the results they get. It seems to me that the spreading floor evidence suggests that the North Pacific was larger in Cretaceous (more than 65 million years ago) than it is now or that several thousand kilometers of *(east-west)* shortening have occurred in Asia since this time. I prefer the first solution. Maybe the actual pole is somewhere between McKenzie's position and yours."

I added in the same letter: "I have a second version of my paper typed now. It has been reviewed within Lamont. My problem is that it has grown out of proportion. I will send you a copy when it is ready, probably before Christmas. I am returning to France before the Spring, so I am rather anxious to have it cleared before I leave." By this time, I had just decided to go back to France and I was to leave in early February 1968. This would have the consequence that I would be cut off from the Lamont data bank and would not be able to work on the development of plate tectonics for the next two years. I had just been offered an associate professorship by Frank Press at Massachusetts Institute of Technology (MIT), which I had seriously considered. However, I thought that if I stayed in the States, I had to be where the action was, and that was clearly at Lamont. Maurice Ewing knew about the MIT offer and wanted to keep me; he tried to push me into heading Bruce Heezen's department, with whom Ewing was now in open conflict. I acted as an intermediary between Ewing and Heezen, who was torn apart by this personal conflict. This experience was so painful that I think it contributed to my decision to go back to France, which I announced to Ewing on December 11.

It is in this context that I had shown my paper to Maurice Ewing and asked him whether he wanted to be an author, as was usually the case in Lamont for papers based on data collected there. He declined and told me: "This is your work; publish it alone." He may have done this because he wanted me so much to stay at Lamont. Alternatively, he may have refused to go against his fixist ideas. In any case, this is how I became the sole author of the most important paper of my career, which was very unusual at Lamont at the time, especially for a young scientist. I waited to submit it until Morgan's paper was accepted, in order to respect his priority. Morgan's paper was delayed three months by Menard's review and could have been published in December 1967 instead of March 1968 if Menard had immediately accepted it, as Wilson and Oliver later did mine. It could also have been published in abbreviated form in June or July, had Morgan decided then to publish a cleaned-up version of his extended outline. I had more luck than he had. My reviewers, Wilson and Oliver, recommended immediate acceptance of my paper. The title would be "Sea Floor Spreading and Continental Drift."[57] The succession of papers that established the plate tectonic model then followed: McKenzie and Parker in *Nature,* in December 1967; Morgan in the *JGR,* in March 1968; and mine, also in the *JGR* in June 1968. Thus, Princeton with Jason Morgan, Cambridge with Dan McKenzie (although the work was done at Scripps and McKenzie insists that he greatly benefited from the environment there), and Lamont with my own contribution were finally united in the definition of the plate tectonic model.

Soon after, Isacks and co-authors demonstrated in their September 1968 paper, "Seismology and the New Global Tectonics," that geophysical data were compatible with my global plate kinematic model.[58] Their paper had a major impact on the geological community. Neither my paper, nor any of the critical ones that followed, would have been possible without the availability of the Lamont sea floor spreading data.[59] I wish here to acknowledge the firm leadership of Jim Heirtzler, who bulldozed us into producing this impressive collection of published work in a short amount of time.

From Ocean- to Space-Based Plate Tectonics

When I reflect with hindsight on what now has become history, it seems clear to me that the elaboration of plate tectonics was mostly the work of a few scientists in continuous interaction with the privilege of rapid access to crucial data and ideas. I wish, however, to highlight the prodigious intuition of Harry Hess and Tuzo Wilson and the immense energy of Maurice Ewing. I also regret the oblivion into which Warren Carey and Bruce Heezen have fallen. I believe, furthermore, that not enough attention has been paid to the privileged relationships existing between a few key laboratories. In a sense, the history of the elaboration of plate tectonics can be read as a concerto for three instruments in which Princeton, Cambridge, and Lamont successively held the soloist role until they joined together in the final chorus.[60] Each of these laboratories was characterized by a grand "multidisciplinarity" and by a strong unity under a charismatic leader. They had, moreover, taken large initiatives toward both continental and oceanic research. Finally, their teams included scientists who did not hesitate to venture outside the strict frames of their specialties, to risk the formulation of very general hypotheses that had to submit to the test of field observation, and to prepare for eventual modifications.

I have been often asked why the Lamont people had such a late conversion to sea floor spreading and why, once they were converted, they were such efficient actors in the development of plate tectonics. In his 1983 letter that I quoted earlier, Dan McKenzie wrote: "Lamont to me has always seemed a very valuable place to test new ideas. But the quantity of data available does not encourage their development. I think that it is no accident that sea floor spreading and plate tectonics were developed elsewhere." On the basis of my experience, I think there is some truth to this. I had devoted my whole research time looking in detail at enormous amounts of data on all aspects of the mid-Atlantic ridge. Any hypothesis that was mentioned on its origin immediately evoked tens of observations that would not fit it. This indeed blocked any progress. In a sense, trees were hiding the forest. On the other hand, once the model became obvious, it was easy to unroll my data bank and to interpret it in terms of this new model. This rather simplistic explanation cannot be the only one. I have insisted earlier in this essay on the quasi-absence of

serious awareness of paleomagnetic results in Lamont and on the preva-
lent fixist culture of the geophysicists there. I know that, as far as I am
concerned, these were two very serious obstacles. When I was a Ph.D. student at Lamont, my supervisors never seriously exposed me to paleomagnetics and continental drift.

But why then were Scripps people, who also had such a large data
bank, slower to jump on the bandwagon, especially in view of the fact
that there was a serious paleomagnetic culture on the west coast? I mentioned earlier as important factors the intensive exchanges between the
different departments and the culture of "multidisciplinarity" imposed
by Ewing. In addition, one should not underestimate the decisive advantage of the computerized data bank put together by Manik Talwani.
Finally, Lamont people had inherited from Ewing a tradition of very
hard work.

It is not my role to comment on the relative importance of my own
contribution in this concerto. All I can say is that my papers, in particular the "Sea Floor Spreading and Continental Drift," were intensively
used by the scientific community in the early years of plate tectonics. I
was the most cited solid earth scientist for the period, 1965–1978 and
one-quarter of the 2,500 citations were assigned to this paper.[61] This
highlights the impact of the demonstration that a global plate kinematic
model could indeed be used as a framework for plate tectonic studies.

In 1973, at my laboratory in Brest, Jean Francheteau, Jean Bonnin,
and I published a book entitled *Plate Tectonics,* which was the first attempt
to present in a coherent fashion the plate tectonic model, from plate
kinematics to processes at plate boundaries, in book form.[62] This kind
of work could be done without having access to a large data bank! Our
book was very well received and widely used – probably because its logic
responded to the needs of the scientific community at the time.

The book also illustrates the remarkable change plate tectonics
brought to the field of geodynamics. Since the beginning of the 20th
century, geophysics and geology had mostly traveled separate ways. What
little interaction there was tended to be tense, if not antagonistic. It is
remarkable that the role of earthquakes was essentially ignored in tectonics, and that mountain-building was considered as a spasmodic
process not directly related to seismic activity. Plate tectonics reconciled
geophysics and geology and showed that earthquakes were the direct
expression of the continuous tectonic activity at the surface of the earth.
From then on, seismologists and tectonic geologists would have to work
together to study geodynamic processes. But the great difficulty of the
study of these processes was that plate kinematics was ocean-based. All

the observations that allowed quantitative plate kinematic determinations, magnetic anomalies, transform faults, and fault-plane solutions at plate boundaries were obtained in the oceans. Continental tectonics, which could be observed and studied in detail, was difficult to relate directly to this quantitative model. Plate tectonics was used more as a help to build a scenario of the genesis of mountain belts than as a quantitative model of earth deformation.

The mutation from an ocean- to a space-based plate tectonics occurred in 1986, after five years of space geodetic measurements between the Westford (United States) and Onsala (Norway) very long baseline interferometry (VLBI) stations. These provided the first space-based direct estimate of the present rate of opening (0.8 inch or 2 centimeters/year) at the rift across the Atlantic Ocean. It was soon followed by the first estimate of the rate of sea floor shortening between Hawaii and Tokyo (3 inches or 8 centimeters/year). A turning point in my scientific journey was reached when I learned about this latter measurement. Of course, I was thrilled to see my 1968 estimate of the amount of subduction in the Japan trench confirmed. But more important, I realized then that we were entering a new era of plate kinematics.

This was rapidly confirmed by increasingly numerous measurements from VLBI techniques, but also satellite laser ranging (SLR) techniques and global positioning systems (GPS). Soon, it was established that the latest ocean-based plate kinematic model (obtained with velocities averaged over 3.5 million years) was in excellent agreement with the space-based model, where all measurements were obtained over a few years on continental sites.[63] Motions of plates could be obtained nearly instantaneously. Moreover, these motions applied to a significant portion of the recent geological past. From then on, the action in plate tectonics moved back to the continents, where one could directly measure not only the instantaneous motions of plates, but the deformations over the plate boundaries, including some complex intracontinental deformation zones. Until then, these had been considered gray areas for plate tectonics, since we had no quantitative knowledge of the kinematics of their deformations. In the same way that seismology had been reconciled to tectonics by the ocean-based plate tectonic model, geodesy became a new and essential partner in the space-based continental plate tectonic model. Today, the integrating power of plate tectonics is becoming more and more evident as tectonics moves from purely kinematic descriptions to dynamic modeling.

In 1990, Jason Morgan, Dan McKenzie and I were awarded the Japan prize for our contributions to the plate tectonic theory. While discussing this together in a hotel in Tokyo, I remember that Dan said: "Never more in our life will we be able to contribute to such a decisive and exciting discovery." This, I thought, was true. We had been involved in a mutation of the whole of earth sciences that occurred within a very short time, and we knew it when we lived it. This explained the extraordinary feeling that carried us through these fascinating years and the sense that never more will we live through something even remotely similar to it.

PART V
FROM THE OCEANS
TO THE CONTINENTS

Continental drift was first proposed on the basis of geological evidence accumulated from fieldwork by geologists on the continents. In contrast, plate tectonics was developed largely on the basis of evidence from the sea floor, or earthquakes under it, collected mostly by geophysicists. When geologists realized what was happening, the most alert among them saw an opportunity for a radical reinterpretation of geological history based on the new model of crustal mobility. Moreover, important geological features that had never been fully understood – like California's great San Andreas Fault – suddenly could be explained, clearly and elegantly, by the new model. With every old understanding up for grabs and new understandings emerging daily, one of the 20th century's greatest scientific revolutions happened.

CHAPTER 14

PLATE TECTONICS AND GEOLOGY, 1965 TO TODAY

John F. Dewey

An accurate portrayal of the events of the 1960s that led to the establishment of the plate tectonic paradigm is a hit and miss affair based upon imperfect personal memories and recollections. Equally, the progressive development of the geological corollaries of plate tectonics has been complicated. The published historical evidence is in the literature but cannot constitute the whole objective truth of who did what and said what, when, where, and to whom. This brief essay constitutes my personal recollections of the events seen from a Cambridge University, and Columbia University's Lamont-Doherty Geological Observatory perspective, and some of my views on the state of related geology from the 1960s until today.

PLATE TECTONICS AND GEOLOGY

My early career from my Ph.D. work through lecturerships in Manchester and Cambridge from 1958 to 1964, in spite of little funding, was a wonderful period of basic geological research, mostly field-based, in Ireland and Newfoundland. They were carefree days of excitement and wonder in the 'preparatory days' to plate tectonics. I was concerned with trying to find out how and when the Lower Paleozoic Grampian Orogeny along the Laurentian continental margin was formed. This orogeny constituted a mountain-building event that stretched from the British Isles throughout the western margin of the northern Appalachians, and has only just been resolved as an early to mid-Ordovician arc-continental margin collisional event that lasted from 472 to 462 million

John Dewey in the southern uplands of Scotland, 1999. Photo courtesy of John Dewey.

years ago. It suggested that the European continent and the North American continent were once joined. Such plate tectonic concepts were, of course, unknown in the late 1950s when I started my Ph.D. mapping on the Ordovician and Silurian rocks (about 490 to 400 million years old) of the South Mayo Trough in western Ireland. Those were the days in which most graduate students in geology were given an area to map, about which little was known. The goal was to produce a detailed geological map of everything from underlying basement to surficial deposits and to follow any or all research leads that emerged from the fieldwork. We did not go into the field to test hypotheses; we were trying to map the planet. We were concerned with process and history but only insofar as it resulted from the fieldwork. This was not an incorrect methodology; it was appropriate for the times before we had the plate tectonics paradigm.

I was profoundly lucky not only in having a marvelous Ph.D. area in County Mayo, Ireland, but in being a graduate student from 1958 to 1960 at Imperial College, London, where the intellectual human resources included Ian Carmichael, Graham Evans, John Ramsay, Doug Sherman, George Walker, and Janet Watson, lecturers and professors, all of whom generously and freely gave their valuable time to helping not only their own but all research students. In addition to intensively detailed fieldwork, my student colleagues and I, including Tony Barber, Peter Kassler, Frank May, and John Spring, spent hours doing basic geological laboratory work: peering at thin sections (actual slices of rock) under a microscope, plotting dozens of stereographic projections with structural data, converting field slips into the final maps, and in reading extensively in our area of geology.[1] Basically, I continued in this mode until 1965, because of a warning I received from an examiner of my Ph.D. dissertation. In my final chapter, I attempted a synthesis of British Ordovician and Silurian paleogeography and paleotectonics, suggesting that the ancient mountain belts seemed the same on either side of the Atlantic. This did not go down well with my internal examiner, who wagged his finger, saying, "This kind of thing gets geology a bad name," and "in any case can only be done properly by very senior people with lots of experience and knowledge." I quietly held my tongue and nodded abjectly.

For the next few years, as a young lecturer in Manchester from 1960 to 1964, I mapped and studied local Irish geology and studied the patterns of kink bands, small angular folds in foliated rocks. This seemed to 'work' because, in 1964, I was appointed to a lecturership in structural geology at Cambridge. Since that time, I have thought a lot about how scientific fashion changes and regiments many of us into particular channels. Professor L. J. Wills was an extremely distinguished and clever geologist of the period between the wars; could it have been that he was not elected to the Royal Society because he published a paleogeographic atlas of the British Isles, the first synthesis of its type in the United Kingdom? It is said that, during the fixist era dominated by Walter Bucher of Columbia University, among others, one could not get a job in a North American university if one embraced continental drift. The reverse is probably true now.

The most critical event in my geological life occurred in 1964. Art Boucot of Oregon State University and Stuart McKerrow of Oxford University invited me to Nova Scotia to help, with Bill Fyson of Carleton University, in the detailed mapping of a large area of Silurian rocks northwest of Antigonish. Marshall Kay of Columbia University, whom I had met earlier that year, heard of my visit and invited me to spend a month in the field with him in Newfoundland immediately afterward. The Canadian Maritimes immediately broadened my geo-vision to a scale even larger than that from which I had been warned. In discussion with Marshall Kay, I realized that there is an extraordinary correspondence between the geology of western Ireland and that of Newfoundland. This marked a sea change for me: since then, I have doggedly stuck to a research course that combines intensely detailed fieldwork with model building, speculation, hypothesis testing, and extensive reading across most of the earth sciences and regional tectonic synthesis, in the firm belief that without detailed field knowledge the rest is useless. As Francis Pettijohn remarked, "the truth resides in the rocks," and "there is nothing as sobering as an outcrop." These truths are extant and even more important today than they were in the 1950s.

In 1965, I was profoundly ignorant of developments taking place from the work of the great oceanographic laboratories of Lamont, Scripps, and the Woods Hole Oceanographic Institution. Although I had read Fred Vine and Drummond Matthews' 1963 paper on sea floor spreading, like many (perhaps most) geologists at the time, the oceans were an irrelevant mystery to me.[2] Although I was dully aware of the work of Harry Hess and Bob Dietz, I was influenced by the fixism of Sir Harold Jeffreys. As an undergraduate I had heard the persuasive mobilist arguments of Arthur Holmes of Durham University, and Lester King and Stanley Hollingworth of University College, London, but for some strange reason the mobilist arguments always seemed irrelevant to my research and thinking.[3] Perhaps it was because I was embedded in the continents.

Three events fundamentally changed my geo-life. First, one early autumn morning in 1964, I was sitting in my room in the Sedgwick Museum in Cambridge trying to fathom the mysteries of kink bands, when Toronto's Tuzo Wilson, on sabbatical leave, sauntered in clearly bursting to tell anyone who would listen about his new ideas. He had discovered that I was the new lecturer in structural geology and said, "Dewey, I have just discovered a new class of fault." "Rubbish," I said, "we know about the geometry and kinematics of every kind of fault known to mankind." Tuzo grinned and produced a simple colored folded-paper version of his now-famous ridge/transform/ridge model and proceeded to open and close, open and close it with that wonderful smile on his face.[4] I was transfixed both by the realization that I was seeing something profoundly new and important and by the fact that I was talking to a very clever and original man. This moment transformed my research life. Although I was embarked irreversibly upon synthesis, it was still part of my intellectual baggage, as a classically trained geologist, that one must always first gather lots of data, then build slowly upward toward models (the inductive or Baconian method). The concept that one could develop a model and then test and falsify it (the deductive or Popperian method) was wholly new and exciting. Some earth scientists have still not grasped the basic concept of how to do science and still believe, mistakenly, that, if one gathers enough random data, the 'answer' will spring, mysteriously, from that data mass.

Second, in 1965, Marshall Kay persuaded Columbia astronomer Bob Jastrow to invite me to the Goddard Institute Conference on continental drift at the Goddard Institute in New York, at which there were spectacular disputes between those like Gordon MacDonald and Art Boucot on the fixist side and Teddy Bullard of Cambridge and others on the mobilist side. Marshall Kay and I rather passively presented a paper showing that, upon closure of the North Atlantic Ocean, the detailed zonal geology of Newfoundland and Ireland fit spectacularly. This could not possibly be fortuitous, and it implied pre-Cenozoic continuity about 200 million years before the development of Appalachian/Caledonian chain.[5] Whether by drift or oceanization (the mysterious mafic metasomatism or transformation of continental into oceanic crust) seemed irrelevant.

Third, immediately following the Goddard Conference, I spent a few days at Lamont chatting with Lynn Sykes, Walter Pitman, and Bill Ryan. Pitman showed me the *Eltanin*-19 profile and told me what it meant for quantifying rates, directions, and timing of sea floor spreading, and that the oceans are all Mesozoic/Cenozoic, that is, younger than about 160 million years. I was mildly impressed and remarked, "Interesting but keep it in the oceans and don't let it onto the continents." Pitman was slightly scathing and muttered darkly about the ignorance and narrow-mindedness of geologists.

Strangely, although my remark was foolish in its intent, sea floor spreading is a solely oceanic phenomenon. There, the boundary conduction layer is generated, cools, subsides, and is subducted as the main engine of global heat loss, in contrast to the weak, old, and largely non-geodegradable continents.[6] Soon Lynn Sykes would publish his masterpiece on the active segments of oceanic transform faults in which he showed that their motions supported precisely the predictions of Tuzo Wilson's 1965 paper that explained oceanic transform faults: namely, that the two sides of the fault were moving exactly as predicted if the ocean crust were splitting apart along the mid-ocean ridges, and the transform faults connected one segment of a ridge to another.[7] Sykes tested Wilson's idea by analyzing the motions on either side of the fault: which side was being compressed, which side was under tension? Lynn Sykes generously spent time explaining it to me, which principle may be explained as follows.

I have a simple imaginary experiment that explains how detailed/compressional first-motion solutions work, which both undergraduate and graduate students seem to enjoy. Imagine a fault, say the San Andreas Fault, along which a segment is 'stuck' and the surrounding rocks are accumulating an elastic shear strain. As the strain accumulates and just before the coefficient of sliding friction is surpassed, and the fault slips to produce an earthquake. Let us set up the following hypothetical scenario. Choose four willing graduate students and supply each with a bag of cement, a sheet of plate glass, sufficient bricks to build a small wall, and a large jar of petroleum jelly. Let each student occupy a quadrant of the fault trace in relation to the expected epicenter, and let each build a wall between a sheet of plate glass cemented to the substrate and the expected epicenter. Then let each student stand on the sheet of plate glass with feet liberally lubricated with petroleum jelly and place his or her hands upon the wall and 'wait for the earthquake.' Where the fault slips, two students are smashed into the wall (compressional quadrants) and two fall backward (dilational quadrants). Hence four quadrants, two compressional and two dilational, define two planes, one parallel with and one orthogonal to the fault through the epicenter, and give the sense of motion on the fault. (Rather than use students in this abusive way, we use seismometers.)

Tuzo Wilson's 1965 paper really was the beginning of plate tectonics, in which the basic concepts of the evolution of boundaries between rigid plates were hammered out. Subsequently, Tuzo told me that, had he known Euler's theorem describing the relative motion of rigid bodies on a spherical surface, he would have nailed plate tectonics cold. Instead, this was done two years later by Dan McKenzie and Bob Parker, and Jason Morgan at Princeton, with important additional contributions by Xavier Le Pichon, Bryan Isacks, Jack Oliver, and Lynn Sykes at Lamont.[8]

There had been many earlier 'sniffs' of some of the elements of plate tectonics, all generated by geologists with great field experience and understanding. Bert Quennell, after many years of working on the geology of the Middle East, far from his native New Zealand, realized that the Dead Sea Fault described a portion of a small circle about a pole in northwest Africa, and that the Arabian "block" had moved north and rotated counterclockwise with respect to an "African block."[9] Earlier, Harry Wellman of Wellington University, New Zealand, thought that the Alpine fault in New Zealand was a connecting "transform" between zones of shortening in the north and south islands.[10] However, although all this "mobilism" was in the background of geology since the days and ideas of Alfred Wegener, Alexander du Toit, Lester King, David Griggs, and Arthur Holmes, it had no proper skeletal structure because the oceans were largely unknown until the postwar beginnings of cruises from the great oceanographic institutions.[11]

A profoundly important contribution to the development of plate tectonics was that of George Plafker, a distinguished geologist working with the U.S. Geological Survey, following the 1964 Anchorage earthquake. Plafker mapped out, over a huge region, those areas that had risen and sunk during the earthquake showing that they were separated by a line running parallel with, and about 160 miles (250 kilometers) inboard of, the oceanic trench.[12] The first motion pattern of the 20-mile (30 kilometer)-deep earthquake indicated a choice between a steeply southeast dipping thrust nodal plane and a gently northwest-dipping thrust nodal plane, both potential fault planes; geophysicists had chosen the former. Plafker pointed out with simple, cold logic that it was impossible to have a 650 mile (1,000-kilometer) long thrust fault that did not break the surface somewhere (blind thrust). He argued that the shallow thrust plane must be the correct choice and, moreover, projected toward the oceanic trench, suggesting that the ocean floor slips beneath Alaska, creating the zones of deep-focus earthquakes described in the 1930s by K. Wadati, and in the 1950s by Hugo Benioff. Such is the nature of simple effective geologic argument. Years before, similar geometric arguments by the Austrian A. Amstutz, from regional structural geology, had led to the development of the idea of limited continental subduction, where one piece of continental crust overrode another in mountain belts. Likewise two Austrians, O. Ampferer and W. Hammer, had argued that the swallowing of continental crust beneath mountains takes place in *Verschluckungzonen*, or zones of swallowing.[13]

Knowing of my developing interests in Newfoundland geology in particular and Appalachian geology in general, Ward Neale, professor of geology and head of the department at Memorial University in St. John's Newfoundland, invited me to spend a field season in Newfoundland in 1967. Luckily, I gained support for this from a Natural Environment Research Council (NERC) three-year research grant. I began work on the Burlington Peninsula in Newfoundland, along the length of which runs a belt of highly deformed mafic rocks (rich in magnesium and iron), which separate continental rocks from an ancient volcanic chain. We now know this zone to be the line of the Ordovician Collision (ca. 470 million years ago) between the ancient North American continent and a contemporaneous volcanic island chain, or "island arc." However, at that early stage, I was unaware of such concepts and was in Newfoundland to begin a project that was to try to measure the amount of shortening across the Appalachians of Newfoundland, a task that we know now to be fruitless and pointless because the closure of an ocean involves massive subduction unrecorded in the rock record. I quickly saw the dead-end nature of my initial plans and switched my attention to mapping and trying to understand the varied rock groups.

In 1967, Marshall Kay of Columbia University organized a Conference in Gander, Newfoundland, which brought together many Appalachian and Caledonian geologists. In spite of some superb geological discussions about how the Appalachians might have once been continuous with the Caledonides prior to the opening of the North Atlantic, there was no mention of Tuzo Wilson's idea that an old ocean might have closed somewhere along the Appalachian/Caledonian Orogen.[14] The arguments from oceanic and seismic data that were leading, inexorably and quickly, to the development of plate tectonics simply were not part of the intellectual equipment of most geologists at the time. We took the somewhat snobbish view that continental geological data for which one sweated was superior to remotely sensed oceanic data. Perhaps this was a defensive counteraction to the physicists' prevailing view that only physics is really science, and the rest is stamp collecting. Both are absurd and extreme positions. We all need each other in the earth sciences if progress is to be made.

In 1967, following a field season in Newfoundland and the Gander Conference, I spent a wonderful sabbatical leave at Lamont. Lamont professor Chuck Drake had been in Cambridge on sabbatical leave during the previous year and, knowing of my interests in synthesizing the geology of the Appalachian/Caledonian chain, he invited me to Columbia University where, he hinted, tectonic things were happening fast and I could avail myself of the opportunity to broaden my education. I was given an office, a large Lamont-style light table, and an inexhaustible supply of tracing linen (my – how computers have changed and vastly improved our drafting capacity) and was encouraged to assemble a cartographic synthesis of the Appalachian/Caledonian belt. I copied and generalized all the state geologic maps of New England and the province maps of the Canadian Maritimes and transposed the results onto a 10 foot long map from New York to Newfoundland. I puzzled and marveled at the zones of Ordovician volcanics, the ultramafic belts, and the zones of Ordovician and Silurian deformation. Like Harry Hess years before, I realized that serpentinites, volcanic zones of particular geochemistry and age, and deformation of various types and ages occur in zones, some of which are continuous or semicontinuous along the length of the mountain belt.

While I was doing all this, Xavier Le Pichon, Bryan Isacks, Jack Oliver, and Lynn Sykes at Lamont, and Jason Morgan at Princeton were writing their profoundly incisive papers on plate tectonics, each with a clearly different slant and approach.[15] Eerily, I felt the restraining influence of Walter Bucher of Columbia University through my conversations with

Maurice Ewing, who was influenced, deeply, by Bucher, probably because in those days geophysicists believed what geologists said. Ewing gave his approval to mobilism only when, in 1971, it was all 'signed, sealed and delivered.' It was during my time in Lamont in 1967–1968 that I realized how very important Keith Runcorn's paleomagnetic work was. Neil Opdyke at Lamont contributed to my geological education by showing me how Runcorn had, years before, demonstrated that the North American and European polar-wandering curves match precisely for a closed North Atlantic.[16] I produced an interpretative tectonic map of the Appalachian/Caledonian chain that was to form the basis of most of my work over the next few years.

My new geological education was completed. I came back to Cambridge in 1968 full of myself, excited and "hot-to-trot" in global tectonics and its geologic expression, and to spread the gospel about the new way to do geology. Not surprisingly, my evangelical zeal led to some resistance. Some resulted from the innate conservatism of classically trained geologists, but some resulted from an ill-founded worry that classical geology was finished. Robin Nicholson of Manchester University said to me, after I had given an enthusiastic lecture, "If you are right, we are all out of jobs." Nothing was further from the truth; we were just embarking on a quite extraordinary voyage of geological discovery from which we would never disembark. I wrote a number of papers on the geology of opening and closing oceans, especially for the Appalachian/Caledonian Orogen, some with Jack Bird of the State University of New York at Albany, whose geological work in and understanding of external zones of the northern Appalachians was, and is, unsurpassed. Jack Bird taught me tectonostratigraphy – how to look at the internal and external stratigraphy of a mountain belt and draw regional conclusions about its timing and kinematics. More recently, my Oxford colleague Maria Mange has taught me that heavy minerals hold the key to understanding the sources of sedimentary materials and the patterns of uplift and erosion in mountain belts.

By early 1969, I was ready and eager to explain the geology of the Appalachian/Caledonian Orogen to anyone who would listen.[17] I had a detailed geotectonic/tectonostratigraphic map of the Orogen that, I believed, was only explicable in terms of an ocean that opened in the late Precambrian, 550 million years ago, and closed during the late Silurian, 400 million years ago. I encountered astonishing resistance from the geological community but huge support from the geophysical community, the result of which being that my allegiance to and interest in geophysics remains. I was floundering around in a no-man's land between geology and geophysics in 1968–1969 when Teddy Bullard, following a seminar that I gave at Madingley Rise in Cambridge, took me out to dinner in College and gave me some profoundly important advice about how to do science, for which I am eternally grateful. It may be summarized as follows. First, do something rather than nothing. Second, have ideas and test them. Third, respect data but remember that 'data' are elusive and observer-controlled. Fourth, think big and do not let senior people dominate you and tell you what to do. I cannot emphasize too strongly what a great man and scientist Teddy Bullard was. He cared deeply about science and truth, and influenced and encouraged a large number of young geoscientists at Cambridge, including myself, Jo Cann, and Dan McKenzie.

In early 1969, I saw an advertisement for the first Penrose Conference on Global Tectonics and Geology organized by Bill Dickinson at the Asilomar Conference Center on the Monterey Peninsula. Dickinson had seen the need for a conference to pull together all the threads of the rapidly developing ideas of plate tectonics and their role in the tectonics of the continents. I phoned Bill from Albany, where I was working with Jack Bird on some Appalachian/Caledonian papers, and pleaded to attend. He quizzed me strongly, demanding to know what intellectual "geo-understanding" I could bring to the conference. I said that I was an admirer of John Steinbeck's novels, especially *Cannery Row* and *East of Eden,* to establish my Californian credentials, and that I believed that I had a modest understanding of both the Appalachian/Caledonian (reasonably justified) and Alpine (partially unjustified) mountain belt systems. Bill very kindly invited me and I arrived in Monterey ready to evangelize and learn. The participant slate was impressive: MIT's Clark Burchfiel, USC's Greg Davis, Jim Gilluly, Steve Oriel, Gary Ernst, Bob Coleman, Warren Hamilton, and Clark Blake, all of the USGS, among many others – all names that I knew from their publications and reputations. Their presentations were all superb, but we were all blown away by one of Bill Menard's graduate students from Scripps, a modest young woman named Tanya Atwater. She showed the geological world, for the first time, how quantitative relative plate motion could be translated into geological predictions that could be tested against the known geology of western North America. The results were spectacular. Tanya showed quietly, with detail and precision, how the migration of two plate triple junctions and the growth of the San Andreas Fault explained the timing of continental margin volcanism in western North America.[18] For me, this was the end of an era of geotectonic speculation and the beginning of a new world of potentially quantitative and testable geotectonic synthesis.

From 1969 to about 1972, the impact of the geological ideas stemming from plate tectonics was muted by the characteristic geological aversion to bold, rational solutions to geological problems; small-scale complexity commonly retards and obscures our understanding of larger-scale simplicity. At the other extreme, the principal characteristics of a major new paradigm took its grip on geology. A reign of terror prevailed for a few years in which blind acceptance was accompanied by the necessity to explain everything in terms of plate tectonics.[19] This was a phase of the often primitive, shallow, and banal casting of all geology in a poorly understood plate tectonics framework. Fortunately, this phase was short-lived and by 1973 geologists were taking plate tectonics more seriously. But the 'reign of terror' phase has left a legacy of poorly constructed and badly understood regional tectonic syntheses.

During the early days of plate tectonics, the words *plate tectonic* were inserted in the title of many papers for which plate tectonics is irrelevant or, at best, of marginal importance. A particular tyranny was the imposition of extant plate tectonic processes on ancient rock systems. This went way beyond the rational interpretation of ancient volcanic belts as former island arcs, ultramatic rocks, or remnants of ancient sea floor, and thick sequences of sedimentary rocks as ancient ocean trenches. Modern plate tectonics was "shoe-horned" into the Precambrian, even back into the Archean, the period at the start of earth history more than 2 billion years ago. Yet even a casual examination of Archean rocks indicates quite clearly that the rocks are different, and arranged in different ways, than in later geological periods. The very early history of the earth is not well explained by the rigid plate model that works so well for most of the rest of earth history. The early earth was probably very hot, and heat loss through convection likely took place at a much faster rate than today. Some form of rapid mobilism, not of large rigid or semi-rigid plates but rather more ductile smaller blocks may have prevailed.

It is also clear that the earth has experienced neither a constant tectonic pattern nor one of simple unidirectional evolution since Archean times. First, there is clearly a global tectonic cycle that assembles continental blocks and then disrupts them, so that episodes of collision, compression and continental amalgamation alternate complexly with periods of continental disruption, breakup, and growth of the ocean floor. Different periods in earth history have been characterized by different dominant tectonic regimes. Therefore, the study of global secular tectonic evolution and comparative tectonics will be an important field, which needs a new generation of ultra-fine field studies that address new questions.

By 1973, we realized that plate tectonics, although a bold, simplifying paradigm, does not simplify geology magically and overnight. Plate tectonics is not only a spectacularly successful and rational framework, but it is a concept that *demands* that geology is as complicated as we know it to be. The evolution of plate boundary mosaics with stable and unstable triple junctions and plate boundaries across which slip directions must change with time generates geological evolution of the greatest complexity.[20] The poor geologist struggles to understand this nightmare of complexity and tries to build a synthetic picture. We are victims of the inversion problem: we can generate the evolution of credible complicated plate boundary mosaics and make precise predictions about the geology that the evolution of such mosaics should create, but we cannot work uniquely the other way around from the result to the model. The value of models is, therefore, not that they provide solutions to the real world, but that they give us ideas about how to proceed.

During the fall quarter of 1998, I was privileged to give a graduate tectonics course at UC Davis. After a few lectures and seminars, several graduate students asked me what geology was like before plate tectonics. I said that there was no unifying theory and that we were subject to a nightmare, necromantic world of geosynclines and tectogenes as portrayed in Umbgrove's books *Symphony of the Earth* and *Pulse of the Earth*. I explained that few geologists really believed in these concepts and that, consequently, we were thrown back into smaller-scale geology to understand history and process. I further explained that, as a result, much of the modern scientific background was laid down, without which little modern process-oriented progress could be made.

My own view of geology changed profoundly from 1969 to 1982. I gave lectures to the New York Academy of Sciences in 1970 and again in 1982. Charlotte Schreiber of Columbia University remarked that, in 1970, I lectured with the conviction that plate tectonics was about to explain with relative simplicity the whole of geology; in 1982, I asseverated, pessimistically, that geology is so complex that we will never explain it. My present position is somewhere between those two extremes. Plate tectonics is, inherently, a simple, beautiful concept, but its smaller-scale results are bewilderingly complicated.

Perhaps the most pressing problem that remains in tectonics, and one that we scarcely understand, is the tectonic and structural evolution of plate boundary zones, particularly in relation to topography. How exactly do mountains grow in areas of crustal compression? What controls the rate of deformation and uplift? A great challenge now is to use seismic data, including data from scarcely detectable earthquakes, to determine the distribution of strain along plate margins and relate this quantitatively to the growth of topography.

The 1970s saw a progressive general acceptance of plate tectonics as the best available model until, by the end of that decade, it became the dominant paradigm. It was used and misused pervasively, as the background basis of regional tectonic analysis. The problem was, and still is, linking continental plate boundary geology with relative plate movements, both quantitatively and convincingly. Atwater's model, linking the Mesozoic/Cenozoic plate boundary of western North America to its resultant geology, remains a unique masterpiece, although similar less successful models have been built for the Alpine system, the India/Asia collisional system, the Andes, and the Caribbean.[21] During the early to mid-1970s, a substantial amount of theoretical quantitative work was done on the evolution of plate boundaries, based on the triple junction analysis of McKenzie and Morgan and the evolution of plate mosaics.[22] This led to a more rational plate tectonic framework for geology, but also to a developing realization that plate tectonics not only can, but must, generate geological patterns and histories of very great complexity for which unique plate tectonic solutions are unlikely to be achieved.

SOME THOUGHTS ON SCIENTIFIC RESEARCH

Science is a strangely addictive beast. I incline strongly toward the Popperian rather than the Baconian view of progress in research, for which I have to thank the University of Houston's Kevin Burke as for much else. I believe that the history of plate tectonics supports Karl Popper's view that science advances by refuting false ideas. When I was a young geologist, money was scarce but one could follow one's nose in research rather than being told what to do to improve the "quality of life," engage in "wealth creation," and be reviewed and assessed to death by bureaucrats. The trust seems to have gone out of modern science; the basic trust that most scientists will work diligently and intelligently to solve what they perceive to be the important and interesting problems in science seems to have vanished. Scientists are generally paid poorly, but most of them love research and work long hours to find things out.[23] Modern bureaucratic obsessions with money and accountability are destroying the trust and the love of science. Young scientists on the postdoctoral treadmill have little or no career structure and little incentive; the best keep going for the excitement of discovery.

When I was a young geologist in Britain in the 1960s and a slightly older geologist in the United States during the 1970s, the funding agencies responded almost wholly to proposals received. The position is now reversed; thematic designed science is geared to perceptions of "social benefit"; buzz words such as "quality of life" and "wealth creation" domi-nate. When will politicians and our funding masters realize that it is impos-sible to predict what science is "important" and "useful" and what is not? The way to promote the best basic science is to have the funding agencies simply judge submitted proposals for their scientific excellence irrespec-tive of their field or topic. It is inappropriate and damaging to the scien-tific effort for bureaucrats, abetted by senior scientists, to solicit proposals in particular thematic areas. Scientists, especially the young, should follow their interests and noses and not be dictated to and told what to do.

Above all, we must not listen to the pronouncements of senior scien-tists that affect the research lives of the young. Senior figures are an equal part of the research community; they may know more but they are not necessarily any wiser or fatidical. For example, in the 1960s, there was a substantial tension among senior figures, such as Maurice Ewing and Walter Bucher, who were committed fixists, and Harry Hess and Bob Dietz, who were enthusiastic mobilists. Walter Bucher had a substantial influence on the thinking of senior researchers at Lamont. Chuck Drake told me that he ridiculed mobilism in his lectures to graduate students and in numerous conversations with colleagues in Columbia University. In some cases his commitment to an immobile crust led to interpreta-tions that were clearly absurd. During my several visits to Lamont in the 1960s, I felt the undercurrent of disapproval and negative background against which the more adventurous younger mobilists were working. It all came right in the end, because clever and original people like Lynn Sykes and Walter Pitman were irrepressible, and because a substantial component of the plate tectonic revolution came from Jason Morgan at Princeton, where Harry Hess was a champion of mobilism, and from Cambridge, where Teddy Bullard encouraged free thinking among the young, especially Dan McKenzie. I sometimes wonder, however, what might have happened if the senior figures of the time had had the power to completely control the funding and research of younger workers. We must beware of fashion in science, particularly when administered by dominant senior figures and funding agencies. It has the power to seri-ously distort and decay our basic research effort.

It is worth briefly examining what I perceive to be a worrying, increas-ingly degraded role for classic field-based geology in the earth sciences. I see the modern earth sciences as rather like a "puff-ball." Since the advent of plate tectonics some 30 years ago, the earth sciences have grown principally by the addition of branches, stand-alone add-ons, and layers, principally of geophysics and geochemistry, commonly driven by new instruments and computer-based numerical modeling. These areas have come to dominate the earth sciences in influence, importance, and funding, to the diminishment and denigration of classic field-based geology. Thus, like the puff-ball, there is a shiny, attractive shell that contains a rotting core of geology. We must redress this gross imbalance by giving our students a solid, basic, geological training and by coming back to the realization that the truth resides ultimately in field geology. Rocks, fossils, and minerals are immensely complicated systems but, with geophysical and geochemical data and the ideas that stem from them in the minds of clever people, they are the substance of our science. A good useful basic rule is, if you don't map it at the appropriate scale, you won't understand it. We must remember that computers do not have ideas; ideas spring from our brains. Geologists especially must not be diminished and subjugated by those who "wield" laboratory instruments and generate numerical models.

Geology is a difficult, field-based science that takes immense amounts of time, patience, and care. The obsession that only fully quantified "solutions" are valid is exceedingly foolish and dangerous. Most of the biological sciences are analogous to geology and not quantified in the way that simple geophysical and geochemical models are. The notion of a fully numerical algorithm to describe the geology and evolution of a large segment of the continental crust is ridiculous and unattainable.

Perhaps it is this impossibility that has led to the fragmentation of geology into small, process-oriented parcels in which a particular modern observable process is studied, described, and quantified as in the methods of physics and chemistry. This, in turn, has led to the idea that history is less important and that we only need to study process to understand the earth. The earth has a complicated history that integrates these process parcels and cannot be understood even by knowing everything about all the individual parcels. The process parcels are essential tools but are not a substitute for understanding the history of the planet. In any case, the modern earth is the result of a long evolution during which quite different processes may have operated. The present is not (fully) the key to the past and the reverse is also true. In my view, it is essential to the future of tectonics that much more intelligently conceived, field-based geology is injected into our thinking if we wish to avoid errors and oversimplification.

CONCLUSION

I am near the end of my formal "employed" career and look back on the way that geology has changed throughout my research life. When I was

young, we thought that geology was done by going into the field, making maps, and describing rocks, minerals, and fossils. Of course that was
fine; we did not have a paradigm and that was all that we could do. Plate
tectonics came along, as a result of oceanographic research and seismic
studies, mainly by geophysicists, and provided a wonderful new paradigm that geologists have explored for 30 years. Tragically, geology
seems to have been, to a large extent, subjugated and ignored as the way
that we find out the truth about the earth. We have become obsessed
with model building and hypothesis testing in the laboratory and by
computer. I am nowhere near the field geologist that John Ramsay of the
ETH Zürich is, nor the physicists that Don Turcotte and Dan McKenzie
of Cambridge are, nor the mathematician that Bob Parker of Scripps is,
but I have the kind of general knowledge and understanding of the earth
sciences that is becoming increasingly rare. It is possessed by people such
as Tanya Atwater, Kevin Burke, Clark Burchfiel, Bill Dickinson, Warren
Hamilton, and Eldridge Moores. If we fail to train the new generations
in geology, the earth sciences will decay to speculative, shallow, model
building. Arthur Holmes, David Griggs, Jim Gilluly, Phil King, Harry
Hess, Bob Dietz, Bert Quennell, Harry Wellman, Tuzo Wilson, Teddy
Bullard, George Plafker, Jason Morgan, Dan McKenzie, Don Turcotte,
Kevin Burke, Bob Stevens, Bill Dickinson, Tanya Atwater, Eldridge
Moores, Clark Burchfiel, Paul Hoffman, Gordon Lister, John Myers,
Cees Van Staal, and Celal Sengör (among others) have generated great
ideas, read eclectically, respected data, communicated wonderfully, and
had a deep influence on my research career. These people, and all my
Cambridge, Albany, Durham, and Oxford graduate students, have taught
me a great deal. I am profoundly grateful to them.

CHAPTER 15

WHEN THE PLATE TECTONIC REVOLUTION MET WESTERN NORTH AMERICA

Tanya Atwater

I WAS IN HIGH SCHOOL IN 1957 WHEN THE RUSSIANS SUCCESSFULLY launched the first man-made satellite, *Sputnik*.[1] It is hard to explain to younger generations just what a profound event that was. To us, it was totally astonishing that we humble humans could put an object into outer space. Until then I had planned to be an artist, but I thought, "Wow! If scientists can do that, they can solve anything (ghettos, hunger, strife . . .)." So began my checkered studies in science.

Various college recruiters came through my high school and I started asking questions about science. When the recruiter from the California Institute of Technology came, he told me straight out that Caltech didn't accept women because they viewed us as a waste of their time – we would just get married, quit, and waste our educations. This was especially ironic since the rhetoric at the time was that no man would marry a woman who was as smart and educated as he was. Luckily I had my brilliant botanist mom and my admiring engineer dad as role models. My brother was a senior at the Massachusetts Institute of Technology (MIT) at the time, so I went to visit him and various eastern schools. At Harvard, they didn't accept women but they proudly told me they would allow me to do the whole Harvard science curriculum if I enrolled at Radcliffe. Meanwhile, they rejected me at Radcliffe because I hadn't studied Greek or Latin. Thank goodness for MIT, where they said, "Sure, come along." They had been accepting a sprinkling of women almost since the institute's inception.[2]

In my junior year at MIT, I was in my fifth major, electrical engineering, when I accidentally took a physical geology course. I was hooked immediately. When they announced summer field camp at the Indiana

Tanya Atwater, at sea with Bill Menard *(right)* and Dick Hey *(left)*, in 1984. This was to be Menard's last cruise. (Photo courtesy of Tanya Atwater.)

University camp in Montana, I was first in line. I loved field camp: the mandate to hike out every day and commune with mountains, discover their secrets. I have always loved hiking, landscapes, and maps, and geometry was my favorite high school subject. The entanglement of geologic structures with land surfaces presented for me an ideal geometric mapping puzzle. I was in heaven.

But I was nervous, too. Everything in geology was so descriptive and detailed. When it came time to discuss the larger forces, we simply drew big arrows at the edges of our maps: the hands of a capricious god shortening or extending our landscapes, willy-nilly. I really didn't want to spend my life adding descriptive observations to the pile, and anyway I wouldn't be very good at it, since I have a terrible memory for isolated facts. The plate tectonic revolution came along just in time to rescue my geo-career.[3]

Once I found geology, I had to move west. The rocks in the east are old, tired, cooked, and spend most of their time covered with green or white stuff. Also, three years in Boston had made me realize what an insufferable Californian I am. I transferred to the University of California at Berkeley and finished my undergraduate degree in geophysics.

During the time that I was studying at Berkeley, my siblings had been roaming around South America having adventures – without me. I needed to get to South America. Cinna Lomnitz, from the University of Chile, was a visitor at the Berkeley Seismology Lab and I asked him about jobs in Chile. He took a long look at my bare feet, beads, and flowers – this was Berkeley, 1965. He laughed, and said, "You'd be good for them." Chile was a very formal place at the time. He gave me addresses and recommendations.

THE MYSTERIES OF THE OCEANIC REALM
AND THEIR REVOLUTIONARY SOLUTION

For the summer of 1965, while I waited to hear from Chile, I applied for and got an internship at the Woods Hole Oceanographic Institution. I was drawn primarily by the romance of the sea and ships. I didn't know enough to realize that the marine scientists were about to unleash a revolution upon the geo-world. When I saw a number of the Woods Hole staff preparing for an Upper Mantle Committee meeting in Ottawa, Canada, I asked my mentor, Brackett Hersey, if I could go too. He said, "Sure. Why not?", and found me travel funds. The meeting was concentrated on the geophysics of the oceans and the various mysteries therein. The list of sessions included all the right things: mid-ocean ridges and rifts, fracture zones, trenches and island arcs, magnetic stripes. They knew what needed explaining, just not quite how to do it.

Most of the major players in this small field were there and I greatly enjoyed meeting them and putting their faces and their quirkinesses to their names. The whole meeting was exciting, but the presentation that made the biggest impression on me was the one about transform faults by J. Tuzo Wilson.[4] Tuzo was a wonderful showman with a great twinkle in his eye. After he had explained his idea, he passed out paper diagrams with two mid-ocean ridges connected by a transform fault. It said "cut here," "fold here," "pull here." We all laughed, and I felt embarrassed (kindergarten games at this august scientific meeting?). But I took the paper back to the privacy of my hotel room and cut and folded and pulled; wow: the light bulbs went on in my brain. The simple geometry of the transform faults with their fracture zones holds the key to the geometry of formation of all the ocean basins – right there in that little piece of paper. I've been handing out versions of that diagram to students ever since, and urging them, after they stop laughing, to cut, fold, and pull.

The revolution caught up with me again in Santiago, Chile, in 1966. I was working as a technician in the Geophysics Institute of the University of Chile when I heard about an international Antarctic meeting to be held there. I did my job reading seismograms in the early mornings and evenings so that I could attend the meetings during the day. One morning session, I was dozing through a series of papers full of Latin names of diatoms and foraminifera (single-celled planktonic organisms) when they announced an extra paper. Jim Heirtzler was passing through from Lamont-Doherty Geological Observatory on the way to meet a ship in Valparaiso and he wanted to present some marine geophysical results. In his talk he put up the *Eltanin*-19 magnetic anomaly profile – still, to this day, the clearest, most beautiful, and symmetrical profile in the world – and with it made the case for sea floor spreading.[5] It was as if a bolt of lightning had struck me. My hair stood on end.

My sisters still remember how crazy I was at dinner that night. I was crazy-excited: this was that big-picture key I had been dreaming of. And I was crazy-disappointed too: there was a revolution going on and I was missing it. I immediately applied to graduate school at the Scripps Institute of Oceanography, but, in the rush of youth, I was sure the excitement would be finished by the time I could get there, six months later. In fact, I was only a few weeks late for the start.

I arrived at Scripps in January 1967 to find the place in chaos. Fred Vine had been there in December 1966 and had presented a collection of magnetic anomaly profiles from various spreading centers around the globe (including *Eltanin*-19).[6] This was extremely compelling evidence for sea floor spreading in all the oceans. Apparently the whole institution attended the talk, most of the scientists going in believing continents were fixed, all coming out believing they moved. In the first meeting of my first class, marine geology, Professor Bill Menard forgot to tell us any of the usual class preliminaries. Instead he just launched into raptures about this "wonderful new idea," scribbling all over the blackboard. I also took a "geosynclines" seminar that spring; in the arrogance of youth, we smirked our way through all that literature with its convoluted explanations and elaborate naming systems (there was a fancy Latin name for every subvariant within the array of geosynclines). Now we realized they were just describing ancient continental margins in their various tectonic situations and combinations.

MY EDUCATION IN PLATE TECTONICS

Since I arrived at Scripps in mid-year, my initiation into graduate school was ad hoc. They sent me to talk to several research groups to try to find a project. My second interview was with John Mudie at the Deep Tow group. His group was developing an instrument package that could be towed very near the ocean floor in order to get a systematic, high-resolution look at various deep-sea features. He was anxiously looking for a student to work up the data to be collected during an upcoming cruise to the Gorda Rift, offshore of northernmost California. It would be the first close-up look at a sea floor spreading center, and they were leaving that spring, just a few months hence. I couldn't believe my good luck. I signed up immediately and never looked back. I heard much later that my little decision set off a long battle over what to do with "the girl" on board the ship – women on ships are bad luck, don't you know? This first battle was fought by Mudie, who had to constantly assert my need to be aboard and, in especially bad moments, my right to be there and to sue if they wouldn't let me go. Apparently there was similar virulent discourse behind the doors each time I went to sea, although I remained happily ignorant of it all.[7]

The results of the cruise were wonderful, showing that most new basaltic sea floor is formed in the narrow rift valley floor of the spreading center, and that the giant rift mountains that flank the valley are built not by volcanism but by uplift of blocks along big normal faults. I wrote up the preliminary results that summer and fall, with lots of help, encouragement, and goading from Mudie, and it was published as a lead article in the journal *Science*.[8] At the time, I had no idea what an honor that was. I presented this work at the American Geophysical Union the following spring, my first professional talk ever, to a full house that came especially to hear me. (When I hear about the miserable first talks of many of my colleagues, I continually marvel at how spoiled I was.) Again, Mudie gave me lots of help and advice and insisted on several rehearsals, so that the presentation went very well.

After the meeting, I heard that some other students were going up to New York for a tour of Lamont Geological Observatory (now known as Lamont-Doherty Earth Observatory). Curious, I joined them. I remember two things from that tour. One is that I was invisible. In every lab we visited, they introduced all the young men and skipped me, every time. I guess our student guide assumed I was someone's tag-along girlfriend and therefore of no account. I introduced myself and tried to establish that I was a scientist, too, but my hints fell on deaf ears. The other thing I remember was the map of earthquake locations that student Muawia Barazangi, working with Jim Dorman, had plotted onto transparent mylar sheets and had overlain on a huge wall map. There they were: the plates of the world all outlined by the earthquakes. It was stunning, awesome, so simple and clear and full of details about the individual plates. It was oh-so-hard to pull myself away from that map.[9]

In those first years we didn't speak about "plate tectonics." The magic phrases were *sea floor spreading* and the *Vine–Matthews hypothesis*. Subduction was a necessary adjunct concept, but one that was much harder to test with marine geophysical techniques. Observations from the field of seismology gave us mantle subduction zones and rigid plates. When the paper by Bryan Isacks, Jack Oliver, and Lynn Sykes, "Seismology and the New Global Tectonics," came out in 1968, we students all read it forward and backward and argued about all its points.[10] It was a seminal paper for me, filling in many vital gaps in my understanding and solidifying my commitment to the whole scenario. The other paper that set me on the rigid plate road was the 1968 paper by Jason Morgan.[11] In this paper, Morgan laid out the mathematical basis for quantifying the displacements of plates on a sphere (they are rotations around "Euler poles") and applied it to the several well-known plate boundaries.

My Education in the Doing of Science

My graduate student years at Scripps were frenetic. All the data ever collected about the solid earth were waiting to be reinterpreted. I got in the habit of dropping in at Bill Menard's lab.[12] It was already known that the magnetic anomalies in the northeast Pacific were exceptionally clear, and that they were well lineated and offset across the fracture zones, but no one had compiled them for a look at the regional pattern.[13] Menard had his draftswoman, Isabel Taylor, transfer all the available magnetic profiles from their paper records to their ship tracks on a big map. She did it all by hand – this was before computer data processing became routine. The result was spectacular. The magnetic anomalies of the northeast Pacific are especially easy to read and the emerging pattern was full of information about sea floor spreading and transform faulting.[14] Every session that we had over the map was full of discovery and excitement. Menard and I began seeking each other out first thing in the morning to share our middle-of-the-night thoughts. Often I couldn't sleep at night, my head was so abuzz with geo-possibilities and implications. Apparently he was having the same problem, because I often arrived in the morning to find his ideas scribbled on my blackboard. "What about this . . . ?"[15]

In Bill Menard, I found a soul mate, a fellow enthusiast for geometric patterns and their implications. He was constantly cutting up pieces of paper and moving them around – "What if such and such happened? How would that play out in the sea floor patterns?" He had a thorough knowledge of the oceanic data sets of the time; we would predict some geometric relationship with our paper cut-outs and he would then recall examples of the same patterns from the real world. Imagine my surprise when, after a few weeks of this, he presented me with a draft manuscript describing our conversations. I was just having fun, playing intellectual games, and it was actually serious science. Indeed, those playful sessions resulted in three early papers in prestigious journals, summarizing the magnetic anomaly patterns in the northeast Pacific and generalizing them to examine the effects of changes in direction of sea floor spreading.[16]

I learned many things from Bill Menard, among them that a new object or phenomenon needs to have a name in order to hold a place in the human mind. For example, some direction changes cause transform faults to pull apart along their lengths, allowing magmas to seep up into the resulting rifts. He dubbed this phenomenon *leaky transform faulting*. I was amazed how often "leaky" transforms appeared in the literature thereafter (although the usage was not always what I would have chosen). In another example, I worked with fellow graduate student John Grow on a paper about the oceanic plate that once lay north of the Pacific plate and that was entirely subducted northward beneath Alaska and the Aleutian island arc. Walter Pitman and Dennis Hayes at Lamont had already pointed out the evidence for this plate, but they had described it and its neighbors as plates I, II, III, and IV, not exactly names that stick in the mind.[17] Plates I, III, and IV were, in fact, the Pacific, North American, and Farallon plates. We needed a name for plate II, the one that had been entirely subducted. We described our need to Donna Hawkins, who had done social work with Native American peoples in Alaska, and she dug out her dictionaries and came up with a possible list of names and their definitions. We chose "Kula," the Athabascan word meaning "all gone."[18] I still blush when I see our paper credited (or sometimes discredited) with the discovery of this plate.[19]

Menard was responsible for naming many of the fracture zones in the North Pacific. He was especially pleased with the fracture zones off Mexico, which had been named after Mexican artists Orozco, Tamayo, Siqueros, and Rivera. Following his lead, I named new fracture zones right and left as they emerged from our patterns, all unknowing that there are weighty rules and procedures concerning the official naming of geographical objects. Happily, he had neglected to teach me about those.

Another rule of Menard's was: when drawing on napkins during a discussion, each individual must have her or his own pencil. Many a joint conversation was put on hold after the first sentence while he went to fetch that second pencil. Our conversations were so geometric that the person without a pencil was rendered voiceless. I am often reminded of this rule when a colleague or student, looking at something I am drawing, starts snatching at my pencil or madly pointing and finger-sketching: ah yes . . . time to implement Menard's multipencil rule.

THE POWER OF TRIPLE JUNCTIONS

Scripps was frequented by visitors from all over the world and they added greatly to the liveliness and depth of this already exciting place. Dan

McKenzie was there during the fall of 1967 and he was thinking hard about many aspects of the new theories. My advisor, John Mudie, had set up a monthly beer party at a local German dance hall to get people together for informal talk. I especially remember one of these sessions during which McKenzie and Bob Parker arrived, bubbling over about some project they were working on.[20] I couldn't figure out what they were talking about and could barely hear them over the loud accordion music, but during a lull I asked, rather timidly, what the fuss was about. Dan took a napkin and sketched out the San Andreas and Queen Charlotte Fault systems and the Aleutian/Alaskan subduction zone. He showed me how all these features lay along the boundary between two large rigid plates, the Pacific and North American plates. "That's all very well, but what about the Mendocino fracture zone? That doesn't line up," I complained, trying to grab his pencil so I could add the offending feature to his tidy sketch. (They were acting so smug, I hoped I could trip them up.) "Easy," said Dan, and he drew a third plate, the Juan de Fuca/Gorda plate, meeting the other two at the Mendocino triple junction. Three plates! Of course. So elegant, so simple, and so powerful. I sat there, agog, my brain zooming around in all directions. Here is what I wrote about this moment a few years later.

It is a wondrous thing to have the random facts in one's head suddenly fall into the slots of an orderly framework. It is like an explosion inside. That is what happened to me that night and that is what I often felt happen to me and to others as I was working out (and talking out) the geometry of the western U.S. I took my ideas to John Crowell [at the University of California at Santa Barbara] one Thanksgiving day. I crept in feeling very self-conscious and embarrassed that I was trying to tell him about land geology starting from ocean geology, using paper and scissors. He was very patient with my long bumbling, but near the end he got terribly excited and I could feel the explosion in his head. He suddenly stopped me and rushed into the other room to show me a map of when and where he had evidence of activity on the San Andreas system. The predicted pattern was all right there. We just stood and stared, stunned.

The best part of the plate business is that it has made us all start communicating. People who squeeze rocks and people who identify deep ocean nannofossils and people who map faults in Montana suddenly all care about each others' work. I think I spend half my time just talking and listening to people from many fields, searching together for how it might all fit together. And when something does fall into place, there is that mental explosion and the wondrous excitement. I think the human brain must love order.[21]

After that evening in the beer hall, I became a McKenzie groupie, attending his seminars, dogging him with questions, making a big nuisance of myself, I'm sure. He was humorously generous and I learned a lot: about tectonics, about the scientific approach, and about tectonic passion and delight.

The magnetic anomaly patterns of the northeast Pacific are different from those in other oceans in that they are almost entirely one-sided. The eastern half of the expected symmetrical pattern was embedded in the Farallon plate and has been subducted beneath North America, along with the spreading center that separated it from the Pacific plate. Only the western half, the half that is embedded in the Pacific plate, remains for us to observe. This one-sided configuration was a hindrance at first, because the lack of symmetry removed one of the most convincing arguments for sea floor spreading. However, once the concept of spreading was demonstrated elsewhere, the one-sidedness revealed a remarkable relationship. The Farallon plate had been completely subducted in the exact regions now occupied by the San Andreas Fault and its relatives, so that the subduction of the Farallon plate and its spreading center holds the key to the origin of the San Andreas system. Dan McKenzie and Jason Morgan first described this geometric relationship (and named the Farallon plate) in their 1969 paper about triple junctions.[22]

The relationship just described is very useful for establishing the timing of events in western North America. Since the San Andreas Fault system forms the boundary between the Pacific and North American plates, it could not have originated until those two plates came into contact. This contact, in turn, could not have occurred until after the complete subduction of the intervening Farallon plate. The offshore magnetic anomalies would constrain when and where that transition occurred, if only we could obtain reliable ages for the magnetic reversals that caused the anomalies. We could not take the next step until we had these dates.

MYSTERIES OF THE CONTINENTAL REALM IN THE SAN ANDREAS FAULT SYSTEM

Meanwhile, I was learning about the San Andreas Fault system. The main fault had been recognized as a major throughgoing structure ever since the 1906 San Francisco earthquake.[23] Mason Hill and Tom Dibblee sharpened our awe of this feature in 1953, when they laid out evidence for at least 300 miles (500 kilometers) of cumulative offset across the fault.[24] By the time I began to study it in the late 1960s, it was clear that the San

Andreas was a profound break, and that it was almost surely a major boundary in the global plate system. But the age of origin and rate of offset were not known. Hill and Dibblee's most convincing evidence for large offset was in the displacement of late Cretaceous granites (80 million years old) from the Tehachapi Mountains in southern California to Bodega Head or beyond along the northern California coast. Had the fault been moving since the late Cretaceous? It seemed likely at the time. We were just realizing, thanks especially to some papers by Warren Hamilton of the U.S. Geological Survey, that the Sierra Nevadan granites were formed in the roots of volcanoes in subduction zones.[25] The Sierran magmatic system had been active during much of the Mesozoic era, implying a major, long-lived subduction zone, but this magmatism had suddenly ceased in the late Cretaceous, about 75 million years ago. This seemed to be just what we were expecting: a cessation of the subduction plate boundary and its replacement by the San Andreas plate boundary. Furthermore, if the fault had been moving steadily since the late Cretaceous, the offset rate would have been less than one-third of an inch per year: quite slow. It all seemed to be coming together, or so we thought.

In 1967 Bill Dickinson hosted a meeting at Stanford to see if the community could solidify the timing and displacement rate along the San Andreas Fault.[26] Some of us students attended this meeting, sitting up high in the back of the big lecture hall, watching with awe as the grand old men presented their work. Dickinson began the meeting by urging the speakers to be wild, to describe any tentative geological correlations that might conceivably bear on the subject. That introduction made a big impression on me. Before that, I had thought all public presentation of science had to be formal and factual and serious; no speculations allowed. How fun to see that the big guys had lots of wild ideas, too.

This august bunch of Californian geologists laid out lots of possible correlations – datable rock bodies or features such as ancient shorelines that occur on one side of the fault and that seem to match with similar bodies or features that occur somewhere on the other side of the fault. If the paired objects started out side-by-side and were later offset, they would help us work out the displacement history. This is always a tricky business, since many different rock bodies are quite similar in their characteristics, and many features, especially shorelines, tend to follow faults, rather than crossing them. Among the many tentative correlations presented, the majority seemed to favor the slow rate described above.

Not everyone agreed with the Cretaceous origin and slow rate, however. A group from U.C. Berkeley, in particular, had evidence for a much faster rate. They presented exceptionally strong evidence for a correlation between the Neenach volcanic rocks in the northwestern Mojave Desert (on the east side of the fault) with volcanic debris in rocks at Pinnacles State Park, near Salinas (on the west side). This match documents an offset of about 200 miles (320 kilometers) sometime after the volcano erupted 23 million years ago. They presented this and related data implying a rate of several inches per year.[27] Thus, we were left with two conflicting scenarios for the San Andreas – a young, fast-slipping fault or an older, slower-moving one. I sat there aching, knowing that the offshore magnetic anomalies would bring an independent voice to this problem, if only they could be reliably dated.

Establishment of the magnetic reversal ages presented a big challenge (as does most geological age dating). During the 1950s and 1960s the paleomagnetic community had honed the ages of the reversals that occurred during the last few million years.[28] They did this using isotopic methods, dating normally and reversely magnetized lava flows on land. These ages were the ones used by Fred Vine in his 1966 compilation to date the youngest magnetic anomalies at each sea floor spreading center and to establish recent spreading rates, a hugely valuable contribution. However, for ages greater than a few million years, the dating errors were too large to distinguish one reversal event from another. So they were useless. We really needed those older ages.

THE LAMONT MIRACLE – FIRST SOLUTIONS FOR THE WORLD'S OCEANS AND THEIR TIMING

Meanwhile, on the global scale, the marine geophysical group at Lamont was busy interpreting the world. For many years their ships had been traversing the global oceans under the somewhat dictatorial direction of Maurice Ewing, Lamont's founder and director. Everywhere these ships sailed, whatever the immediate interests of the shipboard scientists, they collected a coherent set of geophysical and geological data. As part of the routine, they measured magnetic field profiles, even though no one could make sense of the resulting wiggly lines. It was a relatively easy measurement to make, so they made it. The other major U.S. oceangoing research institutions were much more democratic (anarchic?), each scientist following his own agenda and those of his close associates. During traverses between study sites, data collection was somewhat haphazard. When Fred Vine and Drum Matthews finally supplied the key ideas for reading the magnetic anomalies, the Lamont group was in a unique position to interpret the broad histories of most of the world's ocean basins. They presented these interpretations in a series of papers in the March 1968 issue of the *Journal of Geophysical Research*.[29] My copy of this issue is disgustingly grubby and tattered from my constant reference to these papers during the next decade.

The final paper in the March 1968 series was especially important.[30] The preceding papers had presented oceanic histories in terms of magnetic anomaly numbers, using an informal numbering system that Walter Pitman had invented for the purposes of communication within their group. He assigned the central anomaly the number 1 and, working outward, assigned 2 through 32 to distinct bumps in the rest of the known pattern. We still use this numbering system, slightly modified, referring to the numbers as "magnetic isochron numbers," or just "chrons." In that final paper, the Lamont group amassed and compared all their data concerning the distances from the spreading centers out to the various magnetic isochrons. In a series of innovative comparison tests, they concluded that the South Atlantic was the most likely of all the oceans to have spread at a steady rate over the long term. (Of course, no one had any idea if that was even possible.) They then made the leap and extrapolated from the South Atlantic spreading rate, known for the last 4 million years, out to 85 million years – a 20-fold extrapolation. With this audacious extrapolation, they were able to assign tentative dates to magnetic reversal chrons 1–32. The resulting timescale became known as the Heirtzler scale, after the first author, Jim Heirtzler. It has turned out to be surprisingly accurate, good to a few percent in most parts – one of those great strokes of genius or luck or both – but of course, at the time, no one knew if they were even close. Indeed, there was some evidence that the present spreading rates were only good back to about 10 million years (chron 5), and that there may have been a pause in spreading of unknown duration before that.

Meanwhile, back at Scripps I was stewing over our sea floor isochron patterns, yearning for some reliable dates. The one-sided magnetic anomalies nearest the California coast were easily identified as chrons 10–6. These had been formed by the Pacific-Farallon spreading center and had to have preceded the end of subduction and the start of the San Andreas plate boundary. In the Heirtzler scale, the extrapolated ages for chrons 10–6 were about 30 million to 20 million years, implying a quite young San Andreas. But what if there had been a spreading hiatus before chron 5? Then chrons 10–6 could have any older age – maybe even late Cretaceous, seemingly matching the preponderance of evidence from the land. Young? Old? Young? Old? We needed direct dates for these older isochrons.

At first thought, this problem doesn't seem so difficult: just dredge some rocks from the different parts of the sea floor and date them. Unfortunately, all the sea floor is continually being buried in a snowfall of debris (mud and biological remains) so that all the older rocks are buried under a mantle of younger sediments. To get the basement rock ages, we would have to drill through this overlying sedimentary pile.

THE DEEP SEA DRILLING MIRACLE – DIRECT CONFIRMATION AT LAST

The Deep Sea Drilling project came on line at just the right time to give us the gift of age dating that we needed so badly. Various grand schemes to drill through the entire oceanic crust and into the mantle had been around since the Mohole project of the 1950s. By the mid-1960s, these efforts had consolidated into the more modest Deep Sea Drilling project, whose aim was to drill many holes into and through the sea floor sedimentary cover. Quite by lucky chance, the drilling ship, the *Glomar Challenger,* was ready to begin its work just when the community was especially hungry to use it. The ship set sail in the fall of 1968, and after some trials set out for the South Atlantic to test the symmetry of that ocean and to check the proposed constancy of its spreading rate. I have heard that many of the scientists on that expedition boarded the ship in Dakar with considerable skepticism for the whole idea of sea floor spreading. When they got off the ship in Brazil, two months later, they were all avid, noisy believers. They had drilled nine holes along a line across the mid-Atlantic ridge and westward toward South America. By identifying the fossils in the bottom-most sediments, the shipboard scientists had been able to determine the ages at the base of the sedimentary piles in seven of the holes. As each age was determined, they had plotted it on a graph versus its distance from the central ridge. The points formed a perfect straight line (within the errors of the data). To everyone's surprise (including that of its authors), that outrageous Heirtzler scale extrapolation was correct.[31]

WRITING UP THE HISTORY OF THE SAN ANDREAS

With the validation of the Heirtzler timescale, the San Andreas history suddenly became tractable. I don't recall how I first heard about the South Atlantic dating results, but it disrupted my concentration on my thesis work. By summer 1969, I had dropped all pretense at the sea floor work and was struggling along with the San Andreas plate story. There followed the most intense work period of my life. It was almost like a trance that I would be in for many days at a stretch, hardly sleeping or eating.

This brings up one important factor in this story: the nature of funding in the 1950s and 1960s. It was much more general and flexible than the present grant system. Throughout my graduate years I was funded by the U.S. Navy through the Marine Physical Laboratory. Officially, I was working up the Gorda Ridge Deep Tow surveys, but my major excursions into plate tectonics (nine straight months for the San Andreas paper) were accepted, indeed encouraged, by my advisor, John Mudie, and by our Navy funders. The Navy at the time tended to fund productive seagoing groups and individuals in their scientific endeavors, without being too particular about the details. Their view was that any information about the oceans was useful to their mission, so we had a lot of freedom to be productive, wherever our hearts led us. In later years the Navy funding became much more restrictive, so that the pure research community was shifted to the National Science Foundation. Funding from the latter is excellent in many respects, but since it awards money for specific projects there is less flexibility for following up unanticipated avenues as they appear.

That fall (1969) Warren Hamilton came to visit Scripps from the U.S. Geological Survey. He came to learn about the revolution and to present a graduate seminar about continental tectonics. I spent many happy hours in his office, absorbing bits of his vast store of continental geological lore and sharing what I had been learning about marine geophysics. I was still hard at work honing the San Andreas story and loved the chance to try out many of the pieces on him. We had such fun sharing "mind candy" that I remember one middle of the night, about 2 A.M., when I woke up to some mental explosion and just couldn't wait to try it out on him. I called him up and yakked away into his sleepy ear. When he finally managed to get a groggy word in, it was (with patiently humorous undertones), "What time is it, anyway?" I took the hint and let him hang up – promising to repeat it all first thing in the morning.[32] Warren infected me with his passion for big-picture geology – a view of geology that I hadn't really encountered much before.

That winter, Bill Dickinson organized another of his specialty meetings, this one a Penrose Conference at Asilomar, California. It was primarily a land geologists' meeting, but he expressly invited students, so a gang from Scripps went. The meeting was full of good information about the plate tectonic interpretations of many geological phenomena, and it solidified all these for me. It was also very empowering because we oceanography students found ourselves in the role of teachers – about ocean floor features, in particular, and about oceanic plate tectonics, in general. I presented my San Andreas story in a badly crafted talk – I went way over time – but when the moderator decided to cut me off, someone in the audience called out, "Aw, let her go on. This is great stuff." (Bless you, whoever you were.)

Another special memory from that meeting is of a moment at the end of my presentation when someone asked whether he *really* had to accept such young dates for magnetic chrons 10–6 and, thus, a young age for the San Andreas. I was groping in my mind for a convincing description of the South Atlantic drilling results when a voice called out from the audience: "It's true! It's true! Believe it!" The speaker, Ken Hsu, rose up and took over. He had been on that wonderful Deep Sea Drilling expedition and spoke with all the passion of the newly convinced. We all enjoyed his exciting and overwhelming recitation both of the results and of his own personal conversion.

About that time, I ran into Allan Cox from Stanford University at some meeting and told him that I was working on the history of the San Andreas Fault. I could see his eyes rolling up in his head and his struggle to come up with something polite to say to this starry-eyed, impudent student. I asked him if he would read and critique my manuscript. His non-enthusiasm was palpable, but he graciously agreed. I sent it to him a few weeks later and it came back with big letters on the front: "PUBLISH THIS IMMEDIATELY . . ." It was the impetus I needed. I was elated but also scared. This project was my first real solo writing effort. I hit up everyone I knew for reviews and got a lot of excellent advice, including some extremely helpful suggestions from my fellow students. Probably the most useful of all was the reviews from Warren Hamilton. He had a clear understanding of the importance of brevity and clarity, and he didn't hesitate to go after me about it. My original manuscript was dense with "what ifs" and minor possible implications, and so he crossed out whole pages of mine with the simple remarks "FLUFF" and "STUFF." Of course he was also very excited and encouraging. The balance was perfect. It was published as the lead article in the December 1970 volume of the *Bulletin of the Geological Society of America.*[33]

After the Penrose conference, and especially after the San Andreas paper was published, speaking invitations poured in from all over the West. My synthesis was just what many land geologists had been waiting for. They had heard noisy rumblings from the oceangoing community, but it hadn't been clear how the revolution would affect continental work. The San Andreas history is quite unusual in that the oceanic and continental realms are so completely, intricately intertangled. You really can't understand one without the other: the oceanic geophysical record documents the demise of the spreading center while the continental geological record shows the development of the resulting new plate boundary. Although I was officially a marine geophysicist, my passion still held for the mountains and landscapes of the continents. I had a foot in each camp and became a kind of translator, telling each group about the findings of the other side. I suppose it had something to do with being female, too. I knew that a number of those in every audience were there to see the freak. (Indeed, many of the speaking invitations were prefaced with the rationale that their girl students needed to see a real-live female scientist.) I didn't mind. I knew that I had an irresistible tale to tell and was happy to present it for anyone who would listen.

WORK ON PLATE CIRCUIT RECONSTRUCTIONS

At the time of my 1970 paper, I was still missing one important piece of information. In order to work out the details of the plate interactions, I needed to know the long-term history of Pacific–North America relative motions. We had evidence that the Pacific plate is presently moving parallel to the San Andreas Fault about 2 inches (6 centimeters) per year past North America. However, we couldn't be sure how long that had been the case, and a number of lines of geologic evidence suggested that this motion had been slower in the past. In the 1970 paper, I presented two "end member" models: one in which the relative motion had been steady and a second with no relative motion before about 10 million years ago. If we wished to find out the actual history of Pacific–North America motion, we needed to make "plate circuit reconstructions" around the world for a number of past times.

The plate circuit that must be followed in order to calculate a past location of the Pacific plate with respect to North America is one that steps from the Pacific plate to Antarctica to Africa to North America, crossing a spreading center in each step. For example, a reconstruction for 11 million years ago, the time of magnetic chron 5, is based on the following steps. First, we reconstruct the Pacific plate to the Antarctic plate using the chron 5 patterns in the sea floor of the South Pacific. Next, we reconstruct both, together, to the African plate using chron 5 patterns in the southwest Indian Ocean, south of Africa.[34] Finally, we reconstruct those three, all together, to the North American plate using chron 5 patterns in the central-north Atlantic. If we could do similar reconstructions for a number of different chrons, we could work out the approximate track of the Pacific plate past North America through time.

This plate circuit (and every plate circuit that relates Pacific ocean plates to the continents) uses the step from Antarctica to the Pacific plate.

This step is made using the spreading patterns on the Pacific-Antarctic ridge in the South Pacific. In 1970, this ocean was quite poorly known. Geophysicist Peter Molnar came from Lamont to Scripps at about that time on a postdoctoral fellowship. We set out, together with Scripps map maker Jacqueline Mammericx, to quantify the plate motions across this spreading center.[35] Using the results of that study, we were able to construct a track of past Pacific locations with respect to North America for four points in time. The uncertainties on the locations of the points in this first track were quite large, but they did generally support a long-term northwest drift of the Pacific past North America.[36]

FINISHED? (REALLY, JUST WAITING)

In the late 1970s and 1980s, the plate tectonics "revolution" took an interesting turn. It became old-hat for the land geologists. Whole geological meetings were conducted with hardly a mention of plate tectonics. Oceanic work continued apace, deepening and honing the theory, but on the continents, it seemed to have become irrelevant to most new work. The early revelations, of course, had given the community a huge leap forward in general understanding of earth processes, and they definitely set us free: it was suddenly not outrageous to think about terranes or whole continents traveling far distances across the globe. However, the quantitative aspects, so powerful for predicting patterns in the ocean floors, didn't seem helpful on the continents. In western North America, for example, our multistep circuit reconstructions were generally too crude to help with specific geologic problems. (A geologist standing on a hillside outcrop isn't impressed by a prediction that has an uncertainty of hundreds of miles.) At the time, I thought maybe we were done with continental global tectonics. I returned to my ocean floor studies with renewed vigor. It turns out that, rather than being finished, we had simply run through the collected store of relevant information and so had to wait a while for improvements in concepts, techniques, and data sets.

MEASURING THE PRESENT-DAY DRIFTS
AND DEFORMATIONS OF THE PLATES

Several technological advances changed the nature of plate tectonic studies in the 1980s and 1990s. One exciting aspect has been the development and honing of various new systems for measuring the locations of points on the earth's surface. During the mid-20th century, great progress had been made with surveying and laser ranging techniques for characterizing local deformations near and across selected active faults, especially those of the San Andreas system. These had shown us a complex history of ongoing deformations near the plate boundaries, and had given us some understanding of how energy accumulates near plate boundaries and then is released during earthquakes. The real test of plate motions, however, required that we measure the ongoing motions of the plate interiors, far from any plate boundary complications.

Global scale measurements of relative positions on the earth's surface became possible in the latter decades of the 20th century through the "very long baseline interferometry" program, or VLBI. By comparing and timing signals coming to Earth from deep space radio stars, scientists in this project were able to find the locations of their observation points with an accuracy of a few inches. In order to measure the relative displacements of these points through time, they had to measure the locations, then wait years, then measure, then wait again. The results were definitely worth the wait. Repeat measurements of locations over the decades have given us the wonderful (and reassuring) result that the movements of the rigid plate interiors during the past few decades have been the same as the motions over millions of years, that is, motions that we had deduced from the magnetic anomalies!

An especially fun innovation has been the development and democratization of the global positioning system, or GPS.[37] This is the satellite system (originally developed for military navigation) that now allows any citizen to locate herself (or her fancy car) on the earth's surface. A researcher can place a marker, then, with some patience and diligence, can use GPS to determine its position within a few inches. The uncertainty in these measurements is about the same as the displacement of plates in a year, so that one only needs to monitor the location of the marker over a few year's time span to get a quite good estimate of its ongoing movement.[38] Furthermore, the equipment for making GPS measurements is relatively inexpensive so that many groups can make local and regional measurements. Thus, it has become feasible to measure deformations in broad plate boundary zones, both steady motions and the time-dependent deformations around earthquakes and creeping faults. For example, a dense Japanese array of continuous GPS stations is already yielding wonderful images of the ongoing warpings of the land over that major subduction zone.[39] Likewise, the results from periodic measurements across western North America are full of new detail about the way the plate boundary deformation is presently partitioned across the West.[40] I find myself eagerly awaiting each new data set and its revelations.

MEASURING PAST PLATE DISPLACEMENTS AND DEFORMATIONS

Studies of past plate motions have also greatly benefited from technological developments. Our primary data sets for reconstructing the histories of plate motions are oceanic magnetic anomalies and fracture zone trends. These have mostly been gathered aboard oceanographic ships lumbering slowly across the surfaces of the world's oceans. When I began going to sea, our biggest problem was figuring out our position. In the South Pacific and other remote regions, we were proud if we could locate the ship within a few miles twice a day (by measuring the stars at sunrise and sunset, but even then only "if the weather be good").[41] Postcruise data processing often involved Herculean efforts to adjust the navigation record so that the data sets were at least self-consistent, not to mention located well on the earth's surface.[42] The advent of satellite navigation and, eventually, of GPS navigation has changed all this. With this system we can now routinely locate the ship to within a few yards every second. When we tell our students about the bad old days and our navigational labors, they look at us as the poor, deprived, primitive ancients.

Technology has also given us a wonderful gift of ocean topographic coverage with the laser altimetry satellites, *Seasat, Geosat* and *ERS-1*.[43] These satellites measure the height of the top of the water in the oceans with a precision of a few inches, somehow averaging out all the waves and tides. In turn, variations in the water surface height show us gravity variations caused by the topography of the ocean floor. Linear fracture zones show up as some of the most dramatic features on these records and maps, and this is a special boon for us, since our plate tectonic reconstructions of plate motions are based upon fracture zone trends. While the ship-generated sonar records of these features are more detailed and precise than the satellite altimetry records, they are very tedious to collect. In a few years of observations, the satellites filled in the fracture zones in vast regions of the more remote oceans, including those southern oceans so critical for our reconstructions around Antarctica. Combining these with new, well-located, magnetic anomaly data, we are finally able to make round-the-world circuit solutions that have some relevance for land geologic studies. For example, in a recent article Joann Stock at Caltech and I were able to reconstruct the Pacific plate track past North America with some location uncertainties as small as a few miles. This track, in turn, allowed us to formulate a quite precise "deformation budget" for western North America. For example, we predicted that the continent must have stretched more than 150 miles (about 240 kilometers) – that is a lot of extension! – and must have been sheared parallel to the coast as much as 540 miles (860 kilometers) since about 20 million years ago.[44] These budget estimates were made from our chron 6 round-the-world oceanic reconstructions. If they are correct, they should match estimates made from summing the deformations observed across the land.

ESTIMATING THE MAGNITUDES OF PAST LAND DEFORMATIONS

Land deformations are much more difficult to quantify than those in the oceans, because all land deformations are superimposed upon older features. We are spoiled in the oceans, where virtually all of the deformation is accommodated by the creation or destruction of crust. Fortunately, the late 20th century also saw great progress in our ability to quantify continental tectonic deformations.

From the perspective of plate tectonic history, a crucial breakthrough has been the recognition, acceptance, description, and quantification of large magnitude extensional features known as core complexes, or low-angle extensional detachment faults. These deformation systems allow the crust to extend 100 to several 100 percent in a very short time (perhaps less than one million years).[45] These events often bring the ductile middle crust to the surface, laid bare or thinly strewn with fallen-over "dominoes" of the broken, brittle upper crust. The amount of extension represented by one of these features can often be estimated by reerecting the dominoes to reassemble the original upper crust. The timing is often recorded in the lavas that tend to accompany the extensional events.

The Basin and Range Province of interior western North America contains a large number of these extensional features. Many of them date from the Miocene and overlap the San Andreas deformations in time and space. In the east-west corridor near Las Vegas, Brian Wernicke and J. K. Snow of Caltech were able to add up all the extensions between the Colorado Plateau and the Sierra Nevada to estimate for the first time the very large Basin and Range extensional budget.[46] With this piece, we can finally compare the oceanic and continental deformation estimates, and they agree. It took nearly 30 years, but the quantitative power of the plate tectonic theory is finally becoming relevant on land.[47]

BETTER AND BETTER, SO FAR AT LEAST

From time to time, every scientist must step back and reexamine her or his assumptions. I have often done this in my life, sometimes of my own volition and sometimes when under the barrage of some doubter. In plate tectonics work, our most basic assumption is that the aseismic interiors of the plates are rigid, so that we can deduce the motion of every point on each plate using relatively few measurements along the plate edges. It is a pretty outrageous assumption, especially given the array of non-rigid structures that present themselves to the student of continental geology. We must suppose that all these structures were formed when each region lay near a plate boundary – but this supposition often has no independent confirmation.

As the years pass, I have regularly been pleased (and surprised, and relieved, I admit) to see the rigid plate assumption holding true and being reconfirmed with new techniques and data sets. With the passage of time, most scientific ideas are overturned or are greatly modified. Similarly, as the uncertainties in our data sets get smaller and smaller, I fully expect that we will start detecting the non-rigidity of the major plates, but this has yet to happen. So far, with just a few small adjustments, the assumption continues to work. As I tell my students: "Gol' dern! It must be true!"

CHAPTER 16

THE COMING OF PLATE TECTONICS
TO THE PACIFIC RIM

William R. Dickinson

Hanging today on the wall of my study is a dinner plate purloined by colleagues from the Asilomar Conference Center in Pacific Grove, California. Neatly inscribed in its center, in felt-tip pen, are the words "Hero of Plate Tectonics," and around the rim of the plate is written the jocular motto, "In Subduction We Trust." This strange object is my most treasured professional memento, given to me by 95 fellow attendees at the close of a landmark 1969 Penrose Conference that I had convened on behalf of the Geological Society of America (GSA) to consider the implications of the then-fresh concepts of plate tectonics for geologic processes in mountain belts. The collaborative effort of all the participants, each a volunteer attendee, was what made the week-long conference so memorable, but I was proud to accept the honorary "plate" for having been the catalyst for the meeting. The word *subduction* of the motto came into play with its present meaning during the course of the conference itself as the result of a magnificent evening address by Dietrich Roeder, then of Esso (Exxon) Production Research Company. The term was forthwith adopted, by mutual consent of the full company assembled, to describe the leading downward of earth materials by the descent of a tectonic plate from the surface of the earth back into its interior.

Subduction was just one, although the most arresting, of the new words added to the geological vocabulary, or redefined in some significant way, to accommodate the advent of plate tectonics as a guiding intellectual paradigm for geoscience. My own career as a geologist had begun a decade before the Canadian geophysicist Tuzo Wilson (University of Toronto) anointed the concept of moving tectonic plates as the controlling mechanism for driving continental drift and sea floor spreading.[1] In

Bill Dickinson. (Photo courtesy of Bill Dickinson.)

adapting my own point of view to the perspectives that he and other pioneering thinkers offered, I built upon a background of personal knowledge acquired principally in the American West, with an outlook rooted in the challenges presented by geologic relationships around the Pacific rim, the so-called circum-Pacific belt of geologic parlance.

OVERVIEW

In my youth as a geoscientist, I was a casual stabilist, assuming that the continents had maintained their relative positions on the globe throughout geologic time. That outmoded stance stemmed less from informed conviction than from sheer ignorance of how much intellectual ground could be gained with a mobilist outlook embracing the notions of continental drift and sea floor spreading. Conditioned partly by the West Coast locale for my schooling at Stanford University (1948–1958), and partly by a visceral love of field geology in mountainous terrain, my natural focus was on the mountain belts of the continents. I tended to assume naively that orogenic (mountain-building) processes could continue apace within the continents with little or no influence from oceanic affairs, whether or not the continental blocks drifted about. Not being part of the oceanographic or geophysical communities, from which the basic struts of plate theory sprang, I launched into the growing scientific ferment several years after its initiation. My lifelong research interests in relations between tectonics and sedimentation ensured, however, that plate tectonics would be a guiding passion for a number of years.

For a circum-Pacific geologist, the test of any theory is how well it elucidates the configuration and geologic history of the arc-trench systems that adorn the Pacific rim of the globe. The arcs and the trenches are paired geographic features, the trenches being deep oceanic troughs lying close offshore from continental margins, or from associated island chains that fringe continental margins, and the arcs being parallel volcanic chains standing either on the edges of the continents or along the trends of the island chains ("island arcs"). The continental arcs include, among others, the volcanoes of the high Andes in South America, and the island arcs include those of the Aleutians, Japan, the Marianas, Tonga, and their geologic cousins of the western Pacific arena. Other arcs of both kinds adorn the Indonesian and Mediterranean regions. My personal enthusiasm for plate tectonics waxed ever stronger as it became increasingly clear that plate tectonics could foster a fresh analysis of arc-trench systems leading to a deeper understanding of their fundamental geodynamics.

To me, plate tectonics means not only the formal geometric theory for motions of rigid plates, but includes corollary implications for geologic processes within orogenic belts and arc-trench systems, where departures from plate rigidity are inherent during their geologic evolution. Formal plate theory posits plates that are internally undeformed right to crisp plate boundaries, the narrowest of which are transform faults where two plates slide past one another by strike-slip parallel to the plate edges. Where transform faults cross tracts of continental crust, however, not only do broad arrays of parallel fault strands form plate-boundary belts of finite width, but forces transmitted across a plate boundary can induce deformation of the continental crust far into the interiors of the interacting plates. Tanya Atwater, then a graduate student at the Scripps Institution of Oceanography of the University of California (San Diego), pointed out this aspect of plate behavior for the San Andreas Fault system of California at the 1969 Asilomar Penrose Conference.[2] Now a professor at the University of California (Santa Barbara), she undertook her early work almost at the outset of thinking about continental geology in plate tectonic terms.

Arc-trench systems, controlled by plate interactions where one plate slides beneath another, embody internal plate deformation extending into the overriding plates for hundreds of miles away from the plate boundaries involved. The deformation may in some cases involve intraplate extension, to rift apart the arc structures, as first perceived clearly by Dan Karig, now of Cornell University, who also aired his views at the 1969 Asilomar Penrose Conference while still a graduate student at Scripps.[3] In other cases, the deformation may involve intraplate contraction, to thrust the flanks of arc mountain belts over the upper surfaces of continental blocks lying behind the arcs, as the Andes override the Amazonian platform of lowland South America on the other side of the mountains from the trench on their Pacific side. These and other non-rigid aspects of plate behavior are indeed only ancillary derivatives of plate theory, but serve as the crucial springboard for geodynamic analysis of orogenesis, the birth of mountain ranges.

BEFORE PLATES

Prior to the advent of plate tectonics, most students of orogenic belts lived by the precepts that Adolph Knopf, long a professor at Yale University, informally dubbed *geosynclinorial theory* in seminars given in retirement at Stanford during the interval 1955–1957 while I was a graduate student there.[4] By its tenets, rather mystical entities we called *geosynclines*, defined as linear belts of subsidence where thick sediments accumulate, led inexorably to the equally mystical process we called *orogeny*, whereby mountains were built by deformation and uplift of the geosynclinal sediments. I absorbed the intricacies of geosynclinorial theory by learning just which rocks and just which structural features of those rocks gave rise to the notions by which we swore. When plate tectonics entered the picture, it was not a severe challenge to transpose basic information, already in hand, into the stimulating new conceptual format.

On the sedimentary side of my scientific life, I was early inspired by the monograph on geosynclines by Marshall Kay of Columbia University.[5] Chester Longwell, also resident at Stanford in retirement from Yale during my tenure as a graduate student and junior professor, loved to poke fun with his dry wit at the rather esoteric Kay classification, referring to his miogeosynclines as "my" geosynclines and his eugeosynclines as "your" geosynclines. The miogeosynclines and eugeosynclines were envisioned as parallel paleogeographic components of orthogeosynclines, the supposedly "true" geosynclines from which mountain belts sprang. The jocular Longwell usage seems less humorous now that we understand the miogeosynclines of Kay as the miogeoclines of Bob Dietz, long a research scientist with the U.S. Navy Electronics Laboratory in San Diego.[6] His subtle change in nomenclature, from *miogeosyn-* to simply *miogeo-*, was meant to convey the thought that the sedimentary strata in question were not inclined inward, as a *syncline*, from both sides of an imaginary depositional trough, but were inclined in only one direction, seaward as a *cline* along an ancient continental margin. Miogeoclines thus homegrown on "my" continental margins are only later juxtaposed against "your" (foreign) eugeosynclinal rock masses composed of oceanic crustal elements accreted to continental margins by post-depositional plate movements. Through these and other devices, one can rewrite Kay in plate tectonic terms, without changing many of the observations he reported, by simply transposing his geosynclinal nomenclature into a terminology of plate relationships.[7]

My own mind was prepared for the revolutionary new plate vision of geoscience by personal experiences studying a variety of rock masses that were difficult to fit within a stabilist world. Chief among these was the internally disrupted Franciscan assemblage, located along coastal California, now interpreted as a subduction complex related to an ancient trench, but then simply mysterious. Another was a cluster of huge peridotite massifs composed of rock thought to be characteristic of the mantle of the earth lying below the crust, but now exposed at the surface where one can walk around on them in the Klamath Mountains of northwesternmost California and southwestern Oregon. We regard the peridotite bodies now as slices of an oceanic plate emplaced against the edge of the continental block by subduction, but they were then simply mantle rock curiously out of place. Finally, there were the volcanogenic rocks in the Blue Mountains of northeastern Oregon and the Sierra Nevada foothills of Central California that resemble modern oceanic island arcs, yet are now stuck somehow within the continental block.

SHIFTING GROUND

The accustomed ground on which I stood began to tremble during my first sabbatical year (1965) spent mapping geologic relationships in Fiji as a Guggenheim fellow. An extended visit to Australia and New Zealand, while the rainy season made serious fieldwork impossible in the islands, taught me that most alert geoscientists "down-under" took the breakup of the ancient supercontinent of Gondwanaland as a given. They could trace in the field the same rock masses, joined together before continental breakup, from Australia to other southern continents (Africa and Antarctica), and to the Indian subcontinent, which was also born from Gondwanaland. Once those intercontinental correlations are accepted, continental drift is no longer speculation. Working in Fiji taught me also that many of the metamorphosed volcanic rocks in continental orogenic belts are indistinguishable from rocks exposed today in the eroded roots of modern oceanic island arcs, with a degree of similarity too close to be fortuitous. Without continental drift and sea floor spreading, there are no ready means to position oceanic rocks against or within continental blocks.

Even so, I remained for a time quite diffident about the import of sea floor spreading and continental drift for my own research. While in Fiji, away from all libraries, I missed the initial impact of Tuzo Wilson's 1965 paper, which introduced the notion of quasi-rigid plates bounded by coordinated plate boundaries that allow a globally integrated pattern of motions affecting continental blocks and ocean basins alike.[8] With his scheme later in mind, however, it was an exhilarating experience to contemplate on any model globe the interlocking network of mid-ocean spreading ridges, where plates are born from the interior of the earth, and mountain belts, where plates are consumed by subduction. At little more than a glance, it was clear that the marvelously coherent dance of the plates might well explain all the major geodynamic features of the earth, with nothing out of place and nothing left out. No paper I wrote before 1969 gave any hint of mobilism on the scale envisioned by plate tectonics, but everything I wrote thereafter was cast in that mold.

PERSONAL TRANSITION

The change from stabilist to mobilist thinking came quickly for me between mid-1967 and mid-1968 as the cumulative result of a series of conferences and seminars that brushed the cobwebs from my brain. First, a parade of stimulating oral presentations at the 1967 annual meeting of the American Geophysical Union (AGU) in Washington, D.C., late in the spring, codified the concept of sea floor spreading in a plate guise.

Second, a special topical conference on the geologic history of the San Andreas Fault system was convened at Stanford University on September 14–16, 1967, by Arthur Grantz and me, on behalf of the U.S. Geological

Survey (USGS) National Center for Earthquake Research, and the Stanford University School of Earth Sciences.[9] Its sessions removed any remaining doubt about the reality of large cumulative displacement across the San Andreas Fault by indicating the overwhelming geologic evidence for lateral offset of multiple rock masses by dramatic amounts, measured in hundreds of miles for the older rocks offset. In the aftermath of the conclusive evidence for cross-fault movement aired at the conference, the suggestion of Tuzo Wilson that the San Andreas Fault is a transform fault, linked to Pacific plate motion, became an attractive hypothesis supplanting all more stabilist interpretations of San Andreas history.[10] Plate reasoning governed all my own subsequent work in several papers addressing various aspects of the geometrically intricate crustal deformation that affected California and adjacent states in response to evolution of the San Andreas transform system along the coast.

Despite the general success of the plate model for understanding San Andreas evolution, nagging discrepancies between plate predictions and field determinations of fault displacement through geologic time were not fully resolved to my satisfaction until three decades after the San Andreas Conference. Reconciliation of the two independent data sets involved collaborative analysis with Brian Wernicke, a professor at the California Institute of Technology, who was still in knee pants when plates first burst upon the scientific scene. We were able to bring theory and observation into congruence by close attention to complex deformation within the plate edges lying to either side of the San Andreas Fault.[11]

Third, a unique seminar at Stanford, convened jointly by Allan Cox and me as an interdepartmental enterprise involving faculty and students in geology and geophysics, confronted the emerging impact of plate tectonics on geoscience. As the term *plate tectonics* was not yet commonplace, we called our show SPRIFT, an acronym derived from sea floor spreading (SPR) and continental drift (IFT). No participant survived the seminar with his or her previous thinking intact and, in keeping with the mood of the time, it was never clear whether faculty were leading students or students were leading faculty.

During that period, there was also an auxiliary and quite informal seminar held once a week long into the evenings on the floor of my Menlo Park apartment just off the Stanford campus. Students and I sat cross-legged on the floor, much in harmony with the "hippie" craze that swept California in those days, and read aloud to each other, paragraph by paragraph, some selected paper about sea floor spreading or continental drift, with extensive critique and commentary on each paragraph and figure. Some distinguished ears might burn if I could recall all our conversations, but we steeped ourselves thoroughly in the extant lore on spreading and drift, garnering through our efforts a valuable education in depth, with no holds barred.

Finally, stimulating corridor discussions at the week-long International Andesite Conference held in Bend, Oregon, in July 1968 convinced me that the generation of magmas feeding volcanism along both island arcs and continental arcs around the Pacific rim is related to the subduction of oceanic plates at the nearby trenches. The meeting was arranged by Hisashi Kuno of the University of Tokyo and Alexander McBirney of the University of Oregon on behalf, respectively, of the International Upper Mantle Committee and the University of Oregon Center for Volcanology, with logistical support for the conference provided by the State of Oregon Department of Geology and Mineral Industries, headed then by Hollis Dole, an old professional friend from the days of my Ph.D. research in Central Oregon (1956–1957).[12] My paper for that meeting made no mention of plate tectonics, but my paper scarcely a year later for the September 1969 Andesite Symposium of the Volcanic Studies Group of the Geological Society of London, although based on much the same data, was couched explicitly in plate tectonic terms.[13] Although my change in mental orientation appears abrupt in retrospect, at the time it seemed a seamless shift in emphasis as expanding knowledge and insight built naturally upon my past training and experience.

By then it had become clear from field studies that the assemblages of igneous rocks along both island arcs and continental arcs are not restricted to erupted volcanic rocks. Also present are plutonic rocks representing the products of magma batches that are injected into the crustal roots of the arcs beneath the volcanic chains, to be trapped there within subterranean magma chambers where they solidify as granitic batholiths and related intrusions that are in time exposed to view by erosion. With this insight in mind it became attractive, by degrees, to speak of arcs in general as magmatic arcs, rather than simply as volcanic arcs. This currently standard usage embraces both the volcanic and the associated plutonic components of the assemblages of igneous rocks present within circum-Pacific arcs, whether they occur at the edges of continental blocks or along fringing island chains.

INFLUENTIAL PAPERS

Two sets of prescient papers on circum-Pacific geology also encouraged me to join the mobilist bandwagon that was beginning to roll. One was a

string of related papers by Japanese geoscientists correlating geophysical
parameters with geological features and with geochemical data for volcanic rocks in the complex arc-trench systems of the western Pacific
periphery. I had sensed the general flavor of Japanese thinking at the
Tokyo Pacific Science Congress in 1966. Hisashi Kuno, Arata Sugimura,
and Yoshio Katsui had all argued that local variations in the compositions
of arc lavas correlate with depths from individual volcanoes to the inclined
zones of earthquake seismicity that slope deep into the mantle beneath
magmatic arcs from near-surface origins in the vicinities of the paired
trenches offshore.[14] The inclined seismic zones are styled Wadati–Benioff
zones, named in compound fashion in recognition of the Japanese and
American geoscientists who independently established their existence
well before plate tectonics was devised.[15] The zones of seismicity are now
thought, however, to mark the paths of oceanic plates descending into the
mantle beneath arc-trench systems. The very occurrence of earthquakes
at such great depths reflects the subduction of relatively cold and brittle
crustal rocks to subsurface levels, where surrounding mantle rocks are too
hot and ductile to fracture and generate earthquakes.

During his year (1966) at Stanford on sabbatical from the Geophysics
Division of the Department of Scientific and Industrial Research
(DSIRO) in New Zealand, Trevor Hatherton and I tested the reported
Japanese correlation of the composition of arc lavas with depth to the
inclined Wadati–Benioff zone of mantle seismicity. We found that the
most informative variation in composition lies in the content of the element potassium (K) in lavas from different volcanoes, and developed a
global K-h correlation between levels of potassium content (K) and
depths (h) from active volcanic cones to the inclined seismic zones in the
mantle directly beneath.[16] We did not initially perceive that our K-h correlation implies a relationship of arc magmatism to plate descent, but
that connection became obvious in hindsight within just a few years.[17]
In any case, with or without the plate context, looking to the inclined
seismic zone within the mantle as the impetus for arc volcanism, as I did
in 1968 with the K-h correlation in hand, entailed abandoning my own
thoughts of only six years earlier on postulated origins of arc magmas
from strictly crustal processes at much shallower depths.[18]

The paper from the Japanese school most relevant for placing arc-
trench systems as a whole into a plate tectonic framework was the 1967
synthesis of geologic relationships in Japan by Akiho Miyashiro, then at
the Geological Institute of the University of Tokyo, but later affiliated
with the State University of New York (SUNY) at Albany.[19] His key
thoughts were presaged in thumbnail fashion by Hitoshi Takeuchi and

Seiya Uyeda of the Geophysical Institute at the University of Tokyo in 1965.[20]

The explosive insight that the Japanese researchers offered was a shrewd explanation for the so-called paired metamorphic belts of orogenic systems. In many locales around the Pacific rim, sedimentary and volcanic rocks are recrystallized ("metamorphosed") in parallel but separate belts, each hundreds of miles long, into mineral assemblages of the same ages but reflecting different conditions of pressure in relation to temperature within the earth's crust. In both Japan and California, a belt of metamorphic rocks composed of "blueschist" minerals that form at relatively low ratios of temperature to pressure lies parallel to a belt of metamorphic rocks composed of "greenschist" minerals formed at relatively high ratios of temperature to pressure. The connection of paired metamorphic belts to plate tectonics was codified by Miyashiro. He argued cogently that the blueschist belts indicative of low temperature/pressure conditions in the crust formed within subduction zones associated with ancient trenches, where relatively cold surficial rocks were dragged to unusual depths, and consequently subjected to high pressures beneath a thick rock overburden, by plate descent too rapid to allow them to warm up. Conversely, the greenschist belts indicative of high temperature/pressure conditions were identified as regions of crust beneath the volcanic chains of magmatic arcs, where hot magmas rising from the vicinity of the inclined seismic zone within the mantle heated crustal rocks to temperatures above those expected from the normal temperature gradient of crust unaffected by arc magmatism. Armed with these insights, we suddenly had the means to identify the subduction zones and magmatic arcs of ancient arc-trench systems whose geographic configurations have long since been destroyed by the erosion of old orogenic belts.

Two parallel concepts developed by Warren Hamilton, a USGS research geologist, also played a key role in setting the lessons of plate tectonics into a circum-Pacific framework. First was his perception that volcanic rocks of what we used to call eugeosynclinal belts come in two distinct brands, one the basaltic sea floor lavas rafted into subduction zones by plate motion and the other the more andesitic volcanic rocks of magmatic arcs.[21] Distinguishing compositionally between the two kinds of volcanic assemblages provided an independent means for recognition of ancient magmatic arcs apart from their paired subduction zones containing the tectonically transported sea floor lavas. Second was his perception that the intrusive magmas of granitic batholiths commonly are injected into the crustal roots of arc volcanic edifices.[22]

This insight folded the immense circum-Pacific batholiths of the Americas, Japan, and the Pacific margin of Eurasia into an integrated picture for deeply eroded magmatic arcs around the Pacific rim.

ASILOMAR PENROSE

The paradigm of plate tectonics was cemented into place for me by proceedings at the Asilomar Penrose Conference of December 15–20, 1969. As plate tectonics was not then a household term, I gave the conference the rather unwieldy title "The Meaning of the New Global Tectonics for Magmatism, Sedimentation, and Metamorphism in Orogenic Belts." The phrase *new global tectonics* was drawn from the pacesetting 1968 paper by Bryan Isacks, Jack Oliver, and Lynn Sykes of the Lamont-Doherty Geological Observatory at Columbia University setting observations from global seismology into a plate framework (Isacks and Oliver are now at Cornell University).[23] I was especially receptive to their ideas because I had met Sykes, already a faculty member, and Isacks, then still a graduate student, in 1965 while I was mapping geology in Fiji and they were installing research seismometers there for their most critical experiments. Sykes was in attendance at the Asilomar Conference.

My conference proposal of January 25, 1969 spoke of "a comprehensive theory," developed "in recent months," which envisioned "a worldwide movement plan for large, semi-rigid plates of crust and upper mantle combined" with "earlier ideas of continental drift, sea floor spreading, and the deep structure of island arcs incorporated within the theory." The proposal went on to point out that "far-reaching reinterpretations of geologic history are possible with the theory in mind," and further, that "potential consequences are especially severe for the geology of orogenic belts," concluding that the time was "particularly opportune" for a broad-based group of geoscientists to consider "the possible implications" of the then-fresh insights "on an informal basis." The language of the proposal reveals the extent to which plate tectonics had already seized center stage, but also that its applications to the interpretation of geologic history were then still in a nascent state.

There had been an earlier pilot Penrose Conference in Tucson, Arizona, on so-called porphyry coppers, the typical copper deposits of Chile and the American Southwest, but the Asilomar Conference was the first publicly announced Penrose Conference to which any geoscientist in the world could seek invitation. The group in attendance was largely self-selected, apart from a few invited speakers. Only about 150 geoscientists saw fit at the time to apply, and nearly 100 were invited. Among senior researchers, they included legendary figures like Jim Gilluly (a specialist on orogeny) of the USGS, Marshall Kay (the long-time champion of geosynclines) from Columbia University, and John Rodgers (an expert on the Appalachian mountain belt) of Yale University. Incidentally, Kay had by then fully embraced the new mobilistic view of geology, with all that change of viewpoint implied for reinterpretation of geosynclines. Junior scientists in attendance included a number of graduate students who have since made their own marks on global geoscience. A few participants had the kernels of their main contributions to plate tectonics already in print, in press, or in mind before the conference, but many attendees drew direct inspiration from the sessions. In my own case, a paper on plate tectonic models for geosynclines was conceived during the meeting, and given orally on the last day instead of my scheduled presentation, which was by then already outdated. In that paper, I argued that all the traditional lore of geosynclines could be reconfigured to fit within the scheme of plate tectonics, and in a shorter companion paper I argued likewise for the orogenies that build mountain belts.[24]

An objective measure of the aggregate power of the group assembled at Asilomar is the fact that one in five were then, or are now, members of the National Academy of Sciences, with the ratio likely to rise as the work of some of the younger participants continues to mature. All during the conference, our discussions continued night and day at a furious tempo that left us all exhausted by the end of the week. At its close, however, it was clear that we had the "cat by the tail," so to speak, in terms of understanding orogenesis, and I took special care to make my 1970 public reports of the conference as thorough and forward-looking as possible.[25] The figure below illustrates diagrammatically how I came to view the geometry of arc-trench systems in the wake of Asilomar. The broader scientific insights gained at the conference were distilled into a 1971 summary paper on the role of plate tectonics in geologic history.[26] The paper was subtitled "new global tectonic theory leads to revised concepts of geosynclinal deposition and orogenic deformation," and showed how plate tectonics provides a more logical rationale for salient events in geologic history than the older "geosynclinorial" theories had given us. It was as if we had spent decades slowly climbing a high hill to attain a vantage point, and had with a sudden burst of intellectual energy reached the summit from which the whole landscape was at last laid out before us.

There were several direct spin-offs from the Asilomar Conference. The first was a symposium on plate tectonics arranged jointly with Chuck Drake of Dartmouth College, and held on April 28, 1971, at the annual

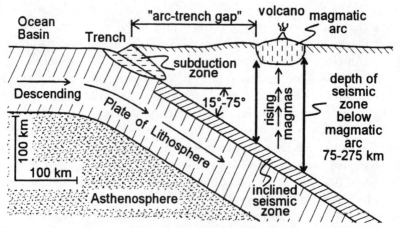

Transverse profile (or "slice") through a typical "arc-trench" system, as I envisioned it in 1970–1971. The profile shows an oceanic plate of "lithosphere" composed of rigid crust and upper mantle descending at an angle into hotter and softer "asthenosphere," a weak zone of partially molten rocks deeper within the mantle, along an inclined zone of seismic activity. A subduction zone marks the locus of plate descent from the surface, and a magmatic arc forms above it as bodies of molten rock rise upward from the mantle into the crust, generating surface uplift and vulcanism.

meeting of the National Academy of Sciences in Washington, D.C., to bring the burgeoning revolution in geoscience to the attention of leading scientists in other fields.[27] The nature of the venue for that session stamped a certain imprimatur of importance to plate phenomena. As the convenor of the Asilomar Penrose Conference, I also arranged a symposium on plate tectonics in geologic history on November 1, 1971, at the annual Geological Society of America (GSA) meeting in Washington, D.C., to bring central insights developed at the conference to the attention of the geoscience community as a whole. The stimulus of that session encouraged the dissemination of information about plate tectonics to an ever-widening circle of geoscientists. Several of the GSA symposium papers, including mine on evidence for past plate tectonic regimes in the rock record, were later published in a special summer 1972 issue of the *American Journal of Science* by invitation of the editor, John Rodgers, who had been in attendance at Asilomar.[28]

Shortly thereafter, I was asked to arrange the annual research symposium of the Society of Economic Paleontologists and Mineralogists under the title of "Tectonics and Sedimentation."[29] Oral presentations were made on November 15, 1973, in Los Angeles at the annual meeting of the American Association of Petroleum Geologists (AAPG), with

Earle McBride (University of Texas, Austin) and me as session co-chairs. The symposium was designed to highlight the significance of plate tectonics for interpretations of sedimentary geology, especially in the service of petroleum exploration, and highlighted my overview paper on relations between plate tectonics and global patterns of sedimentation.[30] Memories of the Asilomar Conference later induced Seiya Uyeda of the University of Tokyo to invite me for a month-long tour of lectures and field trips at seven leading Japanese universities in 1976 under the American Geophysical Union (AGU) auspices.[31] At the time Japanese faculty and students were able to hear directly the implications of plate tectonics for orogenic belts in the spirit of the thinking that had become standard by then for interpreting geologic relationships in California.

CALIFORNIA AFFAIRS

For me personally, the ultimate test of any theory of orogenesis is its success or failure in explaining the geologic evolution of California, a wonderland of orogenic phenomena. The drumbeat that set the pace for my unconditional acceptance of plate tectonics was the growing body of surprising data and mind-warping concepts that emerged right at home in California. The figure below is a sketch map delineating the key rock bodies that define the overall framework and geotectonic patterns of Central California.

Fifty years ago, the ruling notions of California geology included the concept of a paroxysmal Nevadan orogeny, or "revolution" as we often called it, which allegedly occurred during a geologically brief interval of late Jurassic time (ca. 150 million years ago). During that almost magical event, the vast granitic intrusions of the Sierra Nevada batholith were supposedly emplaced, following which rapid erosion of the batholith, and the metamorphosed "wallrocks" that confine it, was thought to have produced voluminous sediment that was quickly deposited west of the batholith to form the extensive strata of the Franciscan assemblage exposed throughout much of the California coast ranges. The Franciscan rocks of the coast ranges and the western flank of the Sierra Nevada were then gradually covered, so we then surmised, by a thick succession of entirely younger sandstones and shales forming the Great Valley sequence, so-called because its principal exposures lie along the Great Valley of California (Sacramento and San Joaquin valleys) between the Sierra Nevada and the coast ranges. We knew from the fossils it contains that deposition of the Great Valley sequence spanned tens of millions of

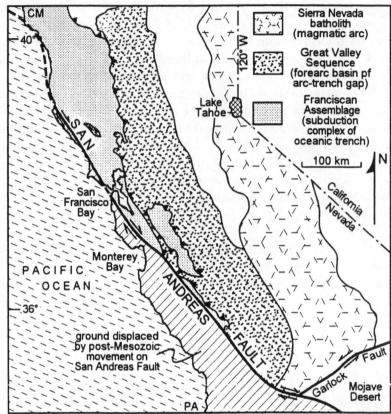

Geologic sketch map of Central California (CM, Cape Mendocino; PA, Point Arguello) showing the geographic arrangement of three great Mesozoic rock assemblages (discussed in the text) formed during the approximate interval from 150 to 50 million years ago. Deformed and displaced counterparts of the three assemblages west of the San Andreas transform fault are not separately delineated for reasons of scale (half arrows beside faults indicate the directions of relative lateral motion of blocks in contact along the faults).

years, from latest Jurassic time through at least all of the Cretaceous period, which did not end until about 65 million years ago. We thought of the Great Valley sequence as strata deposited on a passive continental shelf, much like those surrounding the Atlantic Ocean today. The markedly uneven tempo of geologic events we then envisioned, involving a brief orogenic "revolution" followed by a prolonged period of orogenic quiet, turned out to be all wrong.

Modern geochronology had already dismantled the scheme by 1960, well before the onset of plate reasoning, and cleaned the slate to be rewritten. Potassium-argon (K-Ar) dating, an isotopic method of deter-

mining the ages of rocks, had demonstrated that much of the Sierra
Nevada batholith is Cretaceous in age, and has nothing directly to do
with an older Jurassic Nevadan orogeny.[32] There had also been the revelation that much of the Franciscan assemblage is Cretaceous as well,
covering the same general age span as the supposedly younger Great Valley sequence.[33] Soon thereafter, we learned that the contact between
Franciscan and Great Valley strata is a regional thrust fault, along which
the Great Valley strata were carried above Franciscan strata of the same
general age by tectonic movements.[34] The depositional contact that we
envisaged in our mind's eye between supposedly older Franciscan and
younger Great Valley strata does not actually exist anywhere on the
ground. About the time that the thrust-faulted nature of the contact was
first appreciated, I was mapping the contact in detail within the Diablo
Range, and was able to discern clearly the mutual field relations of Franciscan and Great Valley rocks, thus confirming the interpretation of a
regional thrust fault by independent observations.[35]

A band of altered mantle rock termed *serpentinite* separates the Franciscan assemblage from the Great Valley sequence in many places, and
was at first thought to have been injected along the thrust contact
between the two rock assemblages after they had been placed in contact
by faulting, or perhaps during fault displacement of the two rock masses
adjacent to one another. Later, however, the serpentinite was identified
as part of a so-called ophiolite succession representing oceanic crust and
mantle ("lithosphere") that depositionally underlies the oceanward
flank of the Great Valley sequence.[36] Along the way, Dick Ojakangas, now
a professor at the University of Minnesota (Duluth) but then a Stanford
advisee of mine, had shown that most of the Great Valley sequence is
composed of sediment deposited on the deep sea floor by turbidity currents.[37] That mode of origin befits a succession depositionally overlying
an oceanic ophiolite succession, but rules out the previous interpretation of shelf deposits laid down in shallow marine waters.

We were thus confronted, as plate tectonics loomed over the horizon,
with a grand triad of parallel Cretaceous tectonic elements (see sketch
map above) – Sierra Nevada batholith, Great Valley sequence, Franciscan assemblage – composed of disparate rock masses spanning the same
overall age range. Such an arrangement of tectonic elements fit no pre-
conceived notions, which is to say no pre-plate notions, but was readily
explicable in the context of plate tectonics. In the rush of just a few years,
the Sierra Nevada batholith, encased in greenschist metamorphic rocks,
was interpreted as the eroded roots of a Cretaceous magmatic arc. The
Franciscan assemblage, containing blueschist metamorphic rocks made

from scraps of sea floor sediment intermingled with igneous oceanic crust and mantle, was interpreted as the internally disrupted product of a subduction zone at a paired Cretaceous trench, where an oceanic plate had descended beneath the California continental margin. We perceived then that the Great Valley sequence was deposited during the same time frame as the arc and trench activity within an intervening deep trough, located in the "forearc" region between the Sierra Nevada magmatic arc and the Franciscan subduction complex in a position familiar from the configurations of modern arc-trench systems. Regional underthrusting of the Franciscan subduction complex beneath the Great Valley forearc could then be viewed as the record of plate subduction frozen in geologic time. Only a few months after the Asilomar Penrose Conference, a field trip in the spring of 1970 led by me and the late Ben Page of Stanford University through coastal California, in connection with the annual meeting of the GSA Cordilleran Section in San Francisco, widely aired the sweeping new interpretations.[38]

A rapid-fire series of influential papers codified the new picture. First in print (1969) was a sketch by Warren Hamilton showing the "underflow" (subduction) of Pacific mantle beneath Cretaceous California in a paper that appeared the same month as the Asilomar Penrose Conference.[39] His concepts were outrunning available terminology, and his self-coined term *underflow* to describe the descent of an oceanic plate beneath an arc-trench system was outmoded as soon as the term *subduction* gained favor during the conference proceedings to describe the same process. Next came the identification by Gary Ernst (University of California, Los Angeles), in a 1970 paper in press as the conference convened, of the faulted contact between the Franciscan assemblage and the Great Valley sequence as the crustal expression of a fossil Wadati–Benioff inclined seismic zone.[40] My subsequent review paper of 1970 on arc-trench tectonics was largely put together before the conference but was completed with insights from the conference fresh in mind.[41] A key figure depicted the tectonic triad of Sierra Nevada batholith, Great Valley sequence, and Franciscan assemblage as linked components of the same Cretaceous arc-trench system. The following year (1971), Ken Hsü (Geological Institute, University of Zurich) linked the formation of Franciscan mélanges explicitly to the pervasive effects of underthrusting during subduction.[42] Hsü had himself resurrected the hoary descriptive term *mélange* some years before to denote Franciscan rocks intricately disrupted by closely spaced slip surfaces, but had then favored gravity landsliding on a giant scale as the mechanism that formed them. For him, the concept of plate subduction belatedly came to the interpretive rescue. During the subsequent development of current views on the plate

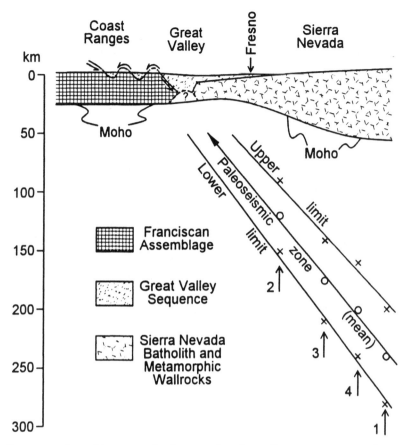

Reconstruction of key tectonic elements forming a Cretaceous arc-trench system in California (transverse profile). The Franciscan assemblage consists of rocks formed under high pressure (blueschists) in the subduction zone; these rocks are now exposed in the California coast ranges. The Great Valley sequence consists of sedimentary and volcanic rocks deposited in a basin formed between the subduction zone and a magmatic arc formed along the edges of the North American continent. These rocks are now exposed near the city of Fresno. The Sierra Nevada batholith represents the exposed root portions of the magmatic arc. See Figure 16.2 for geographic relations in map view. A "paleoseismic zone" marks where the upper surface of the subducted oceanic plate would have lain, and where deep-focus earthquakes would have occurred as the slab sank deeper into the mantle. This zone can be identified in hindsight on the basis of the chemical composition of rocks in the Sierra Nevada batholith; numbers (1–4) denote positions of granitic belts of different ages. Note that the sloping trend of the paleoseismic zone points upward toward the thrust contact between Franciscan and Great Valley rocks marking the intersection of the subduction zone with the surface in Cretaceous time. The "Moho" (Mohorovicic discontinuity) is the contact between the crust (above) and the mantle (below).

geometry of arc-trench systems, my focus centered on the integral place of forearc basins, like that of the Great Valley of California, lying within the elongate belt ("arc-trench gap") between arc and trench.[43]

OVERLOOKED ANTECEDENTS

Interpretive diagrams linking subduction of an oceanic plate to the generation of arc magmatism, stem in part, either consciously or subliminally, from largely forgotten precursors that influenced the outlines of the synthesis achieved in a plate tectonic context. Quite well known was the 1962 scheme of Bob Coats (USGS Alaska Branch) showing oceanic crustal materials dragged down along the seismic zone above a descending slab of oceanic sea floor, akin to a plate, with volatile compounds such as superheated water bleeding upward off the descending slab to foster melting in the mantle between the descending slab and the volcanic chain of the Aleutian island arc.[44] Although his ideas about the critical role of subducted volatiles in generating arc magmas received respectful attention and widespread interest, they also met with considerable skepticism because most geological minds were as yet unprepared to accept dramatic mobility of the ocean floor. Within two decades, however, as modified to conform with improved information about subsurface conditions beneath magmatic arcs, they became a standard part of mainstream interpretations of arc magmatism in a plate tectonic context.[45]

Much less familiar was the even earlier 1955 diagram (next page) of the German tectonicist, Hans Stille, depicting an inclined "megathrust,"

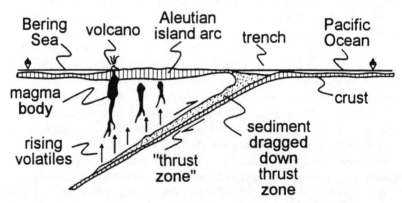

Pre-plate tectonics (1962) scheme of Coats for generation of Aleutian arc magmas (see text for discussion). The half arrows indicate inferred relative motion across a postulated "thrust zone" carrying ocean floor to great depth.

delineated by the Wadati–Benioff seismic zone, with fusion of underthrust crustal materials to produce arc magmas.[46] He termed this suggested process *palingenesis by underthrusting* to contrast it with the then-popular concept of palingenesis, or crustal melting, in the depressed keels of "geosynclines." Just prior to the Second World War, it was Stille who had developed in Europe the very classification of miogeosynclines and eugeosynclines that Marshall Kay further elaborated after the war on this side of the ocean. Few appreciate that the Stille, whose name is so closely linked with geosynclinal theory, had begun to channel his own thoughts into a more mobilistic mode only ten years after the war was over.

Most arc magmas are now thought to derive from melts generated in the mantle wedge above the inclined seismic zone, much as envisioned

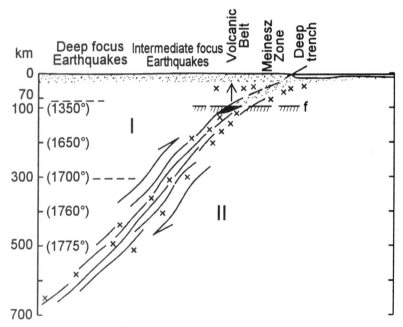

Pre-plate tectonics (1955) scheme for generation of arc magmas by melting along inclined seismic zone (belt of dashed lines, with crosses indicating individual deep-seated seismic events that generate earthquakes) interpreted as a "megathrust" (half arrows denote relative block motions). Block I is a continental part of the Pacific rim and Block II is a part of the Pacific Ocean basin. The letter "f" indicates the "level of fusion [melting]" beneath a volcanic arc. Black indicates a "center of vulcanism" (reservoir of magma at depth below the volcanic chain). Position of subduction zone ("Meinesz Zone") detected then only from its anomalous gravity field. From Stille, 1955. *Geological Society of America Special Paper* 62, p. 184 (reproduced with permission of the Geological Society of America).

by Coats, rather than from direct melting of underthrust sediments as envisioned by Stille. Nevertheless, debates about the detailed geochemical processes giving rise to arc magmas continue today. Arc-trench systems are immensely large and complex phenomena, with much of the geologic action within them hidden from view at great depth. Many years will doubtless pass yet before we understand them fully.

PHILOSOPHICAL ATTITUDES

During the plate revolution, many invoked the religious analogy of "conversion" to describe the change that took place in the minds of geoscientists. Although apt in some respects, I think the religious analogy misses the mark. Devising and developing the plate model required innovation and vision, but it was not plucked by imagination from thin air; it evolved as the logical consequence of crucial new observations and fresh analysis of outstanding problems in the light of those observations. We did not change our way of thinking in response to some spiritual inspiration, or the authority of some admired guru, but because plate tectonics provided far and away the best means to understand otherwise baffling geologic relationships.

In the early going, pioneer thinkers were working against the prevailing intellectual grain and met with widespread skepticism from the geological community. Although the fresh ideas gained general credence within just a few years because of their inherent power, each new conceptual advance was treated with initial suspicion until people had a chance to become comfortable with it. Once convinced that plate tectonics was a much better view of the geological world than any previously available, I spent many lecture hours in many venues, and spilled much ink, helping to persuade geoscientists who were not yet quite "up to speed" that plate theory could improve their approach to almost any problem they faced. For years I "dined out," as the saying goes, in the role of showing how plate tectonics impacted various fields of geoscience or various areas of the world.

Two quite scientific and non-religious lodestars guided my own reasoning as I eased my mind gradually along innovative tracks. One was the recommendation of John Platt (University of Chicago), in a paper published serendipitously just before the plate revolution burst upon us, to use what he termed "strong inference" in scientific analysis.[47] Devise hypotheses inductively, he counseled, test them deductively as rapidly as possible by projecting their implications in as far-ranging a way as possible, reject those found wanting, and move on immediately to alternative hypotheses not precluded by available data. In other words, in the cowboy maxim of the

Old West, "never dally trying to flog a dying pony across a flooding stream." Clinging to failed concepts is only unnecessary delay. This perspective encouraged me to abandon tired old pre-plate concepts, and to embrace substitute plate concepts that better satisfied the data available.

The other perspective goes much deeper and dates from farther back, to a time when I was a graduate student exploring the philosophical underpinnings of science in the stacks of the Stanford library. In 1923, the noted German philosopher Ernst Cassirer, in his book *Substance and Function,* argued that ancient Greek (actually Ionian) science proceeded by identifying the supposedly innate properties or "substance" of objects that control their behavior, whereas modern science seeks to discover the influences that account for observed natural phenomena as a "function" of those governing influences, not as the result of any innate properties.[48] He contrasted the "thing-concepts" (substance) of the former with the "relation-concepts" (function) of the latter.

In that context, geosynclines, and the orogenies to which they supposedly gave rise, were very much "substance" concepts. Geosynclines sank because that was their nature (some wag once remarked that they just "had that sinking feeling"), and climactic orogeny was foreordained because that was what geosynclines inherently fostered. The sequential path from geosyncline to orogeny supposedly defined a geotectonic cycle predetermined by the inherent nature of geosynclines. Plate tectonics wrenched us firmly out of that philosophical dead end, showing us that geosynclines and orogenies, to the extent those terms have any meaning at all, are complex manifestations of plate movements and associated plate evolution.

No one ever dismissed the alleged inevitability of a fixed geotectonic cycle more trenchantly than my University of Arizona colleague, Peter Coney, another veteran of Asilomar. In a paper already in press when the conference convened, he wrote that "saying geosynclinal prisms [of accumulated sediment] lead to mountain systems is a little like saying that fenders lead to automobile accidents."[49] He had in mind, of course, the jostling together of continental blocks when subduction consumes an entire ocean basin, as happened where the Indian subcontinent interacted with Eurasia to raise the Himalayas. The diversity of geosynclinal types and orogenic processes is a function of different styles and sequences of plate interactions.

THE YEARS SINCE

It is no exaggeration to state that the advent of plate tectonics charted a lifelong program of personal research, with few deviations into any avenues unrelated to the marvelous dance of the plates on the surface of the globe. Once the significance of changing plate interactions through geologic time was perceived for the evolution of California, I tried in multiple papers down through the years to discern the implications of plate tectonics for other parts of the American West.[50] The trail is too technical and involved to recount here, but plate tectonics proved the key to puzzle after puzzle buried in the rock record of geologic history. My attention was drawn also to the changing patterns of plate motion that have influenced the evolution of island arcs around other segments of the Pacific periphery, from Alaska down past Japan and through the complex Pacific island chains to New Zealand.[51] Side inquiries involved what happens to subducted plates when they descend into the mantle, and how subduction affects the subterranean migration of petroleum.[52]

Over the years, however, my students and I focused especially on the role of plate tectonics in governing the kinds of sandy detritus derived from highland sources in different plate settings, and in controlling the evolution of sedimentary basins, those regions where subsidence allows the accumulation of thick prisms of sediment.[53] Although the correlation of sandstone composition with paleogeographic patterns controlled by plate tectonics has been a rewarding avenue of research, the results of the exercise are too embedded in the specialist woodwork of sedimentary petrology to discuss in detail here. Throughout our investigations, I tried always to evaluate the relationships of various kinds of sedimentary accumulations to the tectonic evolution of associated orogenic belts.[54] Inquiries into the origins of different types of sand expanded to include tracing the sources of various classes of sand temper in prehistoric ceramics of the Pacific islands. The islands are located in different tectonic realms governed by patterns of plate motion, and the sands available to ancient potters on different islands faithfully reflect the various plate tectonic environments in which the island potters lived and worked.[55]

My plate tectonic classification of sedimentary basins was developed originally for a professional American Association of Petroleum Geologists (AAPG) short course, but was later expanded and codified by Ray Ingersoll (University of California, Los Angeles), and eventually provided the framework for a comprehensive volume on the tectonics of sedimentary basins edited by him and Cathy Busby (University of California, Santa Barbara).[56] The classification scheme delineates different kinds of sedimentary basins as a function of plate setting, but a shadow of "thing-concepts" still attaches to the resulting catalogue of basin types.

The next step in basin analysis is to move still further into the realm of "function-concepts" by acknowledging that each sedimentary basin is unique, with a history that is a function of changing plate settings and sequences of plate interactions that differ in detail through time.[57]

Reconstructing the evolution of sedimentary basins affords special insight into geologic history because understanding patterns of subsidence and subsequent uplift affecting sedimentary basins provides a means to monitor the changing elevations of the earth's surface over geologic time. This is superb leverage for establishing the geodynamic history of the earth. Without a plate framework for analysis, however, the raw data used for basin analysis cannot be amalgamated into an integrated picture. Plate tectonics proved the indispensable tool for evaluating sedimentary basins, and I was immensely fortunate that plates came along during my lifetime of studying basins. Those older or younger cannot possibly imagine the excitement plate theory evoked for those of us who reached scientific maturity before the plate revolution, but were still young enough to reorient our heads and our research programs toward the vistas opened up by plate tectonics. Moreover, working out the new game was sheer fun every step of the way.

CHAPTER 17

FROM PLATE TECTONICS TO CONTINENTAL TECTONICS

An Evolving Perspective of Important Research, from a Graduate Student to an Evolving Curmudgeon

Peter Molnar

I WAS A LUCKY GUY, A GRADUATE STUDENT AT COLUMBIA UNIVERSITY from 1965 to 1970 and able to write a thesis consisting of three papers on different aspects of plate tectonics, one each with Bryan Isacks, Jack Oliver, and Lynn Sykes. Introduced to the forefront of the earth sciences as a student, how could I fail? With an excess of ambition and confidence, and armed with Jack's repeated prodding to think big and to pursue "the next, most important problem," I dismissed plate tectonics as dead in 1970 and sought a new direction. The slow recognition of failure returned me to plate tectonics as a tool (plate reconstructions), as a philosophical approach (to look first at a large scale), and as a battery of methods (mostly seismological in my case) for the study of the continental tectonics of Asia.

Luck struck again, when Paul Tapponnier introduced me to the Landsat imagery and analyzed those from Asia. If plate tectonics were a revolution, and the French one the metaphor, a reign of terror in the early 1970s made many geologists stop saying, "Plate tectonics is fine, but it doesn't work in my area," to avoid the intellectual guillotine. In 1975, Paul and I reached two conclusions that gained us some notoriety: (1) the deformation of Asia seemed to be dominated by strike-slip faulting, which suggested to us that India's penetration was absorbed by eastward extrusion of crust, and (2) Asia, and hence continents in general, behaved like

Peter Molnar, Afghanistan (Photo courtesy of Peter Molnar.)

Peter Molnar, East Africa Rift Basin. (Photo courtesy of Peter Molnar.)

a deformable medium, not like one or a few rigid plates, which rendered plate tectonics, at least in its strictest sense, a poor description of continental tectonics. Ironically, Paul and I now each minimize the significance of one of those two inferences: he the latter and I the former.

Our work might be seen as part of a metaphorical "Thermidorian Reaction"; no wonder field geologists had not discovered plate tectonics, for diffuse deformation and widespread strain makes recognizing rigid plates within continents difficult. By the late 1970s, plate tectonics seemed to have passed into earth science departments as a synonym for ophiolites and paired metamorphic belts, instead of rigid plates and angular velocities. In what may be seen as a counter-revolution, structural geologists defined a new discipline, structure/tectonics, which to a large extent disguised itself as historical geology with a new jargon. As structure/tectonics became the biggest section of the Geological Society of America (GSA), students, when asked how much displacement had occurred on a transform fault, puzzled over piercing points. So, what, after all, were the revolutionary changes brought by plate tectonics? It seems to me that plate tectonics accelerated a transition in the earth sciences from a 19th-century natural science that treated the history of the earth as an end in itself, to a 20th-century physical science focusing on a quantitative understanding of the processes that have shaped the earth.

PLATE TECTONICS (1965–1975)

Timing, as in a good joke, is a vital part of luck, and my experience during the recognition and development of plate tectonics illustrates well the importance of lucky timing. Having studied physics at a college (Oberlin) where teachers were paid to teach and took pride in doing it well, and where students studied more to learn than to get good grades, I was better prepared than most of my fellow students when I entered Columbia University in 1965. I was also lucky to be too young to be intimidated by, let alone aware of, how little I knew.

In 1965, options seemed limited for a physics major like me, physically fit but inept in the laboratory, insecure with mathematics and quantum mechanics, and illiterate compared with his friends. Getting drafted and going to Vietnam was much less attractive than using my training in physics to solve simple problems in the mountains, which had beckoned since my parents introduced me to the Rockies when I was only 9 years old. When I sought advice, however, my physics teachers asked, "But, what is geophysics, anyhow?" to which I responded, "I don't know, but I think it is applying physics to the earth."

The power-forward on the Oberlin faculty's intramural basketball team, Jim Powell, who also taught me a semester of geology while I sought an end-run of chemistry, advised me further, but with the admission that he did not know geophysics well. My father, an experimental physicist/administrator at Bell Labs, gave me the best advice, and on this rare occasion, I took it. His participation on the Berkner panel on nuclear testing had given him the narrow view that geophysics was little more than seismology and had introduced him to two outstanding young seismologists: Frank Press, then at the California Institute of Technology (Caltech), and Jack Oliver, then at Columbia. So I applied to both, hoping to work with one of them. Then, Jim Fisk, my father's boss and a member of the board of directors at Massachusetts Institute of Technology (MIT), leaked to him that Frank Press was moving to MIT to become the department head. Had my father not been a devout atheist, he might have quoted Huckleberry Finn, "I guess I'll go to hell," before breaking the rules of confidentiality he lived by and then confiding to me why I should go to Columbia. Although Caltech, arguably, has maintained the outstanding Earth Sciences Department in the United States since before plate tectonics, during plate tectonics, and subsequently, I consider my father's breach of his ethics, the only example in my experience, to have been another lucky break for me. Despite its pre-eminence, Caltech's direct contributions to plate tectonics lie somewhere between modest and invisible.

When I arrived at Columbia, I was also tired and in need of a rest, as so many students who have worked hard as undergraduates experience when they start graduate school. Before Fred Vine and Drummond Matthews and their hypothesis for how magnetic anomalies over the ocean floor formed were taken seriously, and before many people realized that one of Tuzo Wilson's speculative ideas (transform faulting) was actually right, I audited a course in postimpressionist painting and attended as many concerts and off-off-Broadway theatrical performances as possible, while taking what was reputed to be a full course load. Boredom gave way to more courses in the spring of 1966, but it was the fall that opened the intellectual doors of the earth sciences to me. If in 1965 Lynn Sykes had shown Tuzo to be right and if Walter Pitman and Jim Heirtzler realized earlier that Vine and Matthews had the only sensible explanation for the magnetic anomalies measured by the research vessel *Eltanin* while crossing the South Pacific Rise, I might have missed the fun.

In the fall of 1966, having gleaned more from geophysics than I had dreamed, a summer camping in the mountains of Alaska, I was finally ready to take my field of seismology seriously. Fellow student Bob Liebermann had suggested that deep earthquakes be the subject of the fall semester's two-credit course, Seismology Seminar, led by Jack Oliver and Bryan Isacks. Surely what we discussed was important to Jack and Bryan's recognition of subduction of oceanic lithosphere, although for me, unable to recognize a significant scientific problem, the mere discussion of scientific questions rather than homework problems was most important.[1] One of Jack's tricks was to get us to discuss, among the classics, some decidedly inferior papers, whose shortcomings we were to recognize before he told us what would happen if we had written them.

The seismology group at Lamont Geological Observatory of Columbia University ran another seminar, every Monday night, which was not for credit and not expressly for students, although we all learned quickly that we were expected to know what was discussed. One Monday that fall, Tom Fitch, a fellow student, told me I should stick around, because Jim Heirtzler was going to present evidence supporting "sea floor spreading." Lost, trying to conjure a sensible image of the sea floor framed in the metaphor of an expanding waistline, I stayed and learned more than just what those words meant. It was, however, Sykes' demonstration of transform faulting using seismicity and fault-plane solutions of earthquakes, presented at the Monday night seminar a few weeks later, that made me realize that continents drifted and that something exciting was happening.

When Sykes heard Wilson present his idea of transform faulting, he dropped what he was doing to test the idea; it took me three months to realize that I had better things to do than concentrate on courses. Art McGarr, an advanced student at Lamont, assured me that second-year graduate students could choose their own research problems. After another month, during which Jack Oliver told me daily that my perusal of seismograms (a seemingly aimless search for something interesting) was the best way I could spend my time, I buried my tail between my legs and sought advice from Lynn. He pointed me to a problem that eventually led to our determining the motions of the Caribbean and Cocos plates with respect to each other and to North and South America, just as Jack pointed me toward another, the mapping of lateral heterogeneity in the mantle and hence defining the lateral extent of lithospheric plates.[2] Suddenly I was studying what I wanted, not what teachers in classes expected, as my grades soon reflected. It wasn't long before Jack had to rein me in by pointing out that I seemed to be the kind who could not finish anything. I was a bit afraid of him; after all he was 20 years older than I, and therefore already in his 40s.

Lynn organized a special one-day session at the American Geophysical Union's (AGU) annual meeting in the spring of 1967 to discuss all of the recent developments that related to sea floor spreading, transform faulting, and underthrusting of lithosphere at trenches. Perhaps no single one-day session made the ongoing developments more obvious to the uninitiated than that one, and I felt a special advantage in being already familiar with some of what was presented. Experience had taught us students, however, that when the AGU's morning sessions ended, finding affordable lunch could be hard. To beat the crowds at the sandwich shops, many of us skipped the last talk in the morning. It was to be given by someone who had written a pair of papers interpreting gravity over the Puerto Rico region in terms of dynamic processes in the mantle, and those papers had elicited a critical Letter to the Editor to the *Journal of Geophysical Research* from the big guns at Lamont.[3] I had not heard of the author and remember only by vague innuendo that he didn't know what he was doing. Several years later I read these two gems by Princeton's Jason Morgan, profound and well ahead of their time.[4] By skipping out to minimize time in the queue at the sandwich shop, I missed Jason's presentation of plate tectonics, which was not discussed in his abstract. My impression is that among those who heard his presentation, only Xavier Le Pichon grasped the significance of what he said.

In the summer of 1967, Jason's paper on plate tectonics arrived at Lamont in pre-print form, and Lynn circulated it to all of us who might be interested. Suddenly the jigsaw puzzle fit together. By recognizing the rigid-body movement of lithospheric plates (which he called "crustal blocks"), Morgan gave us the missing glue that united sea floor spreading, transform faulting, and subduction (a word not yet used for what was clearly occurring at island arcs) into "plate tectonics" (more words not used yet).[5] A turning point had obviously been reached, if where next to turn was less obvious to me.

By the fall of 1967, Isacks, Oliver, and Sykes had started meeting regularly to prepare their encyclopedic summary, "Seismology and the New Global Tectonics."[6] With courses still to take and other obstacles to overcome before I could pursue exciting research full-time, I was envious. Poor Jack would have stimulating discussions with Bryan and Lynn, and then I would enter his office with the paper that he and I were writing. At Oberlin, except for a one-semester course in physical geology, I never earned better than a C+ in courses that required writing essays or term papers: English Composition, English Literature, Religion, Modern Painting, Philosophy, and European History. My non-scientist friends called me a "mere technician" and belittled my illiteracy at every opportunity. When done with a page of what would be my first major paper, Jack sometimes would hand it to Judy Healy, his secretary and my English compositional savior, to retype, because the editing had rearranged, if not replaced, most words.

By the time this paper and that with Sykes were approaching submission, what was missing was not the satisfaction of writing clear expository prose; that, like wisdom or humility today, seemed hopelessly out of reach. I sought something more immediately satisfying – my own original idea.

For as long as I can remember, my father, J. P. Molnar, had urged me to think for myself. He also tried repeatedly to impart some of his wisdom to me. Most of what stuck did not do so until after his death in 1973, but when Bryan, Jack, or Lynn reinforced his insight, I took it seriously. Jack could not end a seminar course (or a thesis defense) without asking something like, "What is the next, most important problem to pursue?" Those words tormented me for years, even when I thought I could answer them. Perhaps, if I had understood what made a problem "important," I would have wasted less nervous energy. In any case, whether articulated concisely or merely pervading an atmosphere, the repeated asking of such a question must surely separate institutions that make advances from those that mop up loose ends behind the forefront.

Good ideas at Lamont did not seem to be concepts that related facts and theories, and therefore the kind that often instill possessiveness in myopic egos; rather they focused on topics worth pursuing – like deep earthquake zones or transform faulting.[7] Jack once told me something like, "Yeah. Of course, I want students who can solve problems, but more important, I want students who can choose good problems." Simultaneously, Lynn was constantly suggesting that we students could find ways to test ideas, such as what drove plate motion. This two-pronged attack, of one urging a seemingly blind grope toward a good problem and of the other urging a more orderly analysis of hypotheses followed by tests, kept me alert to "important problems," whatever those words meant. Jack's book, *The Incomplete Guide to the Art of Discovery,* articulates his case more clearly than I had understood and offers advice on how to make such choices.[8] Anyhow, when I got my first good idea, to use fault-plane solutions of intermediate and deep-focus earthquakes to study the "driving mechanism of plate tectonics," I promptly learned that Bryan had already been pursuing this. A kind man, he allowed me to share the study with him, and I think I did contribute to it, although not as much as he did. Twenty-twenty hindsight suggests that Jack and Lynn may have let us share it alone, in part in deference to my palpable ambition.

We showed that fault-plane solutions of intermediate and deep-focus earthquakes defined a simple pattern.[9] First, neither of the possible fault planes determined in a fault-plane solution is oriented parallel to the inclined, deep seismic zone at an island arc. The relatively planar zone of seismicity, therefore, does not define a mega-thrust fault plunging into the mantle, as Caltech's Hugo Benioff had contended.[10] Instead, the approximate orientations of principal stresses causing the earthquakes imply that the earthquakes occur within a strong tabular body, the down-going slab of lithosphere, and result from the stressing of that slab.[11] In areas where no deep earthquakes occur (at depths greater than 200 miles or 300 km), the down-going slabs are stretched, presumably by gravity acting on the excess mass within them to pull the slabs downward in much the same way that a spring hung from the ceiling stretches due to its weight. Essentially all deep-focus earthquakes, however, suggest that the slab encounters resistance (as if the bottom of the spring rested on the floor). We assumed that resistance to reflect increased strength at a depth of 370 to 450 miles (600 to 720 kilometers). In many areas, a gap in earthquakes separates intermediate-depth seismicity from that below 200 miles (300 kilometers), which seems to mark a zone of low stress between down-dip extension above and down-dip compression below, although in some

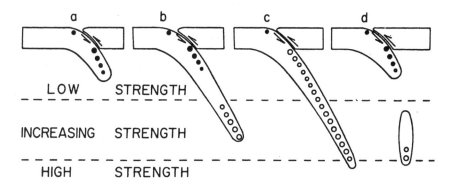

Cartoons illustrating simple interpretations of fault-plane solutions of earthquakes of intermediate and deep-focus earthquakes. Closed and open circles signify earthquakes with down-dip T-axes and P-axes, respectively. (a) Where no deep earthquakes occur, T-axes are down-dip, because gravity acting on the excess mass of the slab pulls the slab down. (b) Where there is a gap in seismicity at depths near 200 miles (300 kilometers), T-axes again are down-dip at intermediate depths (shallower than 200 miles (300 kilometers), and P-axes are down-dip at greater depths, because the slab encounters resistance of some kind. Here, the gap in seismicity marks a zone where stress is low. (c) Where seismicity is continuous to depths of 350 to 450 miles (600 to 700 kilometers), P-axes are down-dip for both intermediate and deep earthquakes, because a rapidly subducting slab has encountered resistance at depth. (d) In some areas, a gap in seismicity might mark a gap in the down-going slab. T-axes are down-dip in the shallower part of the slab, because gravity pulls the slab down, and P-axes are down-dip at great depth because the slab encounters resistance there. Isacks, B. and P. Molnar, Distribution of stresses in the descending lithosphere from a global survey of focal-mechanism solutions of mantle earthquakes, *Rev. Geophys. Space Phys.*, 9, 103–174, 1971.

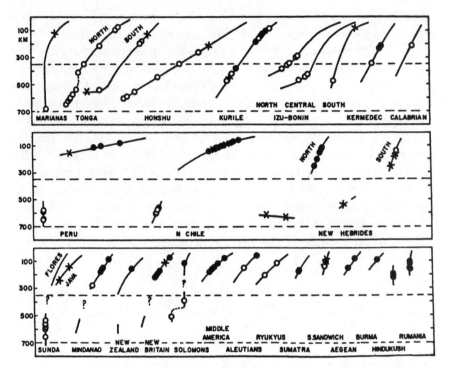

Summary of fault-plane solutions of earthquakes of intermediate and deep-focus earthquakes. Symbols as in Figure 1, with each symbol representing an earthquake that we studied, and with *Xs* showing fault-plane solutions with neither down-dip P- nor down-dip T-axes. A line through symbols denotes the cross-sectional shape and depth range of the seismic zone. Isacks, B. and P. Molnar, Distribution of stresses in the descending lithosphere from a global survey of focal-mechanism solutions of mantle earthquakes, *Rev. Geophys. Space Phys.*, 9, 103–174, 1971.

areas, the slab itself might be discontinuous. The fault-plane solutions, thus, lent support to the idea that gravity acting on the cold, dense mass of the down-going slab played a crucial role in driving plate motion, and also seemed to suggest (perhaps incorrectly) that the slab did not penetrate deeper than approximately 450 miles (720 kilometers).

By 1969, the first of our papers was published, and I felt proud of having done good work. After a presentation of this work in 1969, Seiya Uyeda from the Earthquake Research Institute of Tokyo University gave me what has impressed me as the highest form of compliment one can pay a scientist. He said, "I have to change my ideas."

Workaholism may be hereditary, but still young and untrained in the art of working all the time, I needed a rest. In 1969, I took a seven-month sabbatical to Africa to study earthquakes and intracontinental rifting there, and I failed, as I would time and again. Our results did not enlighten the process of rifting. The Tuesday following a Monday night seismology seminar in which I described my work in Africa, Muawia Barazangi, a fellow student at Lamont, asked me with that ingenuous frankness that has always endeared me to him, "What happened last night? That talk was no good." Although he said nothing after my next seminar, when I discussed what might have become my thesis, he could have said the same. So, with Jack's help in writing a three-page preface, I packaged reprints of three papers published the year before together as a thesis, and in December 1970, I defended a pamphlet whose bound cover was thicker than its pages. Although I felt smug knowing they couldn't fail me with a thesis consisting of papers with each of Isacks, Oliver, and Sykes, something was missing. In 1970, I passed from being a bright young graduate student to just another guy with a Ph.D., a manifest let-down, amplified by the concern that I could not identify "the next, most important problem." Jack had asked during my defense, as I knew he would, but my prepared answer was obviously unsatisfactory. I dropped that topic a year later.

In the late 1960s, the unsolved problem discussed most often surely was "the driving mechanism of plate tectonics." The force moving the plates is obviously gravity. If plate tectonics was a part of thermal convection, as most believed, the energy source must almost surely be radioactivity. No one doubted that subducted oceanic lithosphere played a key role, for the weight of the down-going slab was clearly huge by the standards of likely mass anomalies in the earth. Moreover, the dynamics of flow in a thermally convecting fluid must be expressed in terms of a balance between gradients in stress and *horizontal* gradients of temperature. The large horizontal temperature gradients at subduction zones again pointed to the down-going slab's role in the "driving mechanism." The question seemed and still today seems to me to be not, "What drives plate tectonics?" but "What processes and properties dictate the range of dimensions that plates and the underlying convection take?" In short, "How does mantle convection work?" which comprises more than one well-posed question. In any case, from my point of view in about 1970, the question, "What is the driving mechanism of plate tectonics?" was either already solved or too difficult for me to pursue. I sought a direction that would challenge me to learn something new, but that did not exceed my abilities.

My father had said repeatedly that nine out of ten scientific experiments failed; one had to persevere to get anywhere. I got a post-doc with Jim Brune at the Scripps Institution of Oceanography; Jim is one of those rare individuals who always seemed to find a different way to look at a problem. While reciting conventional views to Jim, and failing to appreciate his approach, I promptly learned that nine out ten ideas of what to pursue might lead to publishable papers, but they need not be important. Jack never defined precisely what "important" meant when applied to the next problem. What I was looking for, however, had gradually become Lynn Sykes' experience with transform faulting, a problem so important that it would cause me to drop what I was doing to pursue it with vigor (almost a love affair with a scientific problem).[12] Lynn did not drop everything when he first read Tuzo Wilson's paper on transform faulting; as he told me, it was not until he heard Wilson, who was renowned for exciting talks, present the idea in a talk given months after its publication.[13] Knowing that I had to be patient did not, however, prevent frustration.

Jason Morgan came to Lamont to give a talk in the summer of 1971, to present his thoughts on "plumes" before he published them.[14] We all went to listen to what was a memorable seminar. He opened with, "Consider plumes rising from the core-mantle boundary."[15] Before uttering the next sentence, of what may have been a rehearsed opening gambit, Xavier Le Pichon blurted out loudly from the front row, "Why?" Somewhat flustered, Jason said, "Well. (pause) Just consider it," but Xavier insisted that Jason give a reason for considering what seemed so preposterous an idea at the time. I sympathized with Xavier and was not inclined to drop everything to pursue this idea. That inclination came a year later, however, when evidence had mounted to suggest first, that "hotspots" – isolated centers of volcanism like Yellowstone or the islands of Hawaii or Iceland, which are the surface manifestations of Jason's plumes – might define a fixed reference frame, and second, that localized anomalies in seismic wave propagation at the core mantle boundary might lie directly beneath some hotspots.

I was delighted to have a problem that might allow me to drop everything, but my efforts to detect anything special at the core-mantle boundary failed. My father would have been happy, for I was learning an important lesson: "most experiments fail." Later, Tanya Atwater and I, then working at Scripps, concluded that the hotspots were not fixed.[16] That conclusion might not have been justified by the data in 1973, but it still stands, in my opinion, with much better data and a more rigorous treatment of uncertainties.[17] If Jason Morgan's paper on plate tectonics had seemed like pieces of a jigsaw puzzle falling out of the sky and into place,[18] resolving plumes and understanding hotspots seemed like assembling a jigsaw puzzle of fluid pieces; they could always be made to fit, but they did not always hang together.

More than once when I groped for a good problem and floundered, Jack Oliver told me, "The worst thing you can do when you can't think of anything to do is to do nothing." After recurring failures to identify a problem that would make me drop everything I was doing to pursue it, I took Jack's advice and returned to what I knew how to do, to determine fault-plane solutions of earthquakes. By the early 1970s, we students and staff at Lamont had studied nearly all regions where fault-plane solutions of earthquakes could resolve plate motions, but what had been ignored were the continents. Dan McKenzie had recognized this omission a couple of years earlier, and during a visit to Lamont in 1968 he began working on the Mediterranean region before returning to Cambridge, England, to complete this study.[19] I chose Asia and promptly learned that Tom Fitch, who had gone to the Australian National University in Canberra for a post-doc, and Francis Wu, of the State University of New York in Binghamton, had also begun work on the same area. When we recognized this, we joined efforts.

My study of Asian tectonics began as a back-burner project and remained so while I was at Scripps, still hoping to find something so important that I could drop everything else. My growing self-doubts about finding such a problem were strengthened when Dan read our paper.[20] We argued that, in contrast to plate boundaries, the consistent parameters in Asia were the approximate orientations of the principal stresses acting in the regions of the earthquakes, or more precisely the axes of principal strains,[21] not the orientations of the two perpendicular planes, which had been so useful in determining the orientation of relative movement at plate boundaries. Having used slip vectors to demonstrate plate tectonics, McKenzie asked how we could regress to old ideas about stress, after plate tectonics had shown the slip vector to be the important parameter.[22] One always takes seriously what Dan says, and he had valid points, which I will not discuss here, but the data showed a regionally coherent strain field (although we foolishly used the word *stress,* not *strain*).

In 1972, John Sclater left Scripps and went to MIT as a professor, in a step to strengthen MIT's program in what we now call geodynamics. As at Scripps, excellent work had been done at MIT in plate tectonics, but as much by graduate students, such as Don Forsyth, Norman Sleep, and Sean Solomon, as by the staff. John persuaded Frank Press that it needed another marine geophysicist, and the outstanding candidate was Tanya Atwater, my partner at the time. In effect, they had to hire me too. (Protestations that I had actually been hired on my own merit were far too emphatic to be believed.) Before going to MIT in January 1974, I wrote two proposals to the National Science Foundation (NSF), one to study large earthquakes in Asia and the other to refine plate reconstructions and develop a method for determining uncertainties. The need to estimate uncertainties was beaten into me unremittingly by my father, and then after a brief respite by Tanya, although she should not be held responsible for the fervor with which I have carried this flag. Anyhow, neither proposal addressed the next important problem, but I reckoned that both would keep me busy for a long time, and they did. Then, Tanya and I went to the Soviet Union for four months.

I had decided, not altogether incorrectly as others showed, that earthquake prediction offered important problems: not only did earthquake prediction seem possible, but precursory processes also could be understood.[23] The Russians were at the forefront, having demonstrated that the ratio of the two waves radiated by earthquakes and passing through the earth, P and S waves, varied over time, but no one had examined their data in any depth. My father would have been proud again, because this experiment failed, too. I could not find the change in the ratio of P and S wave speeds that the Soviet seismologists had reported, that made their work the center of attention in earthquake prediction, and that was subsequently found by others in the United States.[24] During periods mostly lost struggling with and waiting for the Soviet bureaucracy, I resurrected enough of my college Russian to read about the major earthquakes and geology of Soviet Asia and Mongolia and made a few lifelong friends who sympathized with my frustration. Shortly after arriving at MIT in January 1974, I abandoned almost completely earthquake prediction as a direction, with no subsequent regrets.

Satellite Imagery and Large-Scale Asian Tectonics

The intersection of timing and luck greeted me again shortly after starting at MIT. A special student from France, trained in structural geology but with the sound quantitative background typical of French students, had come to MIT to learn rock mechanics from Bill Brace. Already 26 years old, Paul Tapponnier planned to spend one year at MIT and return to France to begin his own laboratory work, based on what he learned from Brace. Anticipating an interest in earthquake prediction, I had asked that my office be on the same floor as Brace's. In the following weeks, the daily ebb and flow between courses and research brought Paul and me together, at first with no thoughts of collaborating. Paul's unusual intellect was soon obvious, and Brace and I thought it would be unfortunate if he left MIT after only one year. We discussed how to make it obvious to Paul that staying at MIT for a Ph.D. would be in his best interest, and Bill said unhesitatingly, "We should find him good research problems." He set Paul going on a careful study of brittle cracks in rocks, and I asked him if he would be interested in looking at the satellite imagery of Asia. The idea was less original than it might seem, because I had never heard of the Earth Resources Technology Satellite (ERTS) imagery, as the Landsat imagery was originally called, before Paul showed me some imagery from Iceland.

By September 1974, Paul had assembled mosaics of images and recognized the prevalence of large strike-slip faults in eastern Tibet. By late January 1975, we had a coherent interpretation of his analysis of the satellite imagery and my India-Eurasia plate reconstructions, fault-plane solutions of earthquakes, and reading of the Soviet geological literature.[25] (He then returned to France, having chosen not to pursue an MIT Ph.D., not live like a student for several more years, and not to pay for a large fraction of the parking tickets issued that year in Boston.) When we presented this work at MIT in January 1975, I could feel Frank Press' relief that the hiring of Tanya Atwater's partner might not have been a mistake after all. More important, I realized not only that Jack had been right about doing something when you couldn't think of anything to do, but also that satisfaction did not require dropping everything else.

Freed (briefly) of self-doubts, my definition of important problems evolved to become those that change the direction that science follows, and therefore the way scientists perceive their fields. A paper could be important if it merely addressed an important problem, without solving it. In 1973, Cambridge's (and Scripps' part-time) Teddy Bullard described to me how his study of gravity anomalies across the East African Rift demonstrated how geophysical methods could attack a geologic problem, like the dynamics of rifting.[26] He was not troubled that his conclusion, that rifting resulted from horizontal shortening of crust, was nonsense.

Paul and I showed that the deformation field over much of Asia bore a coherency that implied, in a loose sense at least, that all of this deformation was part of the same large-scale phenomenon, the collision and subsequent penetration of India into the rest of Eurasia.[27] A dominant feature of this deformation field was strike-slip faulting, in part right-lateral on northwest-southeast planes, but more significantly left-lateral on east-west planes (see figure on next page), from which we inferred that a substantial fraction of India's penetration was absorbed by extrusion of Eurasian crust eastward out of India's northward path. Moreover, active deformation seemed to occur by slip on so many faults, that it was better described as continuous deformation than by the relative movement of plates or blocks. Paul had been taking courses at MIT in plasticity and deformation processing from mechanical engineers and materials scientists and recognized immediately the analogy between faulting in Asia and slip-lines in a deforming plastic solid.[28] In short, although plate tectonics provided boundary conditions on deformation in Asia, the rules of plate tectonics provided little help in understanding, or even describing, continental deformation.

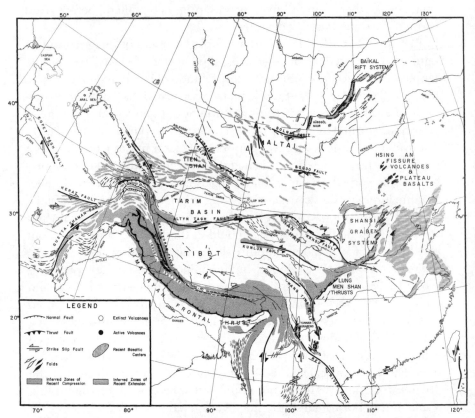

Map of eastern Asia showing large-scale structure inferred from mapped geology, the interpretation of Landsat imagery, published studies of surface faulting associated with great earthquakes, and fault-plane solutions of more moderate earthquakes. Note the prevalence of large strike-slip faults, right-lateral on faults trending NW to NNW-SSE and left-lateral on faults trending E-W. Tapponier, P. and P. Molnar, Active faulting and tectonics of China, *J. Geophys. Res.,* 82, 2905–2930, 1977.

Plate Tectonics and Continental Tectonics

Plate tectonics had an enormous effect on the earth sciences, even though only a small fraction of the field actually worked with the relevant data (magnetic anomalies and seismicity *sensu lato*). It is my own belief, however, that its main effect in the 1960s and 1970s was merely to persuade earth scientists that continental drift had occurred, and not to change how most of them carried out their science in an immediate and profound way. To paraphrase the speech-writer of a famous American president, too many geologists seemed to ask, "What can plate tectonics do for my patch of ground?" instead of "What can my discipline do to enlighten our understanding of tectonics?" Counter-examples of what were lacking do exist. For instance, geologic studies of ophiolites, slices of oceanic crust and upper mantle thrust onto continents, helped bring understanding of how the ocean floor forms.[29] Later, geologic studies of low-angle normal faulting, faults that are nearly horizontal, provided insights into how the earth's crust extends in regions like the Basin and Range Province of the western United States or the Aegean.

In the 1970s, however, a revisionist program developed to reinterpret the geologic history of regions in terms of leitmotifs ("plate-tectonics corollaries"). "Plate tectonics models" that assembled the appropriate leitmotifs for different regions were published frequently for many years. The exception that proves the rule that such studies were merely revisionist is Tanya Atwater's masterpiece that showed how plate tectonics could be used to place quantitative constraints on the history of western North America.[30] As an example of how to use plate tectonics to gain understanding of geological processes on land, her paper went unsurpassed until she revised it almost 20 years later.[32]

Insofar as plate tectonics can be seen as a revolution, its wide acceptance also brought what might be called a reign of terror. Those who quietly eschewed descriptions of the geologic history in terms of plate tectonics corollaries were treated as out of date; those more outspoken were publicly ridiculed. With belated embarrassment, I remember treating with some disdain an eminent geologist, Roye Rutland, who had worked in the Andes and later became the director of the Bureau of Mineral Resources in Australia. At the time, he seemed to know but not appreciate that the Andes formed at an ocean-continent convergent plate boundary. Only with years of hindsight did I realize that he was trying to go beyond such simple classifications to understand the processes that had built the Andes. We all often heard, "Plate tectonics is fine, but it does not work in my area," but only slowly did we realize that marine geologists, not continental geologists, were the only geologists who not only could, but should, see plate tectonics in their data.

Unlike the beheading of Robespierre in 1794 and the Thermidorean Reaction that claimed an end to the Reign of Terror in France, the swing back to treating continental tectonics as something different from plate tectonics was slow, at least from the point of view of a youth 32 years old. Moreover, for me it began with the fiercest of opposition. Paul and I had sent pre-prints out to many people, including Dan McKenzie at Cambridge University. Already a friend for several years, he seemed genuinely pleased that I had done something good, when he visited me in 1975. Yet, while walking from my apartment in Boston to MIT, as we reached the end of the Longfellow Bridge on the Cambridge side of the

Charles River, Dan said with an affirmative nod in reference to Paul's and my work, "But it is still plate tectonics." He had insisted that the active tectonics of the Mediterranean region could be described well with plate tectonics.[32] For at least the next three years, Dan argued with me that what Paul and I were saying about continuous deformation could not be right, because of a bunch of problems, some of which he himself later solved. The following year, 1976, he took advantage of an invitation to write a paper, "Can plate tectonics describe continental deformation?"[33] He seemed to argue Paul's and my case well, but concluded, "Yes, but not in detail." Imagine my surprise in 1980, when a visiting Chinese scientist said over lunch in Cambridge, United Kingdom, "How can we understand the plate tectonics of China?" and Dan replied, "You have to understand that Peter and I do not believe that plate tectonics works in continents." Although it was not long before we again had plenty to disagree about, it seemed like a chapter was finally closed.

But, no. Although no cause-and-effect applies, Dan and Paul rarely agreed, and Paul and I gradually diverged in our views, with each of us abandoning one of the two main inferences of our initial collaborative work. Paul no longer believes that Asia, or any large continental region, is best treated as a continuum; for him thick crust blurs plate tectonics in the uppermost mantle below.[34] Until 1989, I shared his belief that rapid eastward extrusion of China occurs, but no longer. My collaboration with Philip England of Oxford University beginning in 1988, overprinted on latent images articulated by Peter Cobbold and Philippe Davy of the Université de Rennes both orally and in a paper that I had reviewed and promptly forgot, made me realize that South China might move only slowly (less than 1/2 inch or 10 mm/yr) eastward with respect to the rest of Eurasia.[35] To some extent, the same questions that Paul and I addressed in the mid-1970s remain unresolved, and if our papers were "important," it cannot be because we, like Bullard and rifting, had *solved* an important problem.

CONTINENTAL TECTONICS SINCE THE MID-1970S

In the early 1970s Scripps' Bill Menard told me that one key to making progress in science is to choose a subject about which little had been written. Such a choice would obviate the scholarly ethic that requires one to read a pile of literature to get on top of the field. If nothing had been written, you could just write all the papers in the subject and automatically keep on top.[36] Studying continental tectonics was not so easy;

the literature was vast even before plate tectonics was recognized. Fortunately, by the mid-1970s, however, most of it either was irrelevant or had been assimilated into the boilerplate of common knowledge. What seemed weakly developed were a focus on dynamic processes responsible for the tectonics that we could observe and techniques for attacking these processes. Plate tectonics imparted the impetus for the former and offered tools for the latter. (Readers should be warned also that the literature on continental tectonics has grown enormously since the mid-1970s, and the review given here, from my perspective, suffers from both biases and my inability to keep up.)

The beauty of plate tectonics lies, in large part, in the simplicity with which the kinematics can be described – as rigid plates. The relative movements of vast areas can be specified using estimates of relative velocities at only a few points along the plate boundaries. Insofar as continental tectonics is best described by continuous deformation, that simplicity is lost. To describe completely the spatially varying deformation, or velocity field, within continents requires the specification of deformation or velocity at many more points, plus more complex rules than those of rigid-body motion. To appreciate the difficulty here, consider the difference between a boat moving through the ocean and the motion of the ocean itself. If we know how the front of the boat moves over the sea floor, we know quite well how the rest of the boat moves, as we do with one lithospheric plate moving with respect to another. If we know how the water in one part of the ocean moves with respect to the sea floor, however, we must exploit a relatively complicated theory to predict how adjacent water masses move; this difficulty also plagues descriptions of deforming continental regions. Dan McKenzie argued repeatedly in the late 1960s and early 1970s that plate tectonics was easy to accept because the kinematics could be treated separately from the dynamics. For oceanographers or meteorologists, separating the circulation of water or air from the dynamic processes governing that circulation is impossible, except in the simplest of situations or the broadest of generalizations. To a large extent, although many tectonic geologists might not realize it, the "Holy Grail" of continental tectonics has become understanding its governing dynamic processes.

Continental and oceanic lithosphere differ most obviously in the thickness of (low-density) crust atop each: only 4 to 5 miles (6 to 7 kilometers) beneath oceans, compared with 20 to 25 miles (35 to 40 kilometers) of continental crust. Although widely cited as the reason for continental tectonics being so different from plate tectonics, the buoyancy of continental crust may be overrated. Indeed, the negative buoyancy of oceanic lithosphere is more than adequate to carry its thin layer of crust into the asthenosphere once it has been subducted to a depth of approximately 60 miles (100 kilometers), but thick continental crust buoys continental lithosphere up. Even the coldest mantle lithosphere cannot drag with it 20 miles (35 kilometers) of crust into the asthenosphere, although if the upper crust could be scraped off the top of the lithosphere, the mantle part might carry lower crust into the asthenosphere.[37] Being inconsistent with plate tectonics dogma, subduction of continental crust had been dismissed as impossible by many, until Christian Chopin of Ecole Normale Supérieure in Paris found (by looking in his microscope, not with sophisticated instruments) the mineral coesite among what had seemed like garden-variety crustal rock in the Alps.[38] With coesite, a mineral that forms by very high-pressure metamorphism of quartz, Chopin showed that pieces of continental crust had been carried to a depth of approximately 60 miles (100 kilometers). The question of how continental crust could be subducted promptly inspired the next question, how does such crust return to the surface?

Brace–Goetze Strength Profile

In my opinion, the most significant study paving the way to understanding how continental lithosphere and oceanic lithosphere deform differently grew from laboratory measurements of how rocks and minerals deform.[39] At relatively low temperatures, slip occurs on discontinuities (faults) when friction is overcome, and therefore usually during earthquakes.[40] Such resistance increases as confining pressure (depth in the earth) increases.[41] When temperatures are high enough, however, crystals deform by the movement of dislocations within them, and strength (or viscosity) of the crystals, and hence the rock, decreases with increasing temperature. By the mid-1970s, measurements of the temperature dependence of strength (or more precisely, of the dependence of the rate of straining in the crystal on the stress applied to it) of olivine and quartz had shown that quartz deforms much more readily than olivine at the same temperature.[42] Working among scientists at MIT (such as Bill Brace, Jim Byerlee, and David Kohlstedt) who studied thumb-to-fingernail-sized specimens but also posed questions on a larger scale, Chris Goetze recognized that these different temperature dependencies of strength implied that a weak, quartz-rich lower crust underlies the stronger, brittle upper crust and overlies a strong, olivine-rich uppermost mantle.[43] This profile has often been likened to a "jelly-sandwich" (or jam-sandwich in Britain), with the attendant images of the slices of bread sliding with respect to one another and even with jelly being squeezed out between the two stronger bread layers. Where the crust is thin, as in oceanic regions, the jelly layer is absent, and the strong, cold olivine of the mantle makes plates strong enough to behave rigidly. Beneath continents, however, the crust is weak at the same depths where the mantle beneath oceans is strongest.

Goetze, who died of a brain tumor in 1978 at the age of only 38 years, had taught three classes of students at MIT this key to understanding why continental and oceanic lithosphere behaved so differently. Goetze's simple profile called attention to two aspects of the lithosphere that affect how we understand continental deformation at two very different scales. On a large scale, deformation within the mantle should occur as if it were a viscous fluid. On a relatively small scale, the upper crust might deform differently from, and even independently of, the upper mantle.

The rigid-body movement of plates of lithosphere that maintain their integrity as they move over a weak asthenosphere reminds many of us of sheets of ice floating on a lake. The Brace–Goetze strength profile for continental lithosphere, for which the strongest part resides in the mantle but where deformation can occur by plastic flow, suggests a quite different homely analogy: crumbs of bread overlying warm butter, which in turn overlies viscous honey. On a small scale, the crumbs at the top can move past one another as small particles, but on a larger scale their relative movement will be dictated by the flow of the honey below, if decoupled somewhat by the butter beneath them. (Perhaps more poignant, if politically less correct, is Ric Sibson's quote, copied from a stall in the lavatory at the Mayfair station in the London Underground: "The upper crust is just a bunch of crumbs sticking together.")[44] The widespread deformation of continents can be visualized as relative movements of weak bodies, crumbs of upper crust, carried by a stronger, deforming substratum.

The conversion of Goetze's simple image into understanding, replete with quantitative predictions and tests, required a breakthrough that, as usual, came with a simple approximation. What plate tectonics taught us was not to wallow in the complexity of the earth, but to reduce complex interactions to simple processes that could be understood, *before* they were all modeled together in detail. Describing the spatially varying velocity field within continents requires knowing not only the velocity at many more points, but also more complex rules than are needed to describe rigid-body motion. Moreover, continental crust, and presumably the entire continental lithosphere, can thicken or thin when compressed horizontally. Crust thicker than normal stores excess potential energy, and therefore crustal thickening, which requires work to be done against gravity, resists convergence between crustal blocks. Correspondingly, the understanding of continental deformation, in the sense of being able to predict that deformation from basic principles, becomes an integral part of the description of the kinematics.

The breakthrough in understanding of *large-scale* deformation of continents came with the recognition that a thin viscous sheet provides a simple model for continental lithosphere.[45] Only the vertical average of the sheet's strength, and by analogy the lithosphere's, need be considered. A thin viscous sheet deforms as an extremely viscous fluid (the lithosphere is perhaps a sextillion times more viscous than honey), with both viscosity and gravity resisting deformation. Numerical experiments quantify how different proportions of viscous resistance and gravity acting simultaneously affect the distribution and style of deformation on a scale that is large compared both to the thickness of the sheet and to dimensions of crustal blocks separated by faults.

The more important of the two (dimensionless) free parameters dictating the spatial distribution of deformation of such a sheet was officially dubbed the *Argand number* by England and McKenzie in deference to an early giant in large-scale tectonics,[46] Emile Argand, but many of us privately called it the "feta-brie" number. The Argand number scales the relative importance of gravity and strength (both brittle and ductile) in resisting deformation. A large Argand number, corresponding to crust behaving like ripe brie (or, for more Norman tastes, camembert cheese) signifies a weak lithosphere, for which gravity resists deformation most. Conversely, with a small Argand number, the crust behaves more like feta cheese and hence resists deformation with little crustal thickening or thinning and forms rigid blocks. Oceanic lithosphere is much more like feta than brie. Thickened continental crust, however, needs strong surroundings, analogous to the box that contains ripe camembert, with imposed stress to keep it from flowing apart. Such widespread stretching and thinning occurs today in Tibet and the Basin and Range Province of the western United States.

Whereas the thin viscous sheet, like ripe brie, provides a useful analog for deformation of large regions, such as eastern Asia or the western United States, most geologists must work at a smaller scale. As Cambridge's James Jackson and Oxford's Philip England often said, the area that a graduate student maps for a Ph.D. thesis is no larger than that affected by an earthquake with a magnitude of only 5.5 or so. (More such earthquakes occur each year than do students earn Ph.D. theses in field geology!) At such a scale, a weak lower crust allows the upper crust and the upper mantle to deform somewhat differently from one another, and in some regions almost independently.

A weak lower crust provides insight into a number of phenomena previously considered puzzling, if not worthy of much spilled bile. For instance, in regions of widespread horizontal extension, such as the Basin and Range Province of the western United States or the Aegean Sea, inactive normal faults mapped at the earth's surface dip at gentle angles of only a few degrees and separate deeper rock metamorphosed at high temperature from (in some cases) sedimentary rock that has undergone no metamorphism and is barely consolidated into rock.[47] James Jackson and Nicky White, however, showed that faulting associated with moderate-size earthquakes (magnitude between 5.5 and 7) in such settings occurs only on steeply dipping faults, raising the question of how slip occurred to produce the gently dipping inactive faults.[48] Flow in the lower crust, below the layer where earthquakes occur, provides the obvious solution. Many, if not most, of what are mapped as gently dipping, inactive faults at the surface may mark isolated surfaces that evolved from being steeply dipping, active faults to gently dipping, inactive ones. Although they currently separate metamorphic rock that underwent prolonged, extensive deformation in the lower crust, brittle deformation, only in a last gasp, juxtaposed that rock against the unmetamorphosed surface rocks. In many cases, subsequent flow or deformation in the lower crust of the region rotated the material on both sides of the fault, and the fault surface itself making the fault nearly horizontal. (I cite no one here for explicitly stating what the last three sentences say, because my impression is that nearly everyone who has written on this subject now shares this image, but interested readers might consult Block and Royden, Buck, Hamilton, Jackson and McKenzie, Kruse et al., Spencer, and Wernicke and Axen.)[49]

On a somewhat larger scale, a weak lower crust beneath eastern California and Nevada seems to separate very different styles of deformation. Crustal extension in the Death Valley region has thinned the upper crust and created a network of deep valleys, the surface of one of which lies about 250 feet (80 meters) below sea level. To the west, the Sierra Nevada defines a relatively high range (at least by Californian standards). The crust beneath the range, however, is not as thick as we might expect from isostasy.[50] Instead, as the University of Colorado's Craig Jones and colleagues have shown, as the crust beneath the Death Valley region has thinned, the corresponding thinning of the mantle lithosphere apparently occurred in the adjacent region beneath the Sierra Nevada, rather than directly below the area of crustal thinning.[51] The intrusion of hot, low-density asthenosphere into the space where colder, denser mantle lithosphere once lay buoys up the Sierra Nevada.

I have mentioned aspects of continental tectonics not only that have evolved since plate tectonics but also that either addressed processes or provided tools to attack them. Most of us in this field, however, spend our time searching for data to test ideas. As we are now many, our papers have made this field mature, at least as far as Bill Menard would use the word. I suspect that if able to restart again as an earth scientist, he would find the volume of literature for this subject too daunting to merit his attention. Conversely, we have learned a great deal in the past 25 years.

Most tests of ideas for how continents deform have taken one of two approaches: quantification of kinematics of deformation, and imaging of the present-day deep structure of mountain belts. I turn attention to them.

Kinematics of Deformation

Long ago Isaac Newton established a precedent for exploiting kinematic observations to constrain dynamic processes. The apocryphal story of his flash of brilliance when he noticed an apple falling might suggest as much. Less known, however, is his role in discovering Kepler's three laws of orbital motion of planets and satellites. Kepler himself had put forth several "laws," most of which we would today consider New Age fantasy; it took Newton to extract Kepler's brilliant insight from his otherwise muddled thinking.[52] Post-plate tectonics continental tectonics has also relied heavily on advances in measuring the kinematics of continental deformation.

At a small conference in 1976, I tried to call attention to an elegant cross-section of the Canadian Rockies, some 100 miles (160 kilometers) long and richly detailed, by complimenting it as a "beautiful sketch." Bert Bally, then at Shell Oil, now at Rice University, and still a friend, boomed out in an annoyed tone of voice that what I had disrespected as a "sketch" was in fact "high-quality data." Bally, indeed, had had access to excellent seismic reflection profiling across the Rockies, but what his work really showed was how to make a cross-section that was accurate.[53] Although such methods were used in the oil industry, subsequent developments, particularly by John Suppe at Princeton, have transformed the process from one in which each geologist could sketch a different cross-section from the same data to one in which nearly all geologists draw the same one given the same geological data.[54] With balanced cross-sections, total amounts of shortening across (at least some) mountain ranges can be measured.

Concurrently, methods for quantifying rates of active deformation have grown steadily. In the mid-1960s, the rate of slip on the San Andreas

Fault was, at best, poorly known. Today, slip rates for virtually all major faults have been measured by one means or another (although not always with agreement when more than one method has been used). One boon to this enterprise has been the growth of Quaternary geology, once an esoteric field of study, but now a mainstream discipline. Most geologists at one time or another took a course in field geology, where they mapped an area 5 to 10 square miles or more in dimension. Today, Kerry Sieh, one of the leaders in Quaternary faulting, has no qualms about taking his Caltech students into the field to map a region the size of a basketball court, where the uninitiated, as well as the cynical, would describe most of the rock that his students examine as just layers of dirt. The combination of applying classical geological techniques to unconsolidated sediment in small areas with the development of techniques for dating that sediment has enabled geologists to determine rates of slip on faults, rates of recurrence of major earthquakes, and growth of folds. In one of his early studies, Kerry's only fossil was a bottle, although he relied heavily on radiocarbon dates.[55]

In the 19th century much of what we call geophysics was geodesy, or in more mundane terms, surveying. The geodesist's task was to measure the shape of the earth, with the underlying hope that the shape would not change, for that would require redoing the work. Geodesy did not die in the early 20th century, when many earth scientists were unwilling to believe that the earth changed shape and when techniques changed little. The 21st century, however, is witnessing a renaissance in geodesy because of the development of new techniques in the past 30 years: laser ranging to measure distances between benchmarks, Very Long Baseline Interferometry (VLBI) and Global Positioning System (GPS) to measure longer distances, including intercontinental distances, and radar interferometry to map strain at the surface. Measurements with these techniques have determined the slip rates on faults like the San Andreas, demonstrated agreement of rates of plate motion averaged over the past 2 million years with those estimated geodetically for only ten years; and mapped the complete deformation field associated with earthquakes, not just the slip along the surface rupture. It is easy to imagine that in a few years, following the weather forecast, a geodesist will appear on television in front of a map of California in order to summarize the week's strain for the interested public: "We have a had a bit of strain accumulation along the San Andreas Fault east of Los Angeles, which raises the probability of an earthquake of magnitude 6.5 in this region to n percent during the six months, but that patch of accumulation we described last week just north of Los Angeles has relaxed, and the risk of an earthquake there has diminished to *m* percent."

Except for radar interferometry, both geological and geodetic measurements of deformation can be made only at points, and interpolating between such localities need not be straightforward, but in a paper overlooked for many years by nearly everyone except a couple of other New Zealanders, John Haines, at what is now called the Institute for Geological and Nuclear Sciences, solved this problem.[56] When he and Bill Holt, of the State University of New York in Stony Brook, generalized the solution to include measurements of different kinds, they put the determination of the complete strain-rate field of a large area, such as eastern Asia, within reach.[57] The essential constraint, obvious for more than a century (at least, when applied in other contexts), is that strain within adjoining regions must be compatible in the sense that if one sums the deformation along any line connecting two points, one obtains the same relative displacement of the endpoints ("Saint-Venant's compatibility," to the aficionados). For example, if measurable slip occurs on a fault and that fault dies in a region of folding, straining within the folds must accommodate the slip. Even if we cannot measure the rates of deformation everywhere, by knowing the rates in many areas, we can infer it in others.[58]

The ultimate goal of determining the strain-rate field of large regions is to understand the dynamic processes. In my opinion, this goal is within reach, but still not yet attained.[59] Ironically, one of geophysics' venerable techniques, paleomagnetism, has provided a kinematic clue to the underlying dynamics of continental deformation. The uncertainties of several degrees in paleomagnetic measurements, if small enough to allow definitive tests of continental drift, are too large to provide useful bounds on intracontinental deformation across most mountain belts. Paleomagnetism, however, can constrain rotations about vertical axes, and such rotations of material with deforming continental regions can be very large, in some cases greater than 90 degrees.[60] Moreover, the amount of rotation depends on how it is imparted. Consider two situations: (1) a block rubs against its neighbors, which slide past one another so that the block behaves like a ball bearing between the adjacent blocks, and (2) a block floats in a fluid layer undergoing shear (like a twig caught in an eddy in a stream). The rate of rotation of the first is twice that of the second, if all other aspects of the kinematics are the same.[61] Although data may not yet be definitive, they seem to favor the latter mechanism for imparting rotations, as if crustal blocks are carried by continuously deforming substratum.[62]

Imaging Subsurface Structure

Most of us think that the engine that drives continental tectonics lies in the mantle. Inferences about the mechanics of that engine can be made and hypotheses tested using observations of the surface, such as relative movement among plates or the kinematics of deformation in continents. Yet, it is a rare auto mechanic who would not look under the hood of a car to learn why its engine behaved the way it did, and most earth scientists feel the same about the earth's geodynamic engine.

The engine runs by generating lateral density differences so that gravity can move them around with respect to one another, carrying other, arguably more interesting, bits of crust and mantle along. Geophysics has a long tradition of measuring the strength of gravitational acceleration – gravity for short – and then inferring a density structure from the measurements. Because an infinite number of plausible, but very different, density structures can fit the same data exactly, such an approach is doomed to failure. This fact has in no way made measurements of gravity useless, but rather it forces researchers to pose more interesting questions than "What is the structure?" – a question perhaps of interest to geophysical stamp collectors, but only rarely asked, at least without qualification, at the forefront of science.

Were the earth an inviscid fluid, lateral variations in density could not exist, for heavy blobs would sink straight down until they reached a level where the density was the same. Some mechanical process must support heterogeneities. The strength of material can support density heterogeneities, as the walls of a building support its roof. One example is the flexure of the lithosphere due to loads, like mountain ranges, placed atop it. A long tradition of using gravity to constrain properties of an effectively elastic lithosphere preceded the recognition of plate tectonics by decades, but has been developed further as data became more complete. In addition, the flow of a viscous substance requires stress to maintain flow, and that same stress field will support lateral variations in density. The earth's gravity field provides a major constraint on convection in the mantle, particularly in oceanic regions where lateral heterogeneity of the crust is small. Beneath continents, however, the gravity field is blurred by heterogeneity resulting from billions of years of evolution and supported by strength of the lithosphere. To my knowledge, gravity has placed no tight constraint on the dynamic processes of mountain-building since confirming that to a good approximation the crust is in isostatic equilibrium. Depending upon one's point of view, that demonstration occurred in the first third of the 20th century, or was already clear in the mid 19th century when the archdeacon of Calcutta, John Henry Pratt, showed with 45 pages of tedious calculation that mass must be missing beneath Tibet, and George Airy, the Astronomer Royal in Greenwich, England, infuriated Pratt by showing, in four pages of simple argument, that such a state was required by the low strength of rock.[63]

Most techniques for imaging the earth's interior are seismological, but using these images to infer variations in density is risky because seismic wave speeds and density are not uniquely related. For such images to be useful, again the right question must be asked.

In the early 1970s, Jack Oliver and colleagues launched a major program to study the earth's middle and lower crust using seismic reflection techniques, which had been developed in the oil industry to study layered, but sometimes deformed, sedimentary rock. The success of this program, which resolved controversies about dips and depth of faults in the Rocky Mountains and Appalachians,[64] stimulated similar programs abroad. One particularly successful program was the British Institutions Reflection Profiling Syndicate (BIRPS), which took advantage of the ocean surrounding Britain to shoot and record from a homogeneous surface of constant elevation. BIRPS's data are renowned for being especially clear.

Data used to address questions at the forefront of science, however, are only rarely clear (if the data were clear, the question would be answered and would instantly move back from the forefront.) Geoffrey King's 12-year-old daughter Sophia highlighted this difficulty, when she visited her father's office at the Bullard Labs of Cambridge University, where a large display of BIRPS data had been hung on the walls in the seminar room. Transparent mylar sheets were hung in front of the seismic data, lines drawn with colored magic-markers to show the inferred horizons between layers and the inferred faults offsetting them. After a tour around the room, Sophia returned to her father to announce, "Daddy, it's lucky they've drawn the faults. Otherwise you'd never be able to see where they are."

It is my opinion that the most important result that data of this kind have shown us is not the presence of some feature, but rather the absence of one. Profiles that obtain reflected waves from the Moho, the boundary between crust and mantle beneath continents, show, with only one exception to my knowledge, that the Moho is not cut by faults. This continuity of the Moho without faults cutting it may be seen as a test of the hypothesis that the mantle deforms by continuous deformation, not by faulting as occurs at plate boundaries in oceanic regions.

The most popular technique for imaging the upper mantle is seismic tomography, which was pioneered by the University of Southern California's Keiiti Aki, while at MIT, Anders Christoffersson of the University of Uppsala, Sweden, and Eystein Husebye of the Norwegian Seismic Array, but named later for the procedure used for CAT scans of the brain in the medical profession.[65] With elegant color plots, this technique has evolved to assume imperialistic tendencies in seismology and attracted many of its best and brightest young people. The lunar program did the same in the 1960s, by attracting many of the luminaries of the earth sciences at the time to study a tectonically dead moon, allowing others to discover plate tectonics on the tectonically active earth. In my opinion, seismic tomography has hired seismic reflection profiling's tailor, and now dons similar new clothes. Of course, by no means is all tomography devoid of content, but to my knowledge, it has not solved an important problem since Rensselaer Polytechnic Institute's Steve Roecker, when a graduate student at MIT, showed that low-speed material (presumably crust) underlies (and presumably was subducted beneath) the Hindu Kush (mountains) in Afghanistan to a depth of at least 60 miles (100 kilometers).[66] (He did this before Chopin found coesite in the Alps and confirmed subduction to such a depth.)[67]

In my opinion, seismology's most promising techniques for studying dynamic processes occurring within the earth will exploit the anisotropic properties of deformed material in the mantle, as revealed by the difference in speeds of seismic waves propagating in different directions. P and S waves passing through olivine crystals differ by more than 10 percent, depending upon the orientation of the olivine crystal through which the waves propagate. Because olivine also deforms anisotropically, with one set of crystallographic axes being much weaker than the others, the deformation of olivine crystals in mantle rock causes them to align with one another in such a way that measurements of seismic anisotropy can be used to infer finite strain within the mantle.

The existence of seismic anisotropy has been known for a long time. Harry Hess, who described sea floor spreading before most were willing to consider it, was one of the first to demonstrate its existence in the earth.[68] From my perspective, however, anisotropy lingered somewhere between being a nuisance and an interesting, but useless, curiosity until the late 1980s, when papers by Paul Silver of the Carnegie Institute of Washington and Winston Chan of Teledyne Geotech, and by Lev Vinnik of the Institute of Physics of the Earth in Moscow and his French colleagues, convinced me both that it could be measured easily and that it could address important scientific questions.[69] (Many others were convinced of both aspects long before I was.) In terms of understanding continental tectonics, the large magnitude of anisotropy with consistent orientations of the faster of the two quasi-S waves nearly parallel to strikes of major strike-slip faults, recorded at stations as far as 125 miles (200 kilometers) from the faults, suggests (to many, but not to everyone) that the litho-sphere deforms over a broad zone instead of being cut by a fault through its entire thickness.[70] Again, the idea that the lithosphere deforms as a continuous medium passes a test.

Summary

The Chinese commonly look back on the Tang dynasty (seventh to tenth centuries) as the pinnacle of Chinese civilization. They also divide it into four periods, Early, High, Middle, and Late, with an obvious rapid rise and slow decline. I see continental tectonics developing in a similar fashion; following a barbarian period in which students had to study geosynclines and similar woolly thinking, the Early period began in the 1970s. Many of the major unanswered questions were posed. Techniques previously not used to address geologic questions could be borrowed and applied, without (yet) the need for much development. Our field still suffered from the division into subdisciplines, like geophysics, geochemistry, structural geology, stratigraphy, and so on, whose members rarely collaborated or showed interest in the questions of other subdisciplines. To be trained as a geophysicist but to work on geologic problems was an opportunity exploited by only a few and encouraged only rarely. A switch toward a more quantitative study of problems was in progress, but the word *geosyncline* was still often heard and presumably taught.

The High point in the study of continental tectonics came in the late 1970s and early 1980s. The thin, viscous sheet entered geodynamics of continental regions. John Suppe showed how to balance cross-sections with simple rules.[71] New seismological techniques for studying earthquakes, developed a few years before, were put to use.[72] Most good research universities developed programs in Quaternary faulting and earthquake geology. Sedimentary basins became a topic of quantitative analysis. For me, there was a stimulating interaction with geologists, increasingly with Clark Burchfiel and a group of outstanding students working with him at MIT. More important, the training of students in the earth sciences began to focus on studying processes occurring in the earth, not only such that basic mathematics, physics, and chemistry became an integral part of the curriculum, but also with the result that geophysicists and geochemists learned the basics of geology. Many departments renamed themselves Departments of Earth Sciences to eliminate the distinctions between subdisciplines. When I was a student in the late 1960s, the prevailing view was that it was easier to teach physicists geology than geologists physics and mathematics. In the summer of 1996, Dan McKenzie told me that for him times had changed; it had become easier to teach geologists math and physics than the opposite.

Since the early 1990s, what might be called the Middle period of continental tectonics seems to have begun. New techniques and new approaches have been developed steadily, and elegant work has been done, but it seems to me that problems solved have been less significant than those in the earlier periods. More careful analysis with more sophisticated methods must be brought to bear on problems in order to nibble away smaller pieces of them. Some may see this as cynicism, but it seems to me that many of the same old controversies dominate research programs. For instance, to what extent do the rules of plate tectonics apply to continental deformation? Few of us have changed our minds about such issues over the past ten years. Many of us just keep designing studies to prove that what we said ten or 20 years ago is right; few studies cause us to change our minds about fundamental processes. What I have written here illustrates this; I continue to defend what I thought ten years ago. Reader, beware! The examples that I have selected illustrate what I think, not what the many who disagree with me think. Perhaps, continental tectonics will never enter a Late stage analogous to the Late Tang period; some of these controversies will be resolved, but when remains to be seen.

LESSONS FROM PLATE TECTONICS AND ITS AFTERMATH

The recognition of plate tectonics and the subsequent development of continental tectonics illustrate some patterns in the development of science that seem worth noting.

The Importance of a Fresh Point of View

All three of my advisors in graduate school, Bryan Isacks, Jack Oliver, and Lynn Sykes, changed the direction of their research to make their contributions to plate tectonics. When they began their study of deep-focus earthquakes in the Fiji–Tonga region in 1964, Isacks had just finished a thesis in high-frequency instrumentation, Oliver was an established expert in surface-wave seismology, and Sykes had recently written a thesis on short-period surface waves. Pursuing the same problems for too long can cap an open mind.

Many of the key scientists of plate tectonics, not just my advisors, changed directions again shortly after plate tectonics was recognized. Sykes remains a full-time seismologist, but his research has focused on earthquake prediction and the discrimination of underground nuclear explosions from earthquakes. Oliver launched the Consortium for Continental Reflection Profiling (COCORP) to use techniques developed by the oil industry to probe the lower crust. Isacks has become a geomorphologist. Similarly, Dan McKenzie has changed several times, and for 15 years his main focus has been igneous petrology. Much of Jason Morgan's research shifted to mantle plumes. Walter Pitman dropped magnetic anomalies in the early 1970s to pursue sea level changes and, more recently, the geological evidence for a great flood responsible for the legend of Noah's flood. John Sclater virtually abandoned heat flow and turned his attention to the development of continental margins.

Communication at the Forefront

I noted five events that set me forward in the direction of plate tectonics: a seminar course on deep-focus earthquakes, two Monday Night Seismology Seminars at Lamont by Jim Heirtzler and by Lynn Sykes, an American Geophysical Union meeting in 1967, and the arrival of Jason Morgan's pre-print. The AGU meeting was the only truly public affair, and what makes it notable in my experience is not what I learned, but what I missed (Morgan's presentation of plate tectonics). It seems to me that the least blocked channels for communication at the forefront of science ignore the public platform. By the time important developments reach the major scientific meetings, their offspring are well beyond the womb. By the time the funding agencies can respond to them, such developments are somewhere between maturity and senility. Funding agencies should focus less on supporting topics perceived as exciting and more on finding ways to allow individual scientists to create new, exciting topics.

The important developments in the early 1960s involved a small number of people who communicated directly with each other. While a student at Cambridge, Fred Vine benefited immensely from a seminar by Harry Hess and subsequent interactions.[73] Hess' now widely cited paper, however, had no impact; by the time people read it, they already knew its essence.[74] Vine seemed to give direction to Bullard, Matthews, and Wilson, whose open minds were receptive, but Vine and Matthews' paper seemed to have little immediate effect, except perhaps on Tuzo Wilson.[75] By the time Walter Pitman and Jim Heirtzler had recognized the significance of that paper, Vine had already completed his synthesis, in which he not only tied together magnetic anomalies from virtually all oceans, but also corrected the omission of the Jaramillo event from his earlier work.[76] Vine learned of the Jaramillo event from Brent Dalrymple before it was published, an event whose significance historian Bill Glen recounted clearly and insightfully.[77] Moreover, Vine moved to Princeton in 1965, and though he claims no credit, surely he influenced Jason Morgan, his office-mate at that time.[78]

The Earth Sciences as a Modern, Quantitative Physical Science

The beauty of plate tectonics radiated from the ease with which it could be tested quantitatively. The simple description of rigid-body motion on a sphere allowed plate tectonics to exploit magnetic anomalies, orientations of fracture zones, and fault plane solutions of earthquakes in some areas to make predictions of those in others. At first, such data confirmed predictions, and therefore plate tectonics passed these tests. Then with refinements, systematic errors in the data, due for instance to inter-arc spreading or simply to deformation with island arcs, did not refute plate tectonics, but allowed further understanding of processes within the Earth, such as the partitioning of slip into thrust and strike slip at island arcs.[79] Although the testing of hypotheses occurred in the earth sciences before plate tectonics, the development of quantitative analysis grew rapidly afterward.[80]

Continental drift, in its strictest sense, seems to have had little impact on the earth sciences before plate tectonics. "Most geologists could proceed with their research interests without much concern over whether drift theory was right or wrong."[81] My impression is that the interpretation of most geological observations would have been unaffected by confirmation of continental drift.[82] Oreskes has argued quite persuasively that geologists in North America rejected continental drift in part because geophysicists there said it was impossible.[83] Her evidence for the opinions of geophysicists is quite convincing, but it seems to me that her explanation as a whole implicitly requires believing something I do not: that geologists lacked either insight or the ability to think critically. No good scientist accepts uncritically an argument that he or she does not understand, but that is critical to his or her research. Thus, if geologists rejected continental drift because others told them it was nonsense, then either it was not important to their research or they were not good scientists.

Unlike continental drift, plate tectonics, largely through its quantitative implications, affected most subdisciplines of the earth sciences. It seems to me that whereas continental drift offered few solutions to questions asked by sedimentologists and stratigraphers, they now can understand many of their observations in terms of subsidence induced by a cooling lithospheric plate or by flexure of an effectively elastic plate. This analysis can be carried further to estimate maturation of organic material and the potential for petroleum production, because of the simple physical understanding provided by, among other processes, a cooling lithosphere. "Basin analysis" was legitimately born with traditional geology and geophysics already married.

Similarly, continental drift offered little insight into the processes by which igneous rock forms, especially since the vast majority of igneous rock forms at mid-ocean ridges; most of the rest forms at subduction zones, another region largely ignored in continental drift. Subsequent to the discovery of plate tectonics, petrologists recognized that not only the thickness of oceanic crust, but also the composition of magmatic rock could be understood. Essential to plate tectonics, sea floor spreading calls for hot rock in the asthenosphere to rise beneath mid-ocean ridges, but to cool only slightly as it decompresses, for the same reason that air becomes cooler at higher elevation. As pressure decreases, rock can melt at lower temperatures, just as water at high altitude boils at a lower temperature than at sea level. Thus, as hot rock rises beneath a mid-ocean ridge, although it cools slightly, it melts when it reaches sufficiently low pressure without an additional source of heat. Continental drift provided no clue that such processes occur beneath mid-ocean ridges and create most of the igneous rock on the planet.

Measurements of the earth's gravity field provided one source of data used to argue against continental drift.[84] At the time, such analyses treated the earth as static. As no dynamics were considered in most treatments of continental drift, except perhaps those attempting to refute the idea, the scope of problems addressed with gravity anomalies remained limited. With plate tectonics, however, the analysis of the earth's gravity field ceased to be an exercise in choosing one among an infinity of non-unique and very different structures, and became formalized into procedures for constraining processes, such as convective flow within the mantle, which affect density within the earth.[85]

Even profound insights into mountain-building lay fallow. Emile Argand in Neuchatel, Switzerland, recognized that much of Asian deformation could be ascribed to India's penetration into the rest of Eurasia, and on a smaller scale he recognized large folds in crystalline basement, but his analysis was qualitative and his ideas not easily tested.[86] As profound as it was, this work went largely ignored by North Americans for decades; I learned of it only after I had rediscovered in my own data much of what he had described. With balanced cross-sections, however, structural geologists began to quantify amounts of deformation in mountain belts. Similarly, Quaternary geologists appropriated techniques traditionally used to look at vast, ancient sedimentary deposits and applied them to thin layers of historical sediment to place constraints on rates of deformation and even earthquake recurrence. Geosynclines, sketches drawn without even a scale and based on qualitative description, gave way to cross-sections drawn without vertical exaggeration.

The common theme of these post-plate tectonic studies has been quantitative analysis with the goal of understanding. Here "understanding" implies the ability to predict from basic principles, and "quantitative" includes the concept of uncertainty. (Much numerical modeling and seismic tomography remains qualitative.) Although few of the techniques or measurements were invented after plate tectonics was recognized, and many earth scientists had taken a quantitative approach to the earth, plate tectonics accelerated the transformation of the earth sciences from a focus on descriptive classification of phenomena to understanding processes quantitatively.[87]

In an effort to inspire earth scientists, Tuzo Wilson, in what now seems almost clairvoyant, predicted that the days of geology as primarily a historical science were numbered and a new era was dawning.[88] Earth scientists would study processes by trying to develop physical laws that make testable predictions. How Wilson could see this when he did baffles me, for much time elapsed between the recognition of plate tectonics and the permeation of quantitative hypothesizing and testing into the various subdisciplines of the earth sciences.

Although computers have opened new dimensions for experimental science, all too often, numerical modeling seems more to be numerical masturbation, the ejaculation of color plots summarizing simulations of complicated, "realistic" models, perhaps best likened to geophysical Barbie-dolls. My father taught me that "the most beautiful sight to an experimental scientist is a straight line of data points," for data that fit a straight line virtually assure a simple understanding. Although the computer has created a new laboratory for experimentation in all fields, only rarely does numerical modeling follow the tradition of G. I. Taylor, the eminent fluid dynamicist at Cambridge University, whose goal was scaling laws: straight lines of measured data points plotted versus quantities controlled in experiments and scaled to reveal simple relationships.

Sometimes, I think that Wilson will turn out to be wrong, but for the wrong reason.

Plate Tectonics as Revolution, or as the Trigger for Rapid Evolution?

The history of continental drift, as an idea, makes good copy for historians with a journalistic bent, and for a public both sensitive to "revolutions" and sympathetic to underdogs. Le Grand captured this perspective: "The folk-tale of Drift is the stuff of myth and legend in which Cinderella, after years of abuse from her vain step-sisters, is visited by her Fairy Geophysicist, is touched by the Magnetic Wand, goes to the Ball and marries the Prince."[89] No drama is lost by a Greenland winter martyring drift's long-belittled champion, Alfred Wegener.

More important, many who espouse the view that a revolution in the earth sciences occurred concurrently with the recognition of plate tectonics argue that the revolutionary change was the acceptance of continental drift; few seem to see plate tectonics as little more than a version of continental drift.[90] Glen wrote of the "emergent, more complete theory, renamed plate tectonics."[91] In what I consider to be the most insightful historical analysis of changes in the earth sciences brought on by plate tectonics, Le Grand repeatedly refers to the "plate tectonics version of Drift," as well as the earlier "seafloor-spreading version of Drift" and other versions.[92] Wilson wrote that "[t]he acceptance of continental drift has transformed the earth sciences . . . into a unified science," although he had recognized that more than just continental drift was at stake.[93] Menard obviously saw plate tectonics as more than continental drift, but the subject of his book ends in 1968 and looks backward more than forward in his assessment of the impact.[94] Too many observers saw the "revolution" as merely a demonstration that continents had drifted. It seems to me, as I have tried to argue here, that plate tectonics brought a profound change in the way the majority of earth scientists viewed or approached the topic of their study. (To be fair to Homer Le Grand, he recognized that plate tectonics marked a change in how the earth sciences were carried out.)

According to historian of science Thomas Kuhn, scientific revolutions share features that characterize political revolutions.[95] Although no two political revolutions are alike, one can argue that their differences are small compared with their similarities.[96] Among the parallels between continental drift and the French Revolution, one might call the paleomagnetic work of the 1950s an emerging period of enlightenment that prepared the minds of earth scientists, without motivating most of them to pursue continental rift. The "Bastille" fell in the early 1960s, largely with developments in marine geophysics, but it was Vine's verification of the Vine–Matthews hypothesis and Sykes' demonstration of Wilson's transform faulting, by requiring both continental drift and Hess' sea floor spreading, that summarily beheaded fixist ideas (in which positions of continents were assumed forever fixed) of the "establishment."[97] Historical geology, the subject pursued by most earth scientists, was rewritten, as papers presenting "plate tectonics models" of the geologic history of various patches attracted hundreds of reprint requests. A "reign of terror" followed, for those who did not accept plate tectonics were ridiculed as old-fashioned and out-of-date. What was found wanting in such "models," however, rarely included an appreciation for the features of plate tectonics that distinguished it from continental drift. Although no Robespierre lost his head in a Thermidorean Reaction that restored moderation to the earth sciences, it became clear that those "old-fashioned" geologists neither could, nor should, see continental drift, let alone plate tectonics, in their data. Similarly, no geological Napoleon proclaimed an end to revolution and then united earth scientists by giving them an alternative simile to nationalism. Nevertheless, the rise of structure/tectonics as a dominant discipline in geology might be seen as counter-revolutionary. Despite numerous exceptions (two of which are mentioned above), structure/tectonics has focused as much on historical geology, albeit couched in a new jargon, as on discovering new principles or bringing understanding to processes not easily studied with older techniques. Only in the 1990s did the structure/tectonics section of the National Science Foundation begin funding active tectonics, the branch most concerned with understanding the principles. Finally, and gradually, a new approach to the study of the earth is emerging.

Naomi Oreskes has expressed a different view of both hypothesis testing and quantification in the earth sciences.[98] She sees both as in place when plate tectonics was recognized, and she considers the recognition of plate tectonics more as the result of hypothesis testing and quantification than as a stimulus for their growth and development. Perhaps she is right, although if so a revolutionary simile for those earth scientists motivated by hypothesis testing and the need to quantify might be the Bolsheviks in Russia in 1917. As a clear *minority* among Russian revolutionary groups, the Bolsheviks named themselves using a word that meant *majority* and then took over the country. Naomi and I do agree that the earth sciences have become more quantitative since the 1950s, and that quantitative approaches are more common now than before.

These parallels with the French Revolution pose the question: who benefited from the "plate tectonics revolution"? "For most ordinary French subjects, . . . [the French Revolution] had made their lives infinitely more precarious."[99] Beneficiaries included the owners of land, the bourgeoisie, and the soldiers, but not those whose economic plight required the greatest change.[100] "Fat cats got fatter. . . . By contrast, the rural poor gained very little from the Revolution."[101] If the demonstration of continental drift was a revolution, then, similarly, traditional field geologists gained little, for continental drift offered little insight into the solutions to their problems.

One might ask, why do earth scientists tout plate tectonics as a revolution that unified their science? First, plate tectonics is beautifully simple, and most scientists treat simplicity as a prerequisite of scientific profundity. Although no scientific, or epistemological, law guarantees that simple ideas approach truth more closely than complicated ones, simplicity is an expedient, for more people will understand and therefore take interest in a simple idea than one comprehensible only to few. Second, it seems to me that accepting plate tectonics may have been a low hurdle for most scientists. No catharsis was required, or was experienced, by many earth scientists, because the immediate impact on their work was slight. For instance, "in 1978, leaders of the Ministry of Geology of the USSR instructed all field geologists to interpret the results of their 1:25,000 scale survey [i.e., their geologic maps] exclusively in terms of plate tectonics. Almost none of the field geologists was able to do this," not because none understood plate tectonics, but because the sizes of regions mapped could not reveal plate tectonics.[102] At the same time, however, lateral mobility of the crust implied by continental drift and plate tectonics gave mountain belts and other large-scale features a framework into which more detailed studies could be fitted, which surely united earth scientists working at all scales. As the local fans take pride in a winning home team, so perhaps many earth scientists saw plate tectonics as a victory for their team.

If plate tectonics was "revolutionary," how was it so? The significance of "revolution" in history is obscured in part by historians' traditional predilection for considering wars and battles, slaughters and assassinations, and political carnivals to be the grist for their mills, instead of spiritual, intellectual, and technological developments that really have altered the general human condition. Likewise, in part, the "plate tectonics revolution" is obscured by the failure of many historians of science to look beyond the Cinderella story of continental drift. Winston Churchill concluded a book on *The Age of Revolution* by noting that "the principles which had inspired [France in her revolution] lived on . . . to play a notable part in changing the shape of government in every European country."[103] Similarly, as Wilson so precociously recognized, plate tectonics spurred (or validated) a quantitative approach to the earth sciences.[104] If, in general, "history sooner or later takes back her gifts," she seems in this case to have passed them on to a new generation of geological beneficiaries.[105] Plate tectonics was a revolution less because it guillotined existing fixist ideas, and more because it affected the way earth scientists approach the study of the earth. Its impact has been both more gradual and more subtle than most active scientists realized at the time. Perhaps, it is too soon even now to see the impact, Schama wrote: "Asked what he thought was the significance of the French Revolution, the Chinese Premier Zhou En-lai is reported to have answered, 'It is too soon to tell.'"[106]

Looking back on the past 30 years, some of the techniques that have led to quantitative understanding of processes in the earth sciences were in place in the 1960s, but many have been developed subsequently and at a steady rate during those 30 years. For instance, with Stanford's Norman Sleep's analysis of the subsidence of the Atlantic margin (carried out while a graduate student at MIT), much of the formalism was in place to study thermally induced subsidence of sedimentary basins, but nearly 10 years elapsed before this approach became widely exploited.[107] The recognition that the thickness of oceanic crust resulted from a very simple phenomenon apparently was not published until 1988.[108] Although Bally constructed balanced cross-sections before plate tectonics was proposed, their heyday waited another 15 to 20 years.[109] Quantitative methods in metamorphic petrology developed largely in the 1970s and 1980s. In my field of continental tectonics, the most controversial topic 25 years ago was framed by, "Can plate tectonics describe continental tectonics?" and remains unresolved in the minds of many. My impression is that seismic anisotropy, a topic that has blossomed in the past ten years, but is hardly new, will resolve this question.

Revolution was a popular concept in the late 1960s. Presidents John Kennedy and Lyndon Johnson led us Americans into a war we did not want to fight. Anti-Soviet propaganda was too vehement to be believed, at least by idealists. Who could not be revolted by the recurring injustices to African-Americans and other minorities? Although hindsight now clearly exposes previously latent images of the fundamental roles played by established giants in the earth sciences before the 1960s, many of the founding fathers, and the mother, of plate tectonics were young, less than 35 years old, when they wrote their widely cited papers. To many of the youth in the 1960s, revolution seemed like a plausible solution to the world's ills. When Kuhn wrote a book about scientific revolutions, and others deemed plate tectonics an example, it felt good to be part of one, especially a nonviolent revolution.[110]

Harvard's physicist/historian of science Gerald Holton has shown how non-scientific beliefs strongly influence the approach scientists have taken in different eras, and revolution was a theme that pervaded thought in the 1960s.[111] Yet, after the recognition of plate tectonics, most earth scientists returned to what they knew well. Younger scientists, for the most part better trained in basic sciences and mathematics than their mentors, began to push their own subdisciplines forward with rigor inspired by plate tectonics. Meanwhile, however, the mood of the country shifted to the almighty dollar, Americans elected Ronald Reagan with enthusiasm, and political revolutions were associated with their villains, not heroes. No wonder contemporary discussions of "the revolution" employ the past tense. Nevertheless, young scientists should realize that not only do the ideals of the French Revolution remain unattained, but so do those of the plate tectonics revolution.

The Future

One often hears nostalgic discussions of what an exciting time the plate tectonics era was. Today's scientific problems (climate change, landscape evolution, mantle dynamics, etc.) are no less exciting than those 25 to 30 years ago were. What made that time exciting and the present less so is not the nature of the scientific questions, but the number and nature of obstacles that now lie in front of young scientists compared with the unfettered days of plate tectonics. Obviously, difficulties of obtaining funding engender recurring discouragement, compared with what seemed like unlimited freedom in the 1960s. We students at Lamont were encouraged to write proposals to NSF, not to fund our research, but as part of our education; ignorance of the source of money that paid us was bliss. At Columbia, at the beginning of every semester I was required to fill out a form that described my thesis topic and its progress, but no one seemed to notice that for ten consecutive semesters I wrote "Not known at present" on every line except the one containing my name. Freedom to pursue what we wanted seemed to characterize the life of students, post-docs, and anyone eager to seize it. I was a third-year graduate student before I learned what "tenure" meant. Science was fun; that was enough.

Have the pressures to achieve "success," and to feel good about it, supplanted the pleasure of doing scientific research? Has the corporate mentality taken over science? By "reengineering" (sic) students into clients, graduates into products, and imaginative research into oblivion, have universities made funding more important than creativity? (A "creative solution" in the banking world is one so ill conceived that bankers laugh at it.) Young scientists are often encouraged to work on projects that are fun, *only* if funded; the correlation between quality of work and level of funding is immeasurably small. What about the tenure system, once designed to protect academics from the McCarthyism and political correctness of the 1950s, but now the last, highest hurdle in an exhausting race? "Tenure" impedes the development of young scientists less because it maintains departments full of dead wood blocking the paths of a more imaginative youth, and more because it forces young scientists to jump through a sequence of narrowly defined hoops, which, in turn, keep them channeled in research directions that are thought to gain them sufficient fame and funding for promotion, without regard for the pleasure of pursuing what appeals to them. By the time such scientists can relax with a job for life, they may have become so narrow in both scope and self-confidence that most of the field, and its excitement, lies outside their perspective, which now consists of a close-up view of the walls of the rut into which they have dug themselves.

I urge young scientists to assert themselves and to say to their department heads what MIT's John Edmond did, "Leave me alone, Frank. I know what I am doing."[112] Even in the unlikely case that your department head has more vision than Frank Press, if you think you know what you are doing, do it. If you want vision, avoid the ruts. Don't rewrite your Ph.D. thesis. Don't pursue what others want you to do, unless you really want to. Fail, for if you don't fail sometimes, probably you are just polishing the terrestrial monopole. Don't be afraid to be wrong in what you think (although make sure your data are sound). Don't be afraid to flounder at the forefront, at least for a while; it too teaches a lesson. Ask, "What is the next, most important problem?" Change directions to find it, but "when you can't think of anything to do, don't do nothing." Fail again, but most important, find a problem that turns you on. Then, repeat the process, unless you really do want to get old fast.

ACKNOWLEDGMENTS AND DISCLAIMERS

Most of this essay was written while I was a visiting fellow of the Cooperative Institute for Research in Environmental Science at the University of Colorado; support from them was essential in allowing me time to write on a subject that no funding agency would support. I thank R. Bilham, R. and T. J. Fitch, H. Hodder, S. Neustadtl, and especially N. Oreskes, for reading the manuscript and trying to save me from myself.

Three disclaimers. First, my father was not my only, nor my more influential, parent. Without encouragement from my mother, Margaret Molnar, in those few activities that I could actually do better than my father (e.g., sports), I might not have recognized the extent of his insight. Second, this essay is not a study in history of science. I am well aware of Heilbron's admonitions to scientists to take their research of history as seriously as their scientific research.[113] As requested by the editors, I simply offer my personal views (dubbed "unexpurgated" by one editor). Third, I have been aggressive with some general subdisciplines of the earth sciences, for instance, numerical modeling and structure/tectonics. Do not take these personally, as attacks on individuals, but as opinions about popular trends. "Some of my best friends" are modelers and structural geologists who study tectonics, although perhaps for those who know the joke of the 1960s the double entendre applies.

EPILOGUE:
CONTINENTS REALLY DO MOVE

Alfred Wegener died on the Greenland ice cap trying to find proof of continental drift. By proof he meant observations of the continents actually moving today. The geological arguments for drift were all indirect: they were surprising facts that could be explained if the continents had moved, but they were not actual observations of moving continents.

Ironically, plate tectonics was accepted without the evidence that Wegener sought. The geophysical data of plate tectonics – heat flow, seismicity, paleomagnetism – were in their own ways also indirect. They were observations of phenomena that followed from crustal motions or perhaps helped drive them, but they were not actual observations of the motions themselves. It took another decade before such observations could be made through the development of satellite-based global positioning systems. However, because of their military applications, many of the data collected by these satellites remained classified until the 1990s. Finally, almost a century after Alfred Wegener first suggested it and 30 years after earth scientists accepted it, we now have direct evidence that the earth really does move.

CHAPTER 18

PLATE TECTONICS: A MARTIAN VIEW

David T. Sandwell

WHEN WE TEACH PLATE TECTONICS TO YOUNG ADULTS WITH
little or no training in the physical sciences, we tell the "story," but we do
not have enough time to cover the important details. The story is really
quite incredible: rigid crustal slabs sliding thousands of miles across the
mantle of the earth at rates that are too slow to be observed without the
aid of sophisticated instruments; sea floor spreading ridges with submarine
hot springs under thousands of feet of ocean water; deep ocean trenches
where the oceanic plates of the earth literally fall into the mantle; neat
orthogonal ridge/transform patterns, and a magnetic field that reverses at
just the right rate to be recorded by cooling lavas at the spreading ridges.[1]

The only part of the system that can be seen with the naked eye is the
continents, which in fact don't participate in plate recycling and usually

Dave Sandwell, up close and personal with the ocean. (Photo courtesy of Dave Sandwell.)

331

have a long and messy geological history. While plate tectonics describes the motions of the earth's crust, we go further to claim that it is the optimal mechanism for the earth to shed excess radiogenic heat produced in the mantle. Diffusion of heat across a thick lithosphere is less efficient than allowing the oceanic lithosphere to radiate heat as it glides across the slippery asthenosphere and then cools the mantle during subduction.

I encourage my students not to believe any of this without doing a lot more reading. There are always a few religious students who believe literally in the Bible; they question all of these ideas, especially those related to the age of the earth. As a professor, I cannot claim that the story is true just because it is in all the textbooks. Indeed, how do I know the story is true? What are the essential and objective confirmations of plate tectonics, and how do we convey these to our students? Does the earth really behave as described by the theory, or is the theory just a qualitative description of the earth used for instructional purposes?

One of the major difficulties in confirming the theory is that most of the evidence for plate tectonics is covered by 2.5 miles (4 kilometers) of ocean water.[2] The parts of the continents above sea level contain a long and rich history of multiple episodes of collision, drifting, and rifting. Still, one needs a trained geologist who believes in plate tectonic theory to properly interpret the continental geologic record. The continents offer our only means to extend the tectonics of the earth more than about 200 million years into the past, because most of the old sea floor has been subducted. However, data from the oceans provide the primary confirmation of the theory.

Many textbooks, as well as this anthology of essays, are packed with strong evidence for the theory. Today plate tectonics is nearly universally accepted by earth scientists. However, there is always a danger that a prevailing theory will taint observations in a way to further confirm the theory – right or wrong. Take, for example, the construction of bathymetric charts (ocean floor topography) of the southern ocean, where the density of ship soundings is sparse. A reasonable mapmaker would take the available data and the known locations of the ridges and fracture zones nearby to fill in the blanks.[3] If a ship crossed a fracture zone in two locations separated by a great distance, then one could extrapolate the fracture zone along a trend predicted by plate tectonic theory. This approach of filling unknown areas using a guess based on the current understanding of the earth is common to many aspects of geology, and geologists generally mark these areas using dashed lines. The danger of this model-based extrapolation is that it can lead to a too clean and too simple picture of reality. Similarly, to promote learning, most textbooks

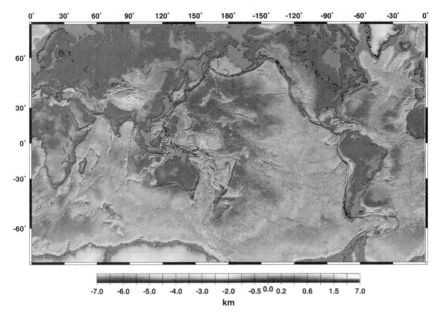

Topography of the earth reveals the sea floor spreading ridge system at a depth of about 7500 feet (2,500 meters). Deep ocean trenches are the sites where the cool and dense plates sink into the earth.

provide an antiseptic view of plate tectonics by selecting examples that reinforce the theory.[4]

Is the earth really that simple? When I was a student of plate tectonics in the late 1970s, I thought the whole theory, although basically correct, was oversimplified. Modern tools have supplied a wealth of new information about the earth and, to my surprise, the plates of the earth behave exactly as described in the early textbooks.[5] For example, they are almost perfectly rigid, the transform faults follow the predictions of Euler's theorem, and the subducted plates penetrate deep into the mantle. Indeed, much of the original evidence for the theory was collected in areas of tectonic complexity, and if one examines the bulk of the ocean basins, an amazingly simple picture emerges.[6]

In this essay, I'll describe a few of the important confirmations of plate tectonic theory provided by satellites and ships. These tools were largely developed to support the Cold War effort,[7] and many are labeled *geodetic* since they are used to make precise measurements of the size and shape of the earth and the spatial variations in the pull of gravity. The tools of satellite geodesy are needed for all aspects of global warfare; precise satellite tracking and gravity field development are needed for precision satellite surveillance as well as for targeting ballistic missiles; the global positioning system is used in all aspects of modern warfare; radar altimetry is used for aiming submarine-launched ballistic missiles as well as for inertial navigation when submerged.[8]

The global seismic networks were developed, primarily, to monitor underground nuclear tests. Marine magnetometers were developed, primarily, for detection of submarines. Multibeam sea floor mapping systems were developed, primarily, for surveying critical and operational areas of the northern oceans.[9] Yet despite these utilitarian origins they have proved fabulously useful for basic science.

EXPLORING THE EARTH FROM MARS

I'll begin by claiming that the reason plate tectonics took so long to become an accepted theory is because the earth was explored backward. Detailed geologic structures on the continents were investigated before the entire planet was properly observed, making it difficult to develop a planetary-scale model. The most efficient way to explore a planet is to start with planetary-scale observations and then design small-scale observational programs to test grand hypotheses. Indeed, this is the current NASA strategy for exploring Mars.

To illustrate this point, let's come up with an earth exploration plan that would lead to the development and confirmation of plate tectonic theory. Suppose that humans evolved on Mars rather than on Earth. The leaders of our great nation decided that Earth may contain life or at least may be a good place to live. Moreover, telescope observations of the land areas of Earth reveal large-scale patterns suggestive of some type of global stress field. The NASA administrator gathers her best engineers and scientists to develop an exploration plan. What is the best observation strategy, and at what point will the hypothesis of plate tectonics become strong enough to pursue more definitive experiments that will lead to confirmation? The exploration plan is designed as a series of hypothetical missions; and I'll highlight the ultimate contribution of each type of observation toward the understanding and confirmation of plate tectonic theory. At the end of the essay, I'll rank the observations in order of importance. Of course, my field of research will come out on top.

Mission 1 is a polar orbiting satellite that will take optical and near-infrared photographs of Earth at 300 feet (100 meter) resolution. A magnetometer is used to measure the external magnetic field, and Doppler tracking of the spacecraft provides a global measure of the gravity field. A radio receiver monitors frequency of the satellite telemetry and compares this with the known carrier frequency on the satellite; a Doppler shift provides an estimate of the velocity of the spacecraft relative to the receiver. Thousands of Doppler observations can be used to establish the precise orbit of the spacecraft and the gravity field perturbations of the planet. The satellite optical imagery reveals the mountains, river channels, and ice-covered areas in great detail. The variety of structures and morphology is overwhelming, and the scientists retire to their labs to try to digest the enormous supply of data. This will take at least five years, and in the end the optical data are not useful for discovering plate tectonics. However, they do reveal something important. The arcuate island chains, first observed through the Mars-based telescopes, are large volcanic structures with central caldera. Many show evidence for young lava flows and a few are actively spewing lava today. Earth is volcanically active; in fact, much more active than Mars. A second major discovery is a prominent magnetic field with north and south poles approximately aligned with the spin axis of Earth. A third major discovery comes from the tracking of the spacecraft orbit that reveals that the core of Earth is quite dense and probably made of iron just like Mars.[10] None of these measurements is unusual for a terrestrial planet (Mercury, Venus, Earth, Earth's moon, and Mars) so they don't provide any hint that Earth has global tectonics.

Scientists are frustrated since the ocean covers two-thirds of Earth's surface. A group of scientists and engineers hold a conference to develop an approach to explore the ocean. How deep is it? What does the bottom look like? Why do the coastlines of some of the continents seem to fit together? Their recommended plan will be implemented in the third mission, since the second mission is ready to launch.

Mission 2 is focused on measurements of land, ice, and ocean topography using a radar altimeter. There is also instrumentation for remotely examining the chemistry of the rocks, as well as a passive microwave radiometer to probe the temperatures of the atmosphere and ocean surface. The radar altimeter reveals linear topographic features on the land, which correlate with the optical imagery from Mission 1. More important, the radar data reveal broad variations in the height of the ocean surface above and below an ideal ellipsoidal shape. What causes these bumps and dips in the ocean surface? To a first approximation, Earth has the shape of an ellipse, where the equatorial radius is about 12 miles (20 kilometers) greater than the polar radius. Over timescales of millions of years, Earth can be thought of as a rotating fluid ball where the equatorial bulge reflects centrifugal force due to rotation. Consider a supertanker full of oil steaming from the north pole to the equator. During its voyage it will move some 12 miles farther from the planet's center. However, since water seeks its own level, the supertanker does not have to do any work to go uphill. The actual ocean surface (the geoid) does not follow the ellipsoidal shape exactly and can bulge outward or inward by up to 300 feet (100 meters). For example, the lowest point in the geoid lies just south of India while the highest point is just north of Australia. These global-scale variations in geoid height reflect the variations in mass inside Earth and are expected. Although this radar has only a 4-inch precision, it shows some prominent lows offshore of the arcuate island chains. These are dominant features of the geoid, but what are they? Based on these initial findings of ocean height, an improved radar altimeter is planned for the third mission.

Mission 3 carries a second radar altimeter, an improved camera, and a magnetometer. After about a year and a half of collecting altimeter profiles, it becomes clear that variations in the height of the ocean surface reflect features on the bottom of the ocean. Researchers present some color-coded maps at scientific meetings that provide direct evidence for the coastline match across the Atlantic Ocean.[11] In the equatorial Atlantic, linear anomalies extend across the ocean basin and seem to connect points on Africa and South America. Moreover, the center of the ocean contains gravity ridges and troughs that are exactly perpendicular to the fracture zones. This planet has some clear and organized surface structures, and the scientists are now working day and night to digest these data and formulate hypotheses. Many puzzles will remain until instruments are sent to the surface of the planet.

Mission 4 carries seismometers that are deployed by parachute to the surface of the continents at six locations. (A similar but much larger set of seismometers was used to locate shallow crustal quakes on Mars.) Moreover, the travel times and shapes of the recorded waves were used to infer the internal structure of Mars in great detail. The mission plan is to monitor Earth in the same fashion. There were two important discoveries derived from the patterns of earthquake locations. First, the earthquakes are not randomly distributed over the planet but are concentrated along discrete zones at the ridges in the center of the oceans and also beneath the island arcs.[12] The biggest surprise is that some of the earthquakes occur up to 450 miles (720 kilometers) beneath the planet's surface.[13] This was completely unexpected. Earthquakes should occur only in brittle material, and since Earth is larger than Mars, it was expected that its interior should be hotter, and grow ductile at a shal-

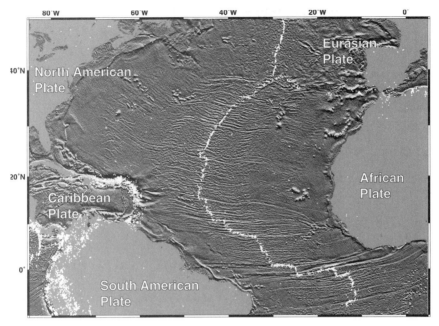

Topography of the ocean surface derived from *Geosat* (U.S. Navy) and *ERS-1* (European Space Agency) satellite altimeter measurements. Fracture zone traces clearly record the opening of the Atlantic ocean basin. Earthquakes mark the plate boundaries. (Sandwell, D. T., and W. H. F. Smith, 1997. Marine gravity anomaly from Geosat and ERS-1 satellite altimetry. *Journal of Geophysical Research* 102: 10,039–10,054, reproduced with permission of the American Geophysical Union.)

lower depth – about 30 miles (50 kilometers). The reason that earthquakes can occur at depths greater than this is still not completely understood.[14] A third major observation is that shallow earthquakes occur precisely on the mid-Atlantic ridge and virtually no earthquakes occur off the ridge.

At this point, all of the elements of plate tectonics are apparent: the altimeter data reveal the opening of the Atlantic Ocean as well as the deep ocean trenches; the seismic data reveal the active plate boundaries; and the deep earthquakes prove that cold slabs plunge into the mantle. The evidence that is still missing is the rate of the tectonic activity and direct measurements of the moving plates.

Mission 5 deploys a robot survey ship to carry out two important experiments. Scientists select a survey site at the Pacific-Antarctic ridge just north of the *Eltanin* fracture zone in the South Pacific Ocean, because this is the simplest structure apparent in the radar altimeter measurements.[15] The basic shipboard instruments are sonar to measure

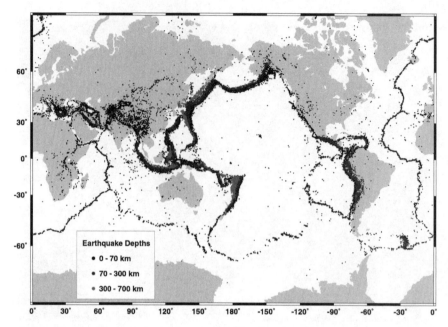

Locations of shallow (black less than 50 miles [70 kilometers] deep), intermediate (medium gray between 50–200 miles [70 and 300 kilometers] deep), and deep (light gray deeper than 200 miles [300 kilometers] deep). Shallow earthquakes occur on the sea floor spreading. (Engdahl, Van der hilst, and Buland, 1998. *Bulletin of Seismological Society of America* 88:722–743. Reproduced with permission of the Geological Society of America).

depth, devices to sample the properties and chemistry of the ocean, and a magnetometer to measure small variations in the magnetic field. The mapping of the magnetic field on the previous missions at satellite altitude did not reveal any unusual crustal anomalies, so the role of the shipboard magnetometer is unclear. Note that the lack of a crustal signal at the altitude of an orbiting satellite is purely a geometric smoothing effect that can only be overcome by moving closer to the surface of the earth. Two shipboard experiments are proposed. Experiment number 1 is a survey of the spreading ridge axis as identified in both the prior altimeter measurements and the earthquake locations. Scientists find a narrow axial ridge 750 to 500 feet (250 to 500 meters) tall that is superimposed on a broad rise where the average depth is 1.5 miles (2,500 meters).[16] The second experiment is a trackline perpendicular to the ridge axis. The sonar readings show a symmetric deepening of the ridge axis as a function of distance from the axial high. The survey is extended far on

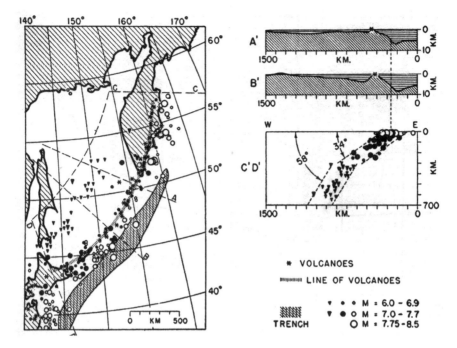

Depth section of earthquakes at the Kurile Trench provides an image of the sub-ducting slab that confirms deep subduction of the lithosphere at ocean trenches. (Figure from Benioff, H., 1954. Orogenesis and deep crustal structure: Additional evidence from seismology. *Bulletin of the Geological Society of America* 65: 385–400. Reproduced with permission of the Geological Society of America).

either side of the ridge to examine this symmetric depth observation. Today, we understand this observation as the signature of the thermal contraction of the oceanic plates as they slide away from the spreading ridges.[17] This cooling of the oceanic plate (the lithosphere) is the primary mechanism for Earth to shed its excess radiogenic heat, so the symmetric deepening of the sea floor as it ages is the planet's primary geodynamic and tectonic signature.

The most surprising result comes from the magnetometer, which shows a square-wave pattern of magnetic highs and lows. A more complete survey reveals that these magnetic anomalies form long stripes parallel to the ridge axis, but, most important, the stripes have spacings that are symmetric on either side of the ridge axis. This observation of symmetric marine magnetic anomalies not only provides direct evidence for symmetric sea floor spreading, but also proves that the global magnetic field of Earth reverses polarity on a timescale that is perfectly recorded in the cooling lava at the sea floor spreading ridges.

SOME RELEVANT ASPECTS OF THE REAL STORY

From this point on, it is impossible to predict when the grand hypothesis of plate tectonics will be proposed. Moreover, the proposition must be forceful enough to prompt further confirmation. At this point, I'll abandon the hypothetical exploration of Earth. There are three issues worth further discussion. First, for the record, I'll provide a brief history of satellite altimetry and the events leading to the declassification of *Geosat* radar altimeter measurements. Then I'll discuss the two remaining important observations of plate tectonics: paleomagnetic intensity variations in sequences of lava flows, and direct measurement of present-day plate motion.

THE HISTORY OF SATELLITE ALTIMETRY

The original altimeters (as in those aboard NASA's *SkyLab* and Defense Mapping Agency's *GEOS-3*) were launched to measure global-scale geoid height variations, but what they discovered were much smaller-scale geoid height variations: 5 to 30 miles (8 to 50 kilometers) horizontally and 1 to 4 inches (2 to 10 centimeter) vertically. The smaller-scale bumps and dips in the ocean surface reflect the gravitational attractions of structures on the ocean floor (spreading ridges, fracture zones, trenches, and volcanoes). When the *GEOS-3* results were first published, it was obvious that radar altimetry was the optimal tool for global mapping of the sea floor.[18] Ship soundings were too sparse to provide a global perspective. What was needed was an improved altimeter to achieve a one-inch (2-centimeter) range precision with dense track coverage. The *Seasat* altimeter launched in 1978 achieved the range precision, but failed after only three months in orbit.

Using data from the *Seasat* altimeter (NASA Jet Propulsion Laboratory), William Haxby and others at Lamont-Doherty Geological Observatory compiled the first completely objective map of the ocean basins in 1983.[19] Unlike other maps of the ocean, where scientists decide where to collect data, how to eliminate bad data, and how to fill in the blank areas, Haxby's map was based on a uniform coverage of the oceans and the data were all treated equally by a single computer algorithm. Most important, another scientist using the same data and the same algorithm could obtain exactly the same answer. This first map had only moderate detail due to the short three-month lifetime of the *Seasat* altimeter. Nevertheless, this map, plus similar maps prepared in our lab (National Geodetic Survey), convinced me that plate tectonics was a fair and accurate description of Earth. In addition to confirming plate tectonic theory, these maps revealed many important geological structures and guided seagoing expeditions for the next 15 years.

It took until July 1995 before better altimeter coverage became available from the *ERS-1* satellite altimeter (European Space Agency) and the *Geosat* altimeter. The *Geosat* satellite, built by the Johns Hopkins University's Applied Physics Laboratory (JHUAPL) and launched in 1985 by the U.S. Navy, collected high-precision sea surface height measurements in a non-repeat orbit for a year and a half and continued in a repeating, unclassified mode for another three years.[20] These data were processed and archived in two locations, the Naval Oceanographic Office (NOO), Stennis Space Center, Mississippi, and the JHUAPL, Maryland. The Navy used the *Geosat*-derived gravity field information to improve the accuracy of sea-launched ballistic missiles for the *Trident* submarine program.[21] The research activities in the classified NOO lab are not yet declassified. The main activity at the JHUAPL and the National Geodetic Survey (NGS) (including work by Bruce Douglas, Robert Cheney, Dave Porter, Dave McAdoo, Laury Miller, Russel Agreen, and David Sandwell) was to extract unclassified altimeter products from the *Geosat* data using a facility at JHUAPL. David McAdoo (NGS) and I (NGS, now at the Scripps Institution of Oceanography) did not have access to the classified data, only the unclassified products. The first unclassified oceanographic data were selected to follow the old *Seasat* track lines so that no significant new gravity information would be revealed.

Understanding the extreme scientific value of these *Geosat* data, Bruce Douglas, Karen Marks, Dave McAdoo (all of NGS), Bernard Minster (Scripps), Walter H. F. Smith (Scripps and NGS), many others, and I sent several requests to the Oceanographer of the Navy asking for release of subsets of data. The declassification of *Geosat* data came in three installments. First, in 1987, the Navy agreed not to classify the *Geosat* altimeter data in Antarctic waters south of 60°S latitude. Second, at the request of the National Research Council Committee on Geodesy (Minster, McAdoo, Sandwell, and others), Rear Admiral Chesbrough, the Oceanographer of the Navy, agreed to declassify all data south of 30°S latitude on June 10, 1992.[22] Finally, at the request of the Medea Committee as well as from the American Geophysical Union,

Admiral Boorda, the Chief of Naval Operations, authorized declassification of all *Geosat* data on July 19, 1995.[23] The most relevant aspect of this final declassification was that the *ERS-1* altimeter had just completed a one-year mapping mission that basically duplicated the still classified *Geosat* data, so there was no longer a reason to keep the *Geosat* data classified.

On a related matter, the Medea Committee assessed the scientific utility of all types of geophysical data collected by the U.S. Navy.[24] The report states:

> During the past 30 years, the Navy's ocean surveys have systematically collected bathymetry, gravity, magnetics, and salinity/temperature data on a global basis. In particular these surveys encompass almost all of the Northern Hemisphere. All together more than 100 ship-years of data acquisitions have been devoted to this effort, making this the most comprehensive surveying activity ever undertaken. It is highly unlikely that such an effort will be repeated, and it is certain that civilian environmental scientific resources could not aspire to an ocean survey program of this magnitude during the next 20 years.

While the *Geosat* altimeter data and the Arctic sea ice thickness data were declassified in 1995, the remaining 100-ship years of bathymetry, gravity, and magnetics data remain classified and discussions of possible declassification have not continued. The main barrier to declassifying the ship data is that the uneven track coverage would reveal which zones are of interest to the Navy. In contrast, the altimeter mapping of the ocean is uniform and unbiased.

MAGNETIC REVERSALS AND DIRECT MEASUREMENT OF PLATE MOTION

According to the textbooks and historical accounts, magnetic reversals at sea, coupled with dated magnetic polarities of lava flows on the continents worldwide, provided the turning point in the real story of plate tectonic acceptance. However, in the scenario just outlined, the magnetic evidence is not absolutely necessary. Indeed, the ability to observe magnetic reversals from a magnetometer towed behind a ship relies on some rather incredible coincidences related to reversal rate, spreading rate, ocean depth, and Earth temperatures.

In the case of marine magnetic anomalies, four scales must match.

First, the temperature at which cooling lava first records the direction of the global magnetic field must lie between the hot temperature of the mantle and the cold temperature of the sea floor. This may not seem that remarkable until one considers that the surface temperature of our sister planet Venus is too high for this to occur. Most of this magnetic field is recorded in the upper mile or two of the oceanic crust. If the thickness of this layer were too great, then as the plate cooled as it moved off the spreading ridge axis, the positive and negative reversals would be juxtaposed in dipping layers; this superposition would smear the pattern observed by a ship. On Earth, the temperatures are just right for creating a thin magnetized layer.

The second scale is related to ocean depth. Consider recording the magnetic field along a track perpendicular to the ridge axis. If the magnetometer is towed close to the bottom of the ocean, just above the magnetized layer, then the square-wave reversal pattern will be sharp and clear. However, most magnetometer measurements are made at the surface of the ocean, which is on average 2.5 miles above the magnetized layer. At this distance, the reversal pattern becomes attenuated and smooth. The result is that anomalies having a spacing of about 2π times the ocean depth will have the strongest signal. As just noted, the crustal anomalies are invisible at the altitude of an orbiting satellite because of the geometric smoothing effect with distance.

The third and fourth scales that must match are the reversal rate and the sea floor spreading rate. Half-spreading rates on Earth vary from 6 to 50 miles (10 to 80 kilometers) per million years. This suggests that for the magnetic anomalies to be most visible on the ocean surface, the reversal rate should be between 2.5 and 0.3 million years. It is astonishing that this is the typical reversal rate observed in sequences of lava flows on land. While most ocean basins display clear reversal patterns, there was a period between 85 and 120 million years ago when the magnetic field polarity of the earth remained positive for a long time, so the ocean surface anomaly is too far from the reversal boundaries to provide timing information. This period of time is called the Cretaceous Quiet Zone and it introduces a large uncertainty in the Cretaceous reconstruction of the plate motions. This lucky convergence of length and timescales makes it very unlikely that magnetic anomalies, due to crustal spreading, will ever be observed on another planet. This is the main reason that I do not believe the recent publication that interprets the Martian field as ancient spreading anomalies – one cannot be this lucky twice.[25]

The final hurdle to be overcome related to the confirmation of plate tectonics is the direct measurement of plate motion using space-geodetic methods. Because atmospheric refraction reduces the speed of light in unpredictable ways, and because plate motions need to be measured between points on opposite sides of Earth, one must use space objects as stable reference points. The typical rate of separation between continents due to plate tectonics is tiny – only 4 inches (10 centimeters) per decade – so a variety of methods needs to be deployed to double check the results.[26] Very long baseline interferometry (VLBI) uses coordinated radio-telescope observations to record microwave emissions from quasi-stellar objects (quasars). Precisely timed tape recordings from two or more VLBI antennas are brought together and correlated in a computer to determine the time delay of signal from the quasar. Because the VLBI antennas are moving on separate plates, this time delay will change over a period of many years. For example, several years of VLBI measurements between Haystack, Massachusetts, and Onsala, Sweden, recorded a plate motion of 0.7 inch (1.7 centimeters per year), which agrees remarkably well with the rate determined from marine magnetic anomalies.[27] Similarly, spacecraft can be simultaneously tracked by a network of lasers (satellite laser ranging – SLR) or a network of antennas (global positioning system – GPS) to establish the relative motions of the plates. The outcome of two decades of these space geodetic measurements is, that to first order, the present-day plate motions agree with a 2-million year average. There are small differences related to deformation of the interiors of the continental plates, but taken as a whole these measurements confirm plate tectonic theory.

In conclusion, the actual path to discovery and confirmation of plate tectonics was slow and painful because many of the Cold War tools were not yet available. The continental drift theory of the 1920s was based on the fit of the continental shorelines, matching fossils and stratigraphies on dispersed continents, and glacial deposits on continents that are now at low latitudes where there is no ice. But there was no *direct* evidence that continents actually moved. Indeed, Alfred Wegener died on the ice cap of Greenland trying to collect such evidence. Forty years later, paleomagnetic evidence and mapping of sea floor anomalies convinced most earth scientists that the sea floor was moving, while earthquake seismology delineated the shallow plate boundaries and deep subducting slabs. Direct proof of moving continents, however, awaited space-age tools.

The path to discovery and confirmation of plate tectonics would have been smoother if we only had the advantage of exploring Earth from another nearby planet. Nevertheless, the outcome would be the same. Today, the most important observations related to plate tectonics are provided by space geodesy, seismology, ship surveys, and geological investigations. I would rank them as follows:

1. radar altimeter measurements of marine gravity, fit of the continents
2. space geodetic measurements of plate motion
3. shallow earthquakes to define plate boundaries
4. deep earthquakes to prove that slabs penetrate into the deep mantle
5. magnetic reversals at sea to provide plate speed
6. mid-ocean ridge axis topography and symmetric deepening about the ridge
7. dating of reversals on land
8. fossil evidence
9. glacial striations
10. matching of rock types on conjugate continental margins

You see: this is exactly the reverse of the order in which things actually occurred.

NOTES

PREFACE

1. Le Goff, Jacques, 1992. *History and Memory*, translated by Steven Randall and Elizabeth Claman. New York: Columbia University Press. I simplify; more precisely, he writes, "the discipline of history . . . enters into the great dialectical process of memory and forgetting experienced by individuals and societies. The historian must be there to render an account of these memories and of what is forgotten, to transform them into something that can be conceived, to make them knowable" (pp. xi–xii). Le Goff notes that the earliest historians were witnesses; over time a shift in focus from oral recollection to post hoc examination of documents created a schism between history as testimony and history as analysis, the latter expanding the timescale of history at the expense of its immediacy. With this analytical expansion came the emergence of "history-as-problem," the resurgent claims of memory to superior authenticity and representativeness, and the counterclaims of historians to greater objectivity and perspective.

2. The notable exception was the Soviet Union, where scientists continued to adhere to a tectonic mode emphasizing vertical tectonics for another decade or so. Moreover, while acceptance among North American and European researchers working on tectonics was very rapid, it took some time for the new ideas to be disseminated. One colleague, who graduated from Caltech in 1972, says he has no recollection of having been taught anything about plate tectonics. Then he came to the Scripps Institution of Oceanography as a graduate student in the fall of 1972, where "it was a different world."

3. Daniel Offer, Marjorie Katz, Kenneth Howard, and Emily S. Bennett, 2000. The altering of reported experiences, *Journal of the American Academy of Child and Adolescent Psychiatry* 39(6): 735–742. In a similar vein, political scientist Gregory B. Markus examined the stability of political beliefs and people's memories of them. Over a nine-year period, he found that individuals' recollections of their earlier opinions were highly unreliable, with most people reporting their beliefs as far more stable than they actually were. Moreover, among those who believed their views had changed, they were often incorrect about the nature of that change and willingly offered plausible but erroneous explanations. For example, they might attribute their shifting beliefs to overall social patterns when in fact their personal shifts did not reflect these social trends, or they invoked simple rules of thumb, such as "people become more conservative as they get

older," whether or not their own beliefs did in fact become more conservative (Markus, Gregory B., 1986. Stability and change in political attitudes: Observed, recalled, and 'explained,' *Political Behavior* 8: 21–44).

Studies have also showed that writing about our beliefs can change them, often without our knowing it (Bem, D. J., and K. H. McConnell, 1970. Testing the self-perception explanation of dissonance phenomena: On the salience of premanipulation attitudes, *Journal of Personality and Social Psychology* 14: 23–31). If so, then scientists who write popular accounts of their work may be unreliable witnesses (caveat emptor!). On the other hand, there is some evidence that the more salient the events in question to the person describing them, the more likely he or she is to have an accurate memory of them (Marcus, 1986, p. 41).

While psychologists and psychiatrists have focused on the instability of memory as held by individual consciousness, historians, anthropologists, and sociologists have emphasized that individual memories are greatly affected by the social dynamics surrounding the individual, including cultural practices that create, reinforce, and alter memory; see, for example, Connerton, Paul, 1989. *How Societies Remember.* New York: Cambridge University Press; Nora, Pierre, 1992. *Realms of Memory: The Construction of the Historical Past.* New York: Columbia University Press; Halbwachs, Maurice, 1992. *On Collective Memory,* edited and translated by Lewis Coser. Chicago: University of Chicago Press; Prager, Jeffrey, 1998. *Presenting the Past: Psychoanalysis and the Sociology of Misremembering.* Cambridge, Mass.: Harvard University Press. Pnina Abir-Am and Clark Elliot note that memory in science is typically commemoration, with its implications of loyalty, gratitude, and appreciation, and therefore often particularly one-sided. Yet, as they note, the ways in which scientists commemorate can be highly informative of the values and beliefs of their communities (Abir-Am, Pnina, and Clark Elliot, eds., 1999. Commemorative practices in science, *Osiris* 14).

Natural scientists have emphasized memory as a complex cognitive process subject to alteration and impairment, most obviously in the various forms of amnesia and dementia but also in more moderate and unusual forms of memory loss and distortion. Recent research has refuted the idea of memory as an unvarnished record of the past, and replaced it with a notion of a complex process of knowledge creation and renewal; memory retrieval is now viewed less as a process of knowledge recapture than as a process of knowledge recreation. This dynamic view of memory seems reconcilable with the sociological perspective that argues for the role of social processes in the creation (and recreation) of memory: the social context of memory retrieval may affect (or even determine) the forms in which the memory is restored (and re-stored). For an excellent introduction to this literature, see Schacter, Daniel L., 1996. *Searching for Memory: The Brain, the Mind, and the Past.* New York: Basic; see also Schacter, Daniel L., ed., 1995. *Memory Distortion: How Mind, Brains, and Societies Reconstruct the Past.* Cambridge, Mass.: Harvard University Press.

4. A number of scientists, historians, and philosophers have written about the scientific developments from continental drift to plate tectonics, among

them the editors of this volume: Le Grand, Homer, 1988. *Drifting Continents and Shifting Theories.* Cambridge: Cambridge University Press; and Oreskes, Naomi, 1999. *The Rejection of Continental Drift: Theory and Method in American Earth Science.* New York: Oxford University Press. Other relevant works include Allegre, Claude, 1988. *The Behavior of the Earth.* Cambridge, Mass.: Harvard University Press; Frankel, Henry 1987. The continental drift debate. In *Resolution of Scientific Controversies: Case Studies in the Resolution and Closure of Disputes in Science and Technology,* H. T. Engelhardt, Jr., and A. L. Caplan, eds. Cambridge: Cambridge University Press, pp. 203–248; Glen, William, 1982. *The Road to Jaramillo: Critical Years of the Revolution in Earth Science.* Stanford, Calif.: Stanford University Press; Laudan, Rachel, 1980. The method of multiple working hypotheses and the discovery of plate tectonic theory. In *Scientific Discovery: Case Studies, Boston Studies in the Philosophy of Science* 60. Dordrecht: Reidel, pp. 331–343; Laudan, Rachel, and Larry Laudan, 1989. Dominance and the disunity of method: Solving the problem of innovation and consensus, *Philosophy of Science* 56(2): 221–237; Le Pichon, Xavier, Jean Francheteau and Jean Bonnin, 1973. *Plate Tectonics.* Amsterdam: Elsevier Scientific; Marvin, Ursula B., 1973. *Continental Drift: The Evolution of a Concept.* Washington, D.C.: Smithsonian Institution; Menard, H. W., 1986. *The Ocean of Truth: A Personal History of Global Tectonics.* Princeton: Princeton University Press; Stewart, John A. 1990. *Drifting Continents and Colliding Paradigms.* Bloomington: Indiana University Press; Wood, Robert Muir, 1985. *The Dark Side of the Earth.* London: Allen and Unwin. The best compendium of benchmark scientific papers related to plate tectonics is Cox, Allan, 1973. *Plate Tectonics and Geomagnetic Reversals.* San Francisco: W. H. Freeman. On the life and work of Alfred Wegener, the originator of continental drift theory, see Greene, Mott T., forthcoming. *Alfred Wegener and the Origins of Modern Earth Science.* Baltimore: Johns Hopkins University Press.

5. Brush, Stephen G., 1974. Should the history of science be rated X? *Science* 183: 1164–1172.

6. Reichenbach, Hans, 1938. *Experience and Prediction: An Analysis of the Foundations and Structure of Knowledge.* Chicago: University of Chicago Press, see especially pp. 3–16.

7. Except sometimes, as one of my colleagues notes, the stupidity of their opponents.

8. There is a large literature on how the rhetoric of detachment serves to create an aura, or perhaps even to instill an actual stance, of objectivity. The classic statement on scientific disinterestedness is Merton, Robert K., 1942. The normative structure of science. Reprinted in *The Sociology of Science,* Norman W. Storer, ed. Chicago: University of Chicago Press, 1973, pp. 267–278. For recent work on objectivity in science, see Porter, T. M., 1992. The accounting ideal in science, *Social Studies of Science* 22: 633–652, and idem, 1992. Objectivity as standardization: The rhetoric of impersonality of measurement, statistics, and cost-benefit analysis, *Annals of Scholarship* 9: 19–61; Daston, Lorraine, 1992.

Objectivity and the escape from perspective, *Social Studies of Science* 22: 597–618; Daston, Lorraine, and Peter Galison, 1992. The image of objectivity, *Representations* 40: 81–128; and Dear, Peter, 1992. From truth to disinterestedness in the seventeenth century, *Social Studies of Science* 22: 619–631.

9. Offer et al., 2000, note 2.

10. Of 17 essayists in this volume, all but two had major or primary affiliations with one or more of these four institutions. The Woods Hole Oceanographic Institution (WHOI), Caltech, Imperial College, London, and the U.S. Geological Survey also figure, but to a much lesser degree. On the question of what makes scientific research institutions productive, see Hollingsworth, J. Rogers, and Ellen Jane Hollingsworth, 2000. Major discoveries and biomedical research organizations: Perspectives on interdisciplinarity, nurturing leadership, and integrated structure and cultures. In *Practising Interdisciplinarity*, Peter Weingart and Nico Stehr, eds. Toronto: University of Toronto Press, pp. 215–244; and Hollingsworth, J. Rogers, Ellen Hollingsworth, and Jerald Hage, eds., in press. *Fostering Scientific Creativity: Institutions, Organizations, and Major Discoveries*. Cambridge: Cambridge University Press.

11. It also helps to explain why it has proved difficult for women and minorities to gain a stronger presence in science, even once they are welcome: the odds of making it to one of these few institutions without informed advice are not great.

12. Nor does this include individuals who in retrospect deserve more credit for their work than they have generally received. One example is R. R. Coats, who in 1962 published a paper connecting subduction of the ocean crust, deep focus earthquakes, and island-arc vulcanism. (Coats, R. R., 1962. Magma type and crustal structure in the Aleutian Arc. In *The Crust of the Pacific Basin*, Gordon A. MacDonald and Hisashi Kuno, eds. *Geophysical Monograph* 6. Washington, D.C.: American Geophysical Union, pp. 92–109.) While Coats' remarkable cross-sections through the crust and upper mantle of a generalized island arc were republished in Allan Cox's 1973 compendium of benchmark papers (Cox, 1973, note 3), Cox did not reprint the paper itself and Coats' work remains little known and rarely cited. There are also nationalistic issues in the assignment of credit: American students are often taught about the work of Cox and co-workers on the establishment of geomagnetic reversals, but may not know about the comparable work done in Australia by Ian McDougall and D. H. Tarling (cf. Cox, Allan, Richard R. Doell, and G. Brent Dalrymple, 1963. Geomagnetic polarity epochs and Pleistocene geochronometry, *Nature* 198: 1049–1051, and McDougall, Ian, and Don H. Tarling, 1963. Dating of polarity zones in the Hawaiian Islands, *Nature* 200: 54–56. Finally, if we were to add those scientists who contributed to serious discussion of continental drift before the 1950s, we should have to add at least the names of Otto Ampferer, Warren Carey, Reginald Daly, Alexander du Toit, Arthur Holmes, John Joly, Philip Keunen, Felix Vening Meinesz, and, of course, Alfred Wegener.

13. Price, Derek De Solla, 1963. *Little Science, Big Science.* New York: Columbia University Press.

14. On U.S. military funding of science and its effect on the size, subject, and style of scientific research, see Dennis, M. A., 1991. A change of state: The political cultures of technical practice at the MIT Instrumentation Laboratory and the Johns Hopkins University, Ph.D. dissertation, Johns Hopkins University; idem, 1994. Our first line of defense: Two university laboratories in the postwar American state, *Isis* 85: 427–455; Forman, Paul, 1987. Behind quantum electronics: National security as basis for physical research in the United States, 1940–1960, *Historical Studies in the Physical and Biological Sciences* 18: 149–229; Forman, Paul, and José Sanchez-Ron, 1996. *National Military Establishments and the Advancement of Science and Technology.* Dordrecht: Kluwer Academic; Galison, Peter, and Bruce Hevly, eds., 1992. *Big Science: The Growth of Large-scale Research.* Stanford, Calif.: Stanford University Press; Kevles, Daniel J., 1990. Cold war and hot physics: Science, security, and the American state, *Historical Studies in the Physical and Biological Sciences* 20: 239–264; Leslie, S. W., 1993. *The Cold War and American Science: The Military-Industrial-Academic Complex at MIT and Stanford.* New York: Columbia University Press; and Mendelsohn, E., M. R. Smith, and P. Weingart, eds., 1988. *Science, Technology, and the Military.* Dordrecht: Kluwer Academic.

15. On the diversity of theoretical opinions in the earth sciences in the 19th century, see Greene, Mott T., 1982. *Geology in the Nineteenth Century: Changing Views of a Changing World.* Ithaca: Cornell University Press; for the persistence of this theoretical diversity into the 20th century, see Le Grand, 1988, note 3, and Oreskes, 1999, note 3.

16. Hodgson, J. H., 1957. Nature of faulting in large earthquakes, *Geological Society of America Bulletin* 68: 611–644; see also Hodgson, J. H. and W. Milne, 1951. Direction of faulting in certain earthquakes of the North Pacific, *Seismological Society of America Bulletin* 41: 221–242; Hodgson, J. H. and R. S. Storey, 1953. Tables extending Byerly's fault-plane technique to earthquakes of any focal depth, *Seismological Society of America Bulletin* 43: 49–61; Hodgson, J. H., and J. I. Cock, 1956. Direction of faulting in the deep focus Spanish earthquake of March 29, 1954, *Tellus* 8: 321–328; Hodgson, J. H., and W. M. Adams, 1958. A study of inconsistent observations in the fault-plane project, *Seismological Society of America Bulletin* 48: 17–31; Hodgson, J. H., and A. E. Stevens, 1964. Seismicity and earthquake mechanism. In *Research in Geophysics* 2, H. Odishaw, ed. Cambridge, Mass.: MIT, pp. 27–50.

17. In their 1964 review article, Hodgson and Stevens acknowledge both the analytical and instrumental difficulties: "Because of high background noise, weak first motion, and, sometimes, reversed galvanometer wires, about 15% of reporting stations usually provide incorrect data. The selection of the correct solution is normally something of an art; as an art of course it is scientifically suspect." In discussing stereographic projections of fault-plane solutions, which generate two possible solutions for the orientation of the fault plane, they continue: "The diagrams

which we have been looking at have come to be called 'fault-plane solutions.' They suffer from two limitations: first that they do not tell us which plane represents the fault, second that a substantial body of seismologists are of the opinion that they do not represent the fault at all." (Hodson and Stevens, 1964, note 14, first quote on p. 33, second on p. 35.) After more than a decade of work in this area, Hodgson came to admit that the results obtained were fundamentally ambiguous.

Despite these acknowledged ambiguities, many scientists have told me that "everyone" in the 1950s and early 1960s believed that Hodgson had shown that deep-focus earthquakes were strike-slip. For example, in their 1963 review article on ocean trenches, Robert Fisher and Harry Hess wrote: "Hodgson (1957) studied the first motion of large deep focus earthquakes. . . . [F]rom these he deduced that the movement causing the earthquakes occurred on nearly vertical planes. This type of analysis results in defining two planes at right angles to each other. Which is the real fault plane is indeterminate. In either case, Benioff's postulated 40–45° thrust motion is ruled out" (Fisher, R. L., and H. H. Hess, 1963. Trenches. In *The Sea*, vol. 3 *The Earth Beneath the Sea*, M. N. Hill, ed. New York: Wiley, pp. 411–433, on p. 432). No doubt Hess was eager to accept Hodgson's result because it permitted him to retain his tectogene concept, which he had first proposed in the 1930s, but Hess was not alone; in the absence of other evidence on fault-plane solutions, most people accepted the strike-slip model. Why Benioff's evidence was not taken more seriously is not clear; perhaps it is because Benioff himself was ambiguous. His now-classic 1954 work clearly outlined the moderately dipping zones of deep-focus earthquakes beneath ocean trenches, which now bear his name, but he did not interpret them as subducting ocean crust. Rather, he saw them as defining the margins of continental blocks, which were moving downward en masse, effectively forming an inverted graben (Benioff, Hugo, 1954. Orogenesis and deep crustal structure: Additional evidence from seismology, *Geological Society of America Bulletin* 65: 385–400).

In 1964, a New Zealand seismologist, R. D. Adams, realized what was wrong with Hodgson's analysis (and the answer is banal): his conclusions were an artifact of his plotting technique, which made points from distant sources appear to lie on steeply dipping planes. Adams explained:

"When the number of observations is limited there has been a tendency to infer a quandrantal distribution, although various other patterns would fit equally well, if not better. . . . One of the nodal planes of the quadrantal solution is usually assumed to be a fault plane, and it has become common to refer to first-motion investigations as "fault-plane" studies.

In many first-motion studies most of the available readings are from P phases at large distances, and from KPP phases, for both of which the ray paths leave the focus downwards at a very steep angle. For shallow earthquakes all waves recorded as PKP phases are contained within less than 3% of the total solid angle surrounding the focus. The "extended distance" projection generally used by North American workers to display their analyses is such that the area representing these distant phases is greatly exaggerated. This increases the probability that their nodal lines will be drawn through this area, and hence favors solutions with steeply dipping nodal planes. On the fault hypothesis [that is, that one of the nodal planes is actually the fault plane] this implies that the faults are transcurrent. Thus the fact that most first-motion analyses have been interpreted as indicating transcurrent faults could be due to the method used to display the results." (Adams, R. D., 1963. Source characteristics of some deep New Zealand earthquakes, *New Zealand Journal of Geology and Geophysics* 6(2): 209–220, quote on pp. 216–217)

18. Bill Menard, Dietz's close colleague at Scripps, vigorously denied this. See Menard, 1986, note 4, chap.13.

19. Dietz , Robert S., 1961. Continent and ocean basin evolution by the spreading of the sea floor, *Nature* 190: 854–857; idem, 1962. Ocean-basin evolution by sea floor spreading. In *The Crust of the Pacific Basin*, Gordon A. MacDonald and Hisashi Kuno, eds. *Geophysical Monograph* 6. Washington, D.C.: American Geophysical Union, pp. 11–12; and Hess, Harry H., 1962. History of the ocean basins. In *Petrologic Studies: A Volume to Honor A.F. Buddington*, A. E. J. Engel, H. L. James, and B. F. Leonard, eds. Denver: Geological Society of America, pp. 599–620.

Allan Cox conspicuously omitted Dietz's 1961 *Nature* article in his compendium of benchmark papers, presumably because he believed that Dietz had stolen the idea from Hess. Yet the two versions of sea floor spreading were not identical. Hess interpreted the ocean crust as a hydration rind on exposed mantle; in his model, there was effectively no oceanic crust (which, ironically, is what Alfred Wegener thought). In contrast, Dietz (1961) argued for an ocean floor generated by submarine basalt eruptions. One might conclude that both Hess and Dietz stole the idea of sea floor spreading from Arthur Holmes, an accusation that was lodged in 1968 by geologist Arthur Meyerhoff (see Meyerhoff, A. A., 1968. Arthur Holmes: Originator of the spreading ocean floor hypothesis, *Journal of Geophysical Research* 73: 6563–6565; Dietz, Robert S., 1968. Reply to comment by A.A. Meyerhoff, *Journal of Geophysical Research* 73: 6567; and Hess, H. H., 1968. Reply to comment by A.A. Meyerhoff, *Journal of Geophysical Research* 73: 6569). My own view is that scientists often hear about things without knowing where they originated, and often read articles but forget they have read them. This is consistent with research in cognitive science on "implicit memory" – things we remember without realizing that we are remembering, and therefore think we thought of ourselves (Schacter, 1996, note 2) – or what sociologist Robert Merton called *cryptomesia*. On the other hand, sometimes scientists steal ideas.

20. McKenzie, D. P., and R. L. Parker, 1967. The North Pacific: An example of tectonics on a sphere, *Nature* 216: 1276–1280, on p. 1276.

21. Morgan, W. Jason, 1968. Rises, trenches, great faults, and crustal blocks, *Journal of Geophysical Research* 73: 1959–1982.

22. Le Pichon, this volume.

23. Bullard, Sir Edward, J. E. Everett, and A. Gilbert Smith, 1965. The fit of the continents around the Atlantic. In *A Symposium on Continental Drift*, P. M. S. Blackett, Sir Edward Bullard, and S. K. Runcorn, eds. London: Royal Society, pp. 41–51; see also McKenzie, D. P. 1987. Edward Crisp Bullard 1907–1980, *Biographical Memoirs of Fellows of the Royal Society* 33: 67–98.

24. Backus, George E., 1964. Magnetic anomalies over oceanic ridges, *Nature* 201: 591–592. Backus later wrote an NSF proposal to pursue the idea, but it was rejected, and like Robert Parker he moved on to work on geophysical inverse theory. Jason Morgan says he was unfamiliar with Backus's paper, having moved into geophysics from physics, and only found out about it many years later, when Backus gave him a reprint (W. Jason Morgan, pers. comm., March 7, 2001.) Dan McKenzie knew about Backus's paper, but at the time did not find it convincing. It is important to realize that while we speak of magnetic stripes as if they were something you could see – visually – by looking at the sea floor, this is not at all the case. Magnetic anomalies are geophysical signals, measured by instruments, which require considerable data processing to convert into something that can be drawn on a piece of paper and rendered visible. Backus' argument involved trying to decipher the power spectra of magnetic anomalies as a function of the distance from the rotation pole. McKenzie recalls: "I did not find what he did at all convincing, and the central idea, of using Euler's theorem, was already common knowledge at Madingley at the time because of Teddy's fits. This is why George's work did not make a deeper impression." McKenzie notes that later attempts to use sophisticated signal processing to understand the pattern of magnetic anomalies has not really paid off (McKenzie, pers. comm., July 15, 2001, and pers. comm. July 18, 2001).

25. Wilson, J. Tuzo, 1965. A new class of faults and their bearing on continental drift, *Nature* 207: 343–347. See Vine, this volume, for discussion of Wilson's visits to Cambridge.

26. Wilson, 1965, note 21, p. 343. In a recently published textbook, Geoffrey Davies credits Wilson for being the first to introduce the term *plate* to describe crustal segments, and the first to "see plates in simple and complete form" (Davies, Geoffrey, 1999. *Dynamic Earth: Plates, Plumes, and Mantle Convection.* Cambridge: Cambridge University Press, p. 42). Davies is probably right, although the term *plate* had a history of usage in geodesy, perhaps derived from engineering, to describe regions that were responding uniformly to isostatic load, as in the "Gulf of Mexico plate" described by Felix Vening Meinesz and Frederick E. Wright in the 1930s; see Oreskes, 1999, note 3, p. 241.

27. McKenzie, 1967, note 18, p. 1277; and Morgan, 1968, note 19, p. 1959.

28. Le Pichon, this volume.

29. Coode, A. M., 1965. A note on oceanic transcurrent faults, *Canadian Journal of Earth Science* 2: 400–401, ref. cit. Davies, 1999, note 26; and Menard, 1986, note 4, pp. 238 and 242–243.

CHAPTER 1

1. The earliest commonly cited example is Abraham Ortelius, a Dutch cartographer (Romm, J. 1994. A new forerunner for continental drift, *Nature* 367: 207; see also Carozzi, A. V., 1969. A propos de l'origine de l'théories des dérives continentales: Francis Bacon (1620), Francois Placet (1668), A. von

Humboldt (1801), et A. Snider [1858], *Comptes Rendus et Séances de la Société des Physiques et de la Histoire Naturelle,* n.s., 4(3): 171–179). The first developed scientific theories of continental splitting, and perhaps drifting, date from the 19th century. They can be roughly categorized into two types. One group, led by British geophysicists George Darwin and Osmond Fisher and American geodesist William Bowie, viewed continental breakup as a unique event of early earth history, which created the morphology of the continents and oceans for the remainder of geological time (Darwin, G. H., 1879. The precession of a viscous spheroid and the remote history of the Earth, *Philosophical Transactions of the Royal Society* 170: 447–538; Fisher, Osmond, 1881. *Physics of the Earth's Crust.* London: Macmillan; Bowie, William, 1935. The origin of continents and oceans, *Scientific Monthly* 41: 444–449). Another group, led by Frank Taylor, Howard Baker, and, more famously, Alfred Wegener, viewed continental drift as a continuous feature of earth history (Taylor, F. B., 1910. Bearing of the Tertiary mountain belt on the origin of the Earth's plan, *Geological Society of America Bulletin* 21: 179–226; Baker, Howard B., 1914. The origin of continental forms, *Annual Report of the Michigan Academy of Sciences,* 99–103; Wegener, Alfred L., 1912. Die Entstehung der Kontinente, *Geologische Rundschau* 3: 276–292; idem, *The Origin of Continents and Oceans,* 3rd edition translated into English by J. G. A. Skerl. London: Methuen). Somewhere in between these two camps was the American geologist Joseph Barrell, who believed that basalt intrusion created ocean basins by fragmentation of the continents, effectively driving them apart. Barrell used the same paleontological evidence as Wegener to place the opening of the current North Atlantic in the early Tertiary period, and coined the term *asthenosphere* to refer to a weak zone, beneath the earth's surface layer, in which isostatic compensation would occur and from which basaltic magmas would arise (Barrell, Joseph, 1927. On continental fragmentation and the geologic bearing of the Moon's surface features, *American Journal of Science* 213: 283–314).

   2. Suess, Eduard, 1904–1924. *The Face of the Earth,* translated by Hertha B. C. Sollas. Oxford: Clarendon; see also Leviton, Alan, and Michele Aldrich, 1991. The origin of the notion of Gondwanaland: The Indian Geological Survey, *Geological Society of America Abstracts with Programs* 23(5): A 47. On the early recognition of faunal homologies, see Marvin, Ursula B., 1973. *Continental Drift: The Evolution of a Concept.* Washington, D.C.: Smithsonian Institution.

   3. Dana, James D., 1846. On the volcanoes of the moon, *American Journal of Science* 52 (2nd series, vol. 2): 335–355; idem, 1847. On the origin of continents, *American Journal of Science* 53 (2nd series, vol. 3): 94–100; idem, 1847. Geological results of the Earth's contraction in consequence of cooling, *American Journal of Science* 53 (2nd series, vol. 3): 176–188; idem, 1847. Origin of the grand outline features of the Earth, *American Journal of Science* 53 (2nd series, vol. 3): 381–398; idem, 1873. On the origin of mountains, *American Journal of Science* 105 (3nd series, vol. 5): 347–350; idem, 1873. On some results of the Earth's contraction

from cooling, including a discussion of the origin of mountains, and the nature of the Earth's interior, *American Journal of Science* 105 (3nd series, vol. 5): 423–443.

4. Hall, James, Jr., 1859. *Paleontology, v. III, Geological Survey of New York.* Albany: Van Benthuysen; idem, 1882. Contributions to the geological history of the North American Continents, *Proceedings of the American Association for the Advancement of Science* 31: 29–71 (presidential address of 1857).

5. Markham, Clements R., 1878. *A Memoir on the Indian Surveys.* 2nd ed. London: W. H. Allen, reprinted Amsterdam: Meridian, 1969.

6. While the term *earth science* was not common in the early 20th century, there was a group of sciences – geology, geodesy, geophysics, meteorology, and oceanography – that were commonly understood as having shared interests, and frequently involved the same individuals. In this sense, meteorology was a branch of earth science, and Wegener was an earth scientist. (On the history of scientific meteorology, see Fleming, James R., 1990. *Meteorology in America, 1800–1870.* Baltimore: Johns Hopkins University Press; and idem, ed., 2001. *Weathering the Storm: Sverre Petterssen, the D-Day Forecast and the Rise of Modern Meteorology.* Boston, Mass.: American Meteorological Society; and Friedman, Robert Marc, 1989. *Appropriating the Weather: Vilhelm Bjerknes and the Construction of a Modern Meteorology.* Ithaca, N.Y.: Cornell University Press.)

On the other hand, of all the earth sciences, geology was the oldest and most institutionally well-established, and Wegener was not a geologist. Therefore, some questioned his ability to interpret geological evidence (Oreskes, Naomi, 1999. *The Rejection of Continental Drift: Theory and Method in American Earth Science.* New York: Oxford University Press, ch. 5). An early example of this mind-set comes from the pen of German Leopold von Buch, who, in 1846, described the speculations of a young Englishman: "The spirited [young man], with all his remarkable vivacity of mind, is for me no Geologist. . . . [He] could never make a tolerable geological map." The man in question? Charles Darwin (von Buch to Roderick Murchison, April 20, 1846, cited in Laudan, Rachel, 1987. *From Mineralogy to Geology: The Foundations of a Science 1650–1830.* Chicago: University of Chicago Press).

7. Daly, Reginald A., 1926. *Our Mobile Earth.* New York: Charles Scribner's Sons.

8. Joly, John, 1925. *The Surface-History of the Earth.* Oxford: Clarendon.

9. Holmes, Arthur, 1925. Radioactivity and the earth's thermal history, *Geological Magazine* 62: 404–528 and 529–544; idem, 1926. Contribution to the theory of magmatic cycles, *Geological Magazine* 63: 306–329; idem, 1928. Radioactivity and continental drift, *Geological Magazine* 65: 236–238; idem, 1929. A review of the continental drift hypothesis, *Mining Magazine* 40: 205–209, 286–288, and 340–347; idem, 1929. Radioactivity and earth movements, *Transactions of the Geological Society of Glasgow* 18: 559–606; idem, 1933. The thermal history of the Earth, *Journal of the Washington Academy of Science* 23: 169–195; and idem, 1944. *Principles of Physical Geology.* London: T. Nelson and Sons, first American edition 1945, New York: Ronald, reprinted 1965.

10. Meredith, Margaret, 2002. A noble commerce: American interpretations of fossil bones in a trans-Atlantic context," Ph.D. dissertation, University of California, San Diego.

11. On the methodological commitments of earth scientists in relation to the drift debate, see Le Grand, H. E., 1996. Steady as a rock: Methodology and moving continents. In *The Politics and Rhetoric of Scientific Method: Historical Studies*, J. Schuster and R. Yeo, eds. Dordrecht: Reidel, pp. 97–138; and Oreskes, Naomi, 1999. *The Rejection of Continental Drift: Theory and Method in American Earth Science*. New York: Oxford University Press, pp. 123–156.

12. McKenzie, D. P., 1987. Edward Crisp Bullard, 1907–1980, *Biographical Memoirs of Fellows of the Royal Society* 33: 67–98.

13. Vening Meinesz, Felix, J. H. F. Umbrove, and P. H. Kuenen, 1934. *Gravity Expeditions at Sea, 1923–1932*. Vol. 2. Delft: Netherlands Geodetic Commission; see also idem, 1929. Gravity expeditions of the U.S. Navy, *Nature* 123: 473–475; and idem, 1952. *Convection Currents in the Earth and the Origins of Continents* I. Paper read at Koninklÿke Nederlandse Akademische van Wetenschappen, Amsterdam.

14. Hess, H. H., 1933. Interpretation of geological and geophysical observations (with figures). In *The Navy-Princeton Gravity Expedition in the West Indies*, R. M. Field, ed. Washington, D.C.: U.S. Government Printing Office, pp. 27–54; see also Hess, H. H., 1932. Interpretation of gravity anomalies and sounding profiles obtained in the West Indies by the International Expedition to the West Indies in 1932, *Transactions of the American Geophysical Union* 13: 26–33; idem, 1937. Geological interpretation of data collected on cruise of *U.S.S. Barracuda* in the West Indies – preliminary report, *Transactions of the American Geophysical Union* 18: 69–77; idem, 1938. Gravity anomalies and island arc structure with particular reference to the West Indies, *Proceedings of the American Philosophical Society* 79(1): 71–96. See also Allwardt, Alan O., 1990. The roles of Arthur Holmes and Harry Hess in the development of modern global tectonics, Ph.D. dissertation, University of California, Santa Cruz.

15. Kuenen, Philip H., 1936. The negative isostatic anomalies in the East Indies (with experiments), *Leidsche Geologische Mededeelingen* 8(2): 327–351; Griggs, D. T., 1938. Convection currents and mountain building (abstract), *Geological Society of America Bulletin* 49: 1884; idem, 1939. A theory of mountain building, *American Journal of Science* 237(9): 611–650; see also Griggs, D. T., 1974. Presentation of the Arthur L. Day award to David T. Griggs: Response by David Griggs, *Geological Society of America Bulletin* 85: 1342–1343.

16. Griggs, 1939, note 15, p. 647.

17. Ibid., p. 612.

18. Hess, H. H., undated manuscript, circa 1939. Recent advances in the interpretation of gravity anomalies and island-arc structures, *Advanced Report of the Commission on Continental and Oceanic Structure*, pp. 46–48; Papers of Harry H. Hess, Princeton University Archives.

19. Ewing, Maurice, and J. Lamar Worzel, 1945. *Long range sound transmis-*

sion, Contract Nobs-2083, Report No. 9 (Woods Hole, Mass.: The Woods Hole Oceanographic Institution), declassified March 12, 1946; see also Weir, Gary, 1993. *Forged in War: The Naval-Industrial Complex and American Submarine Construction 1940–1961.* Washington, D.C.: U.S. Government Printing Office.

20. Hess, H. H., 1946. Drowned ancient islands of the Pacific Basin, *American Journal of Science* 244: 772–791.

21. On the history of ONR, see Office of Naval Research, 1987. *Forty Years of Excellence in Support of Naval Science: 40th Anniversary, 1946–1986.* Arlington, Va.: Office of Naval Research, Department of the Navy, and Sapolsky, Harvey M., 1990. *Science and the Navy: The History of the Office of Naval Research.* Princeton: Princeton University Press.

22. Munk, Walter, 1967. *Dedication.* In *Topics in Nonlinear Dynamics: A Tribute to Sir Edward Bulllard,* Siebe Jorna, ed. New York: American Institute of Physics, pp. v–viii; and also McKenzie, 1987, note 12.

23. Blackett, P. M. S., 1949. *Fear, War, and the Bomb: Military and Political Consequences of Atomic Energy.* New York: Whittlesey House.

24. Blackett, P. M. S., 1947. The magnetic field of massive rotating bodies, *Nature* 159: 658–666; idem, 1952. A negative experiment relating to magnetism and the earth's rotation, *Transactions of the Royal Society* A245: 309–370; see also Le Grand, H. E., 1988. *Drifting Continents and Shifting Theories.* Cambridge: Cambridge University Press, pp. 142–143, and Nye, Mary Jo, 1999. Temptations of theory, strategies of evidence: P. M. S. Blackett and the Earth's magnetism, 1947–1952, *British Journal for the History of Science* 32: 69–92.

25. Elsasser, W. M., 1946. Induction effects in terrestrial magnetism, Part I, *Reviews in Physics* 69: 106; idem, 1946. Induction effects in terrestrial magnetism, Part II, *Reviews in Physics* 70: 202; idem, 1947. Induction effects in terrestrial magnetism, Part III, *Reviews in Physics* 72: 821; Bullard, E. C., 1949. The magnetic field within the earth, *Proceedings of the Royal Society* A197: 433–453; idem, 1955. The stability of a homopolar dynamo, *Proceedings of the Cambridge Philosophical Society* 51: 744–760; and Bullard, E. C., and H. Gellman, 1954. Homogeneous dynamos and terrestrial magnetism, *Philosophical Transactions of the Royal Society* A247: 213–278.

26. Le Grand, H. E., 1994. Chopping and changing at the DTM 1946–1958: Merle Tuve, rock magnetism, and isotope dating. In *The Earth, the Heavens, and the Carnegie Institution of Washington,* Gregory Good, ed. Washington, D.C.: American Geophysical Union, *History of Geophysics* 5: 173–184.

27. Creer, K. M., E. Irving, and S. K. Runcorn, 1954. The direction of the geomagnetic field in remote epochs in Great Britain, *Journal of Geomagnetism and Geoelectricity* 6: 163–168; Clegg, J. A., M. Almond, and P. H. S. Stubbs, 1954. The remanent magnetism of some sedimentary rocks in Britain, *Philosophical Magazine* 45 (series 7): 583–598; idem, 1954. Some recent studies of the pre-history of the earth's magnetic field, *Journal of Geomagnetism and Geoelectricity* 6: 194–199.

28. Clegg, J. A., E. R. Deutsch, and D. H. Griffiths, 1956. Rock magnetism in India, *Philosophical Magazine* 1 (series 8): 419–431; Blackett, P. M. S., 1956. *Lectures on Rock Magnetism.* Jerusalem: Weizmann Science, and idem, 1961. Comparison of ancient climates with the ancient latitudes deduced from rock

magnetic measurements, *Proceedings of the Royal Society* A263: 1–30; Irving, E. 1956. Paleomagnetic and paleoclimatological aspects of polar wandering, *Geofisica pura e Applicata* 33: 23–41; idem, 1958. Rock magnetism: A new approach to the problem of polar wandering and continental drift. In *Continental Drift: A Symposium,* S. W. Carey, ed. Hobart: University of Tasmania, pp. 24–61; Runcorn, S. K., 1959. Rock magnetism, *Science* 129: 1002–1011.

29. Bullard, E. C., A. E. Maxwell, and Roger Revelle, 1956. Heat flow through the deep sea floor, *Advances in Geophysics* 3: 153–181.

30. Hess, H. H., 1962. History of ocean basins. In *Petrologic Studies: A Volume in Honor of A. F. Buddington,* A. E. J. Engel, H. L. James, and B. F. Leonard, eds. Denver: The Geological Society of America, pp. 599–620.

31. Allwardt, 1990, note 14.

32. For the stalemate comment, see J. H. Taylor, 1965. Discussion of Bullard, Sir Edward, J. E. Everett, and A. Gilbert Smith, The fit of the continents around the Atlantic, *Philosophical Transactions of the Royal Society* A258, *A Symposium on Continental Drift,* P. M. S. Blackett, Sir Edward Bullard, and S. K. Runcorn, eds. London: Royal Society, pp. 52–58, on p. 52.

33. Brunhes, B., 1906. *Recherches sur la direction d'aimentation des roches volcaniques* (1), *Journale Physique* 4e sér. 5: 705–724; Mercanton, P. L., 1926. Inversion de l'inclinaison magnétique terrestre aux âges géologique, *Journal of Geophysical Research* 31: 187–190; and idem, 1926. Magnétisme terrestre – Aimentation de basaltes groenlandais, *Comptes Rendus de l'Académie des Sciences* (Paris), 182: 859–860; see also Glen, William, 1982, *The Road to Jaramillo: Critical Years of the Revolution in Earth Science.* Stanford, Calif.: Stanford University Press.

34. Matuyama, M., 1929. On the direction of magnetization of basalt in Japan, Tyôsen and Manchuria, *Proceedings of the Japanese Academy* 5: 203–205; see discussion in Cox, Allan, 1973. *Plate Tectonics and Geomagnetic Reversals.* San Francisco: W. H. Freeman, pp. 138–139.

35. Hospers, J., 1951. Remanent magnetism of rocks and the history of the geomagnetic field, *Nature* 168: 1111-1112; idem, 1953. Reversals of the main magnetic field, *Koninklÿke Nederlandse Akademische van Wetenschappen* 56 (series B): 467–477; idem, 1954. Magnetic correlation in volcanic districts, *Geological Magazine* 91: 352–360; see also Frankel, Henry, 1987. Jan Hospers and the rise of paleomagnetism, *EOS* 68(24): 156–160.

36. Cox, Allan, R. R. Doell, and G. B. Dalrymple, 1963. Geomagnetic polarity epochs and Pleistocene geochronometry, *Nature* 198: 1049–1051; idem, 1963. Geomagnetic polarity epochs: Sierra Nevada II, *Science* 142: 382–385; idem, 1964. Geomagnetic polarity epochs, *Science* 143: 351–352; and idem, 1965. Reversals of the earth's magnetic field, *Science* 144: 1537–1543; Tarling, D. H., 1962. Tentative correlation of Samoan and Hawaiian Islands using "reversals" of magnetization, *Nature* 196: 882–883; McDougall, I., and D. H. Tarling, 1963. Dating of polarity zones in the Hawaiian Islands, *Nature* 200: 54–56; idem, 1964. Dating geomagnetic polarity zones, *Nature* 202: 171–172. For a summary of the early geomagnetic chronologies, see Cox, 1973, note 35, on p. 145.

37. Heirtzler, J. R., and Xavier Le Pichon, 1965. Crustal structure of the mid-ocean ridges, 3. Magnetic anomalies over the mid-Atlantic ridge, *Journal of Geophysical Research* 70: 4013–4033; Heirtzler, J. R., Xavier Le Pichon, and J. G. Baron, 1966. Magnetic anomalies over the Reykjanes ridge, *Deep Sea Research* 13: 427–443; Pitman, W. C., III, and J. R Heirtzler, 1966. Magnetic anomalies over the Pacific-Antarctic ridge, *Science* 154: 1164–1171; Herron, E. M. and J. R. Heirtzler, 1967. Sea-floor spreading near the Galapagos, *Science* 158: 775–580; Heirtzler, J. R., 1968. Evidence for ocean floor spreading across the ocean basins. In *The History of the Earth's Crust*, R. A. Phinney, ed. Princeton: Princeton University Press, pp. 90–100; Pitman, W. C., III, E. M. Herron, and J. R. Heirtzler, 1968. Magnetic anomalies in the Pacific and sea floor spreading, *Journal of Geophysical Research* 73: 2069–2085; Heirtzler, J. R., G. O. Dickson, E. M. Herron, W. C. Pitman III, and Xavier Le Pichon, 1968. Marine magnetic anomalies, geomagnetic field reversals, and motions of the ocean floor and continents, *Journal of Geophysical Research* 73: 2119–2136; see also Vine, F. J. 1966. Spreading of the ocean floor: New evidence, *Science* 145: 1405–1415, and idem, 1968. Magnetic anomalies associated with mid-ocean ridges. In Phinney, 1968, pp. 73–89.

38. Pitman, this volume.

39. Opdyke, N. D., B. P. Glass, J. D. Hays, and J. H. Foster, 1966. Paleomagnetic study of Antarctic deep-sea cores, *Science* 154: 349–357; Opdyke, N. D., and B. P., Glass, 1969. The paleomagnetism of sediment cores from the Indian Ocean, *Deep Sea Research* 16: 249–261.

40. Wilson, J. T., 1965. Evidence from ocean islands suggesting movement in the earth. In Blackett et al., 1965, note 33, pp. 145–167.

41. Ibid, on pp. 163 and 165.

42. See Vine, this volume.

43. Wilson, J. T. 1965. A new class of faults and their bearing on continental drift, *Nature* 207: 343–347; see also Wilson, J. T., 1965. Transform faults, oceanic ridges and magnetic anomalies southwest of Vancouver Island, *Science* 150: 482–485.

44. Oliver J., and B. Isacks, 1967. Deep earthquakes zones, anomalous structures in the upper mantle, and the lithosphere, *Journal of Geophysical Research* 72: 4259–4275.

45. Sykes, L. R., 1968. Seismological evidence for transform faults, sea-floor spreading, and continental drift. In Phinney, 1968, note 37, pp. 120–150; Isacks, B., L. R. Sykes, and J. Oliver, 1968. Seismology and the new global tectonics, *Journal of Geophysical Research* 73: 5855–5899.

46. McKenzie, D. P., and R. L. Parker, 1967. The North Pacific: An example of tectonics on a sphere, *Nature* 216: 1276–1280; Morgan, W. Jason, 1968. Rises, trenches, great faults, and crustal blocks, *Journal of Geophysical Research* 73: 1959–1982.

47. Le Pichon, Xavier, 1968. Sea floor spreading and continental drift, *Journal of Geophysical Research* 73: 3661–3697.

48. See Dewey, Atwater, Dickinson, and Molnar, this volume.

CHAPTER 2

1. Alldredge, L. R., and F. Keller, 1949. Preliminary report on magnetic anomalies between Adak, Alaska and Kwajalein, Marshall Islands, *Transactions of the American Geophysical Union* 30: 494–500.

2. Heezen, B. C., M. Ewing, and E. T. Miller, 1953. Trans-Atlantic profile of total magnetic field and topography, Dakar to Barbados, *Deep Sea Research* 1: 25.

3. Mason, R. G., 1958. A magnetic survey off the west coast of the United States between latitudes 32° and 36° N, longitudes 121° and 128° W, *Geophysical Journal* 1: 320–329.

4. Mason, R. G., and A. D. Raff, 1961. Magnetic survey off the west coast of North America 32° N latitude to 42° N latitude, *Geological Society of America Bulletin* 72: 1259–1266; Raff, A. D., and R. G. Mason, 1961. Magnetic survey off the west coast of North America 40° N latitude to 52° N latitude, *Geological Society of America Bulletin* 72: 1267–1270.

5. Mason, A. D., and Martin Vitousek, 1959. Some geomagnetic phenomena associated with nuclear explosions, *Nature* 185: 52–54.

6. Vacquier, V., A. D. Raff, and R. E. Warren, 1961. Horizontal displacements on the floor of the Pacific Ocean, *Geological Society of America Bulletin* 72: 1251–1258.

7. Vacquier, V., 1965. Transcurrent faulting in the ocean floor, *Philosophical Transactions of the Royal Society* A258: 77–81; Menard, H. W., 1960. The East Pacific Rise, *Science* 132: 1737–1746.

8. Wilson, J. T., 1965. A new class of faults and their bearing on continental drift, *Nature* 207: 343–347.

9. Hess, H. H., 1962. History of ocean basins. In *Petrologic Studies*. A. E. J. Engel, H. L. James, and B. F. Leonard, eds. Denver: The Geological Society of America, pp. 599–620; Dietz, R. S., 1961. Continent and ocean basin evolution by spreading of the seafloor, *Nature* 190: 854–857.

10. Sykes, L. R., 1967. Mechanism of earthquakes and nature of faulting on the mid-oceanic ridges, *Journal of Geophysical Research* 72: 2131–2153.

11. Vine, F. J., and D. H. Matthews, 1963. Magnetic anomalies over ocean ridges, *Nature* 99: 947–999.

12. Cox, A., R. R. Doell, and G. B. Dalrymple, 1963. Geomagnetic polarity epochs and Pleistocene geochronometry, *Nature* 198: 1049–1051.

13. Wilson, J. T., 1965. Transform faults, ocean ridges and magnetic anomalies southwest of Vancouver Island, *Science* 150: 482–485; Vine, F. J., and J. T. Wilson, 1965. Magnetic anomalies over a young ocean ridge off Vancouver Island, *Science* 150: 485–489.

14. Heirtzler, J. R., X. Le Pichon, and J. G. Baron, 1966. Magnetic anomalies over the Reykjanes Ridge, *Deep-sea Research* 13: 427–443.

15. Vine, F. J., 1968. Magnetic anomalies associated with mid-ocean ridges.

In *The History of the Earth's Crust,* R. A. Phinney, ed. Princeton: Princeton University Press, pp. 73–79.

16. Pitman, W. C., and J. R Heirtzler, 1966. Magnetic anomalies over the Pacific-Antarctic ridge, *Science* 154: 1164–1171.

17. Bullard, E. C., 1964. Continental drift, *Quarterly Journal of the Geological Society* (London) 120: 1–33.

18. Blackett, P. M. S., J. A. Clegg, and P. H. S. Stubbs, 1960. An analysis of rock magnetic data, *Proceedings of the Royal Society* (London) A256: 291–322.

CHAPTER 3

1. Vine, F. J., and D. H. Matthews, 1963. Magnetic anomalies over oceanic ridges, *Nature* 199: 947–949.

2. Hess, H. H., 1962. History of ocean basins. In *Petrologic Studies: A Volume to Honor A. F. Buddington,* A. E. J. Engel, H. L. James, and B. F. Leonard, eds. Denver: The Geological Society of America, pp. 599–620.

3. Wilson, J. T., 1965. Transform faults, oceanic ridges, and magnetic anomalies southwest of Vancouver Island, *Science* 150: 482–485; Doell, R. R., and G. B. Dalrymple, 1966. Geomagnetic polarity epochs: A new polarity event and the age of the Brunhes-Matuyama boundary, *Science* 152:1060–1061.

4. Pitman, W. C., and J. R. Heirtzler, 1966. Magnetic anomalies over the Pacific Antarctic Ridge, *Science* 154: 1164–1171; Heirtzler, J. R., X. Le Pichon, and J. G. Baron, 1966. Magnetic anomalies over the Reykjanes Ridge, *Deep-Sea Research* 13: 427–443.

5. Heirtzler, J. R., 1968. Evidence for ocean floor spreading across the oceans. In *History of the Earth's Crust,* R. A. Phinney, ed. Princeton: Princeton University Press, pp. 90–100; Vine, F. J., 1968. Magnetic anomalies associated with mid-ocean ridges. In *History of the Earth's Crust,* R. A. Phinney, ed. Princeton: Princeton University Press, pp. 73–89.

6. Pitman and Heirtzler, 1966, note 4; Vine, F. J., 1966. Spreading of the ocean floor: New evidence, *Science* 154: 1405–1415.

7. Bullard, E. C., A. E. Maxwell, and R. Revelle, 1956. Heat flow through the deep sea floor, *Advances in Geophysics* 3: 153–181; Hill, M. N., 1959. A shipborne nuclear-spin magnetometer, *Deep-Sea Research* 5: 309–311.

8. Hess, 1962, note 2; Dietz, R. S., 1961. Continent and ocean basin evolution by spreading of the sea floor, *Nature* 190: 854–857.

9. Hess, H. H., 1946. Drowned ancient islands of the Pacific Basin, *American Journal of Science* 244: 772–791.

10. Vine, F. J., 1977. The continental drift debate, *Nature* 266: 19–22.

11. Mason, R. G., 1958. A magnetic survey off the west coast of the United States: 32°–36° N latitude, 121°–128° W longitude, *Geophysical Journal* 1: 320–329; Mason, R. G., and A. D. Raff, 1961. Magnetic survey off the west coast of North America, 32° N latitude to 42° N latitude, *Bulletin of the Geological Society of America* 72: 1259–1265; Raff, A. D., and R. G. Mason, 1961. Magnetic

survey off the west coast of North America, 40° N latitude to 52° N latitude, *Bulletin of the Geological Society of America* 72: 1267–1270.

12. Miller, E. T., and M. Ewing, 1956. Geomagnetic measurements in the Gulf of Mexico and in the vicinity of Caryn Peak, *Geophysics* 21: 406–432.

13. Girdler, R. W., and G. Peter, 1960. An example of the importance of natural remanent magnetisation in the interpretation of magnetic anomalies, *Geophysical Prospecting* 8: 474–483.

14. Runcorn, S. K., ed., 1962. *Continental Drift.* New York, Academic.

15. Laughton, A. S., M. N. Hill, and T. D. Allan, 1960. Geophysical investigation of a seamount 150 miles north of Madeira, *Deep-Sea Research* 7: 117–141; Matthews, D. H., 1961. Lavas from an abyssal hill on the floor of the North Atlantic Ocean, *Nature* 190: 158–159.

16. Raitt, R. W., R. L. Fisher, and R. G. Mason, 1955. Tonga trench. In *Crust of the Earth,* A. Poldervaart, ed. Geological Society of America Special Paper 62: 237–254; Mason, 1958, note 11.

17. Heiland, C. A., 1940. *Geophysical Exploration.* Englewood Cliffs, N.J.: Prentice Hall.

18. Matthews, D. H., F. J. Vine, and J. R. Cann, 1965. Geology of an area of the Carlsberg Ridge, Indian Ocean, *Bulletin of the Geological Society of America* 76: 675–682; Cann, J. R., and F. J. Vine, 1966. An area on the crest of the Carlsberg Ridge: Petrology and magnetic survey, *Philosophical Transactions of the Royal Society* A259: 198–217.

19. Kunaratnam, K., 1963. Applications of digital electronic computers to gravity and magnetic interpretation. Ph.D. thesis, University of London.

20. Mason, 1958; Mason and Raff, 1961; Raff and Mason, 1961, note 11.

21. Emiliani, C., ed., 1981. The oceanic lithosphere. In *The Sea,* vol. 7. New York: Wiley. (The letter submitted to *Nature* by L. W. Morley in February 1963 is reproduced in Appendix I, pp. 1717–1719.)

22. Vacquier, V., 1965. Transcurrent faulting in the ocean floor. In *Symposium on Continental Drift,* P. M. S. Blackett, E. C. Bullard, and S. K. Runcorn, eds. *Philosophical Transactions of the Royal Society* A258: 77–81.

23. Talwani, M., 1964. A review of marine geophysics, *Marine Geology* 2: 29–80.

24. Peter, G., and H. B. Stewart, 1965. Ocean surveys: The systematic approach, *Nature* 206: 1017–1018; Heirtzler, J. R., and X. Le Pichon, 1965. Crustal structure of the mid-ocean ridges 3: Magnetic anomalies over the Mid-Atlantic Ridge, *Journal of Geophysical Research* 70: 4013–4034.

25. Hess, H. H., 1964. Seismic anisotropy of the uppermost mantle under the oceans, *Nature* 203: 629–631; Backus, G. E., 1964. Magnetic anomalies over oceanic ridges, *Nature* 201: 591–592.

26. Wilson, J. T., 1965. A new class of faults and their bearing upon continental drift, *Nature* 207: 343–347.

27. Raff and Mason, 1961, note 11.

28. Menard, H. W., 1986. *The Ocean of Truth.* Princeton: Princeton University Press.

29. Cox, A., R. R. Doell, and G. B. Dalrymple, 1964. Reversals of the earth's magnetic field, *Science* 144: 1537–1543.

30. Wilson, 1965, note 3; Vine, F. J., and J. T. Wilson, 1965. Magnetic anomalies over a young oceanic ridge off Vancouver Island, *Science* 150: 485–489.

31. Wilson, 1965, note 3.

32. Heirtzler, Le Pichon, and Baron, 1966, note 4.

33. Pitman and Heirtzler, 1966, note 4; Vine, 1966, note 6.

34. Sykes, L. R., 1967. Mechanism of earthquakes and nature of faulting on the mid-ocean ridges, *Journal of Geophysical Research* 72: 2131–2153.

CHAPTER 4

1. Mason, R. G., and A. D. Raff, 1961. Magnetic survey off the west coast of North America, 40°N latitude to 52°N latitude, *Geological Society of America Bulletin* 72: 1267–1270.

2. The old magnetic balance type of magnetometer for measuring anomalies in the strength of the geomagnetic field was impossible to use in a moving vehicle. The airborne magnetometer, being electronic, is insensitive to accelerations of the vehicle in which it is carried. It could therefore be used as a towed instrument behind an aircraft or ship as long as the towline was long enough to escape the effect of magnetic interference caused by the ship or aircraft.

3. Sedimentary rocks and igneous rocks achieve remanent magnetization under the influence of the earth's field in different ways. The mineral in sedimentary rocks that gives them the ability to retain remanence or permanent magnetism is most often hematite (Fe_2O_3). As the particles settle out from turbid water under quiet conditions, each fine grain of hematite behaves like a small compass needle and aligns itself in the direction of the earth's field. When it settles and becomes compacted with the other non-ferromagnetic minerals, the whole rock takes on the direction of the prevailing geomagnetic field and retains this magnetization indefinitely, unless it is remagnetized by being heated to high temperature and cooling or by being subjected to a large magnetic field.

4. Igneous rocks receive their magnetizations in the direction of the earth's field when cooled below the Curie point of the mineral magnetite, which is 582°C. Magnetite (Fe_3O_4) is by far the most common ferromagnetic mineral in igneous rocks. It was the physicist husband of Mme. Curie, Pierre, who discovered that when a ferromagnetic material is heated to a certain high temperature (now called the Curie point), it totally loses its remanence. Conversely, when it is cooled from a temperature above the Curie point, it achieves a very hard or stable remanence in the direction of the earth's field. This is called *thermoremanence*. Physicists have found that ferromagnetic crystals of iron or iron oxide are divided into ferromagnetic 'domains' and when heated they lose the domain structure. Generally, hematite in sedimentary rocks retains its

remanence better than magnetite in igneous rocks, largely because of the fineness of the grains. Hematite also has a higher Curie point than magnetite (680°C). That is why sedimentary rocks are much more stable paleomagnetically than igneous rocks. However, the strength of their magnetization is much weaker. They require a much more sensitive remanent magnetometer than do most igneous rocks.

5. Graham, J. W., 1949. The stability and significance of magnetism in sedimentary rocks, *Journal of Geophysical Research* 54: 131–167.

6. Johnson, R. A., 1938. The limiting sensitivity of an A-C method for measuring small magnetic moments, *Review of Scientific Instruments* 9: 263–266.

7. Wegener, A., 1912. Die Entstehung der Kontinente, *Petermanns Mitteilungen* 58: 185–195, 253–56, 305–08.

8. Ferromagnetic materials have two kinds of magnetization: induced and remanent (or permanent). The common bar magnet has remanent magnetism, which it retains unless demagnetized. An ordinary piece of iron has very little remanent magnetization, but it is strongly attracted to a permanent magnet, and when it does so, it has magnetism induced due to the field of the permanent magnet, giving it the ability to attract another piece of iron. When the permanent magnet is released, the induced magnetism in the other two pieces is lost and they immediately fall apart. So it is in the case of rocks: they all contain both remanent and permanent magnetizations. The direction of the permanent magnetization is achieved when the rock is consolidated and is the direction of the earth's field prevailing at the time. The induced magnetism is always in the direction of the *present* earth's field. The field that creates the anomaly as measured above the surface of the earth is a combination of both kinds of magnetization. All theories developed for interpreting magnetic anomalies assume that all the magnetization is induced and is therefore in the direction of the *present* earth's field. When the effect of a negatively polarized rock is observed from the surface, the remanent magnetization must be stronger than the induced magnetization or it cannot be observed from the surface. This is what happened in the zebra maps. The inverse remanence was much stronger than the induced. That is why most scientists were not able to interpret the zebra pattern at first.

9. A Helmholtz coil (named after its inventor), consists of two equal coils of wire connected in series. The two coils are fastened parallel to each other and spaced apart at a distance roughly equal to their diameters. When an electric current is sent through the wire, it sets up a magnetic field at the center and between the two coils that is very spatially uniform in strength and direction over quite a large volume. If a paleomagnetic sample is placed at the center of such a coil and a strong alternating current is passed through the coils, the resulting strong magnetic field at the center will demagnetize the sample. For paleomagnetic purposes, the intensity of the current is adjusted to wash out the soft component of magnetization, leaving only the hard paleomagnetic component.

10. Dubois, P. M., 1957. Comparison of palaeomagnetic results for selected rocks of Great Britain and North America, *Advances in Physics* 6: 177.

11. The "polar-wandering school" refers to those scientists who believed that the age of a rock could be determined by calculating the location of the north magnetic pole from a rock sample's direction of magnetization and comparing this pole position with an established polar-wandering curve as the pole migrated throughout geological history.

12. Einarsson, Tr., and T. Sigugeirsson, 1955. Rock magnetism in Iceland, *Nature* 197: 892.

13. Hess, H. H., 1965. Mid-ocean ridges and tectonics of the sea floor. In *Submarine Geology and Geophysics*, W. F. Whittardand and R. Bradshaw, eds. London: Butterworth, pp. 317–332; Dietz, R. S., 1961. Continent and ocean basin evolution by spreading of the sea floor, *Nature* 190: 854–857.

14. Wegener, A., 1912, note 7.

15. Refer to note 2.

16. Hess, H. H., 1965. Mid-ocean ridges and tectonics of the sea floor. In Whittardand and Bradshaw, eds. 1965, note 13, London: Butterworth, pp. 317–320.

17. Dietz, R. S., 1961, note 13.

18. K values refer to the magnetic susceptibility of a sample.

19. Vine, F. J., and D. H. Matthews, 1963. Magnetic anomalies over ocean ridges, *Nature* 99: 947.

CHAPTER 5

1. Blackett, P. M. S., E. Bullard, and S. K. Runcorn, eds. 1965. A symposium on continental drift, *Philosophical Transactions of the Royal Society* (London) A258: 323.

2. Hess, H. H., 1962. History of ocean basins. In *Petrologic Studies: A Volume in Honor of A.F. Buddington*, A. E. J. Engle, ed. Denver: The Geological Society of America, pp. 599–620; Dietz, R. S., 1961. Continent and ocean basin evolution by spreading of the sea floor, *Nature* 190: 854-857.

3. Holmes, A., 1945. *Principles of Physical Geology*. New York: Ronald.

4. Dietz, 1961, note 2; Wilson, J. T., 1963. Hypothesis of the earth's behavior, *Nature* 198: 925–929.

5. Vine, F. J., and D. H. Matthews, 1963. Magnetic anomalies over oceanic ridges, *Nature* 199: 947–949.

6. Vine, F. J., and J. T. Wilson, 1965. Magnetic anomalies over a young oceanic ridge off Vancouver Island, *Science* 150: 485–489.

7. Cox, A., R. R. Doell, and G. B. Dalrymple, 1963. Geomagnetic polarity epochs and Pleistocene geochronology, *Nature* 198: 1049–1051; Cox, A., R. R. Doell, and G. B. Dalrymple, 1964. Reversals of the earth's magnetic field, *Science* 144: 1537–1543; Doell, R. R., and G. B. Dalrymple, 1966. Geomagnetic

polarity epochs: A new polarity event and the age of the Brunhes-Matuyama boundary, *Science* 152: 1060–1061.

8. Vine and Wilson, 1965, note 6.

9. Raff, A. D., and R. G. Mason, 1961. Magnetic survey off the west coast of North America, 40°N latitude to 52°N latitude, *Geological Society of America Bulletin* 72: 1267–1270; Mason, R. G., and A. D. Raff, 1961. Magnetic survey off the west coast of North America, 32°N latitude to 42°N latitude, *Geological Society of America Bulletin* 72: 1259–1265.

10. Vine and Wilson, 1965, note 6.

11. Vine and Matthews, 1963, note 5.

12. Cox et al., 1963, 1964, 1966, note 7.

13. Pitman, W. C., and J. R. Heirtzler, 1966. Magnetic anomalies over the Pacific-Antarctic ridge, *Science* 154: 1164–1171.

14. Cox, Doell, and Dalrymple, 1963, note 7; Cox, Doell, and Dalrymple, 1964, note 7; Doell and Dalrymple, 1966, note 7.

15. Opdyke, N. D., B. P. Glass, J. D. Hays, and J. H. Foster, 1966. Paleomagnetic study of Antarctic deep sea cores, *Science* 154: 349–357.

16. Opdyke, N. D., 1972. Paleomagnetism of deep sea cores, *Reviews of Geophysics and Space Physics* 10: 213–249.

17. Opdyke, N. D., 1972, note 16.

18. Sykes, L. R., 1967. Mechanism of earthquakes and nature of faulting on the mid-oceanic ridges, *Journal of Geophysical Research* 72: 2131–2153.

19. Isacks, B., J. Oliver, and L. R. Sykes, 1968. Seismology and the new global tectonics, *Journal of Geophysical Research* 73: 5855–5899; *Editor's note:* see also Oliver, this volume.

20. Vine, F. J., 1966. Spreading of the ocean floor: New evidence, *Science* 154: 1405–1415.

21. Heirtzler, J. R., G. O. Dickson, E. M. Herron, W. C. Pitman, and X. Le Pichon, 1968. Marine magnetic anomalies, geomagnetic field reversals and motions of the ocean floor and continents, *Journal of Geophysical Research* 73: 2119–2136.

22. Morgan, W. J., 1968. Rises, trenches, great faults and crustal blocks, *Journal of Geophysical Research* 72: 1959–1982.

CHAPTER 6

1. Matuyama, M., 1929. On the direction of magnetization of basalts in Japan, Tyôsen and Manchuria, *Proceedings of the Imperial Academy* (Japan) 5: 203–205; Mercanton, P. L., 1926. Inversion de l'inclinacion magnétique terrestre aux âges géologiques, *Journal of Geophysical Research* 31: 187–190.

2. Roche, A., 1951. Sur les inversion de l'aimentation rémanente des roches d'Auvergne, *Comptes Rendus de l'Académie des Sciences* (Paris) 236: 107–109.

3. Nagata, J., 1952. Reverse thermo-remanent magnetism, *Nature* 169: 704.

4. Hospers, J., 1951. Remanent magnetization of rocks and the history of the geomagnetic field, *Nature* 168: 1111–1112; Hospers, J., 1953–1954. Reversals of the main geomagnetic field I, II and III, *Koninklijke Nederlandse Akademische van Wetenschappen B.,* 56: 467–476 and 477–491, and 57: 112–121.

5. Irving, E., and S. K. Runcorn, 1957. Analysis of the paleomagnetism of the Torridonian Sandstone Series of Northwest Scotland I, *Philosophical Transactions of the Royal Society* A250: 83–99.

6. Nagata, 1952, note 3.

7. Uyeda, S., 1958. Thermoremanent magnetism as a medium of palaeomagnetism with special reference to reverse thermoremanent magnetism, *Japanese Journal of Geophysics* 2: 1–123.

8. Uyeda, pers. com., 1998.

9. Balsley, J. R. and A. F. Buddington, 1954. Correlation of reverse remanent magnetism and negative anomalies with certain minerals, *Journal of Geomagnetism and Geoelectricity* 6: 176–181; Balsley, J. R. and A. F. Buddington, 1958. Iron-titanium oxide minerals, rocks and aeromagnetic anomalies of the Adirondack area, New York, *Economic Geology* 53: 777–805.

10. Opdyke, N. D., and S. K. Runcorn, 1956. New evidence for reversal of the geomagnetic field near the Plio-Pleistocene boundary, *Science* 123: 1126–1127.

11. Mason, R. G. and A. D. Raff, 1961. A magnetic survey off the west coast of North America 32°N latitude to 42°N latitude, *Geological Society of America Bulletin* 72: 1259–1265.

12. *Editor's note:* it is now recognized as a transform fault on what was originally the mid-Pacific ridge, although now located well to the east of the middle of the Pacific basin.

13. Khramov, A. N, 1958. Paleomagnetism and stratigraphic correlation, *Gostop Techizdat Leningrad*: 218.

14. Hess, H. H., 1962. History of ocean basins. In *Petrologic Studies: A Volume in Honor of A.F. Buddington,* A. E. J. Engel, H. L. James, and B. F. Leonard, eds. Denver: The Geological Society of America, pp. 599–620; Dietz, R. S., 1961. Continent and ocean basin evolution by spreading of the sea floor, *Nature* 190: 854–857.

15. Vine, F. J., and D. H. Matthews, 1963. Magnetic anomalies over oceanic ridges, *Nature* 199: 947–949; *Editor's note:* see also Vine, F. J., and J. T. Wilson, 1965. Magnetic anomalies over a young ocean ridge off Vancouver Island, *Science* 150: 485–489; and Vine, this volume.

16. *Editor's note:* see Morley, this volume.

17. Cox, A., and R. R. Doell, 1960. Review of paleomagnetism, *Geological Society of America* 71: 645–768; Cox, A., R. R. Doell, and G. B. Dalrymple, 1963. Geomagnetic polarity epochs and Pleistocene geochronology, *Nature* 198: 1049–1051; Cox, A., R. R. Doell, and G. B. Dalrymple, 1963. Radiometric dating of geomagnetic field reversals, *Science* 140: 1021–1023; Doell, R. R., and G. B. Dalrymple, 1966. Geomagnetic polarity epochs: A new polarity event and

the age of the Brunhes–Matuyama boundary, *Science* 152: 854–857; McDougall, I., and D. H. Tarling, 1963. Dating of reversals of the earth's magnetic fields, *Nature* 198: 1012–1013.

18. Harrison, C. G. A., and B. M. Funnell, 1964. Relationship of palaeomagnetic reversals and micropalaeontology in two late Cenozoic cores from the Pacific ocean, *Nature* 204: 566.

19. Wensink, H., 1964. Paleomagnetic stratigraphy of younger basalts and intercalated Plio-Pleistocene tillites in Iceland, *Geologische Rundschau* 54: 364–384.

20. Foster, J. H., 1966. A paleomagnetic spinner magnetometer using a fuxgate gradiometer, *Earth and Planetary Science Letters* 1: 463–466.

21. Heirtzler, J. R., X. Le Pichon, and J. G. Baron, 1966. Magnetic anomalies over the Reykjanes Ridge, *Deep Sea Research* 13: 427–443.

22. Ade-Hall, J. M., 1964. The magnetic properties of some submarine oceanic lavas, *Geophysical Journal of the Royal Astronomical Society* 9: 85–92.

23. *Editor's note:* see Pitman, this volume.

24. *Editor's note:* the proceedings of this meeting, along with many comments from the audience, was published by Princeton University Press in 1968, *The History of the Earth's Crust*, edited by Robert Phinney; for his version of events, see MacDonald, this volume.

CHAPTER 7

1. Munk, W. H., and G. J. F. MacDonald, 1960. *The Rotation of the Earth.* Cambridge: Cambridge University Press.

2. Darwin, G., 1887. On the influence of geological change on the Earth's axis of rotation, *Philosophical Transactions of the Royal Society* 167: 271.

3. Daly, R. A., 1926. *Our Mobile Earth.* New York: Scribner.

4. *Editor's note:* see discussion in ch. 1.

5. Jeffreys, H., 1959. *The Earth.* 4th ed. Cambridge: Cambridge University Press.

6. Jeffreys, H., 1939. *The Theory of Probability.* Oxford: Clarendon.

7. Birch, F., 1952. Elasticity and constitution of the earth's interior, *Journal of Geophysical Research* 57: 227–286.

8. Harry Hess was a longtime (1934–1969) professor of geology at Princeton. His interests ranged from mineralogy to the structure of the ocean basins. His wartime duties in the early 1940s provided Hess with the opportunity to use his ship's echo sounder to collect ocean floor profiles across the North Pacific Ocean. Using these unplanned-for profiles, Hess discovered flat-topped submarine volcanoes, which he labeled *guyots* after the name of the Princeton geology building and first geology professor. Hess' rediscovery of drowned ocean structures earlier described by Charles Darwin was the first of his several major contributions to geology. Robert Dietz, like Hess, had broad interests and worked on the nature and origins of the moon's surface features, as well as on

meteorites and the craters they left when they struck the earth and the moon. His oceanographic work was extensive; he supervised the oceanographic research on Admiral Richard Byrd's final Antarctic expedition (1941–1947). Together Hess and Dietz are generally credited with the idea of sea floor spreading.

9. MacDonald, G. J., 1963. The deep structure of continents, *Reviews of Geophysics* 1: 587–665; MacDonald, G. J., 1964. The deep structure of continents, *Science* 143: 921–929.

10. The theory of isostasy is based on the findings of two American geodesists working for the U.S. Coast and Geodetic Survey in the early 20th century, John Hayford and William Bowie. See Hayford, J. F., 1909. *The Figure of the Earth and Isostasy from Measurements in the United States.* Washington, D.C.: U.S. Government Printing Office; Hayford, J. F., and William Bowie, 1912. The effect of topography and isostatic compensation upon the intensity of gravity, *U.S. Coast and Geodetic Survey Publication 10: Geodesy.* Washington, D.C.: U.S. Government Printing Office. For a historical discussion of their work and its impact on the debate over continental drift, see Oreskes, Naomi, 1999. *The Rejection of Continental Drift: Theory and Method in American Earth Science.* New York: Oxford University Press, pp. 39–48.

11. *Editor's note:* this requirement holds only if heat flow is entirely (or almost entirely) vertical. Given a mechanism for distributing heat laterally, this requirement is weakened or obviated. Similarly, isostatic balance can be partly maintained by distributing stresses laterally.

12. Jeffreys, H., 1959, note 5.

13. Munk, W. H., and G. J. F. MacDonald, 1960. Continentality and the gravitational field of the earth, *Journal of Geophysical Research* 65: 2169–2172.

14. Revelle, R., and A. E. Maxwell, 1952. Heat flow through the floor of the eastern North Pacific Ocean, *Nature* 170: 199–200; Bullard, E. C., 1954. The flow of heat through the floor of the Atlantic Ocean, *Proceedings of the Royal Society* (London) A222: 408–429.

15. Lee, W. H. K., and G. J. F. MacDonald, 1963. The global variation of terrestrial heat flow, *Journal of Geophysical Research* 68: 6481–6492.

16. Jeffreys, H., 1939. The times of P, S, and SKS, *Monthly Notices of the Royal Astronomical Society, Geophysics Supplement* 4: 498–533; Gutenberg, B., and C. Richter, 1935. On seismic waves (second paper), *Gerlands Beitrage Geophysik* 45: 280–360; Gutenberg, B., and C. F. Richter, 1936. On seismic waves (third paper), *Gerlands Beitrage Geophysik* 47: 73–131.

17. Pekeris, C. L., Z. Alterman, and H. Jarosch, 1952. Comparison of theoretical with observed values of the periods of the free oscillations of the earth, *Proceedings of the National Academy of Sciences, U.S.* 47: 91–98; MacDonald, G. J. F., and N. F. Ness, 1961. A study of the free oscillations of the earth, *Journal of Geophysical Research* 66: 1865–1911.

18. Gutenberg, B., and C. F. Richter, 1954. *Seismicity of the Earth and Associated Phenomena.* Princeton: Princeton University Press; Gutenberg, B., and C. F.

Richter, 1956a. Magnitude and energy of earthquakes, *Annales Geophysicae* 9: 1–15; Gutenberg, B., and C. F. Richter, 1956. Earthquake magnitude, intensity, energy and acceleration (second paper), *Bulletin of the Seismological Society of America* 46: 105–145.

19. MacDonald, G. J., 1963. The deep structure of continents, *Reviews of Geophysics* 1: 587–665.

20. Munk and MacDonald, 1960, note 1.

21. Jeffreys, H., 1959, note 5.

22. Jeffreys, H., 1959, note 5.

23. Menard, H. W., 1986, *The Ocean of Truth*. Princeton: Princeton University Press, p. 227.

24. Worzel, J. L., 1965. Discussion, *Philosophical Transactions of the Royal Society* 258: 167.

25. Phinney, R. A., ed., 1968. *The History of the Earth's Crust*. Princeton: Princeton University Press.

26. Gold, T., 1999. *The Deep Hot Biosphere*. New York: Springer-Verlag.

27. Muller, R. A., and G. J. MacDonald, 1997. Glacial cycles and astronomical forcing, *Science* 277: 215–218; idem, 2000. *Ice Ages and Astronomical Causes*. Chichester, U.K.: Springer–Praxis.

28. Cox, Allan, ed., 1973. *Plate Tectonics and Geomagnetic Reversals*. San Francisco: W. H. Freeman, p. 3.

29. Fischer, K. M., and R. D. van der Hilst, 1999. Enhanced: A seismic look under continents, *Science* 285: 1365–1366.

CHAPTER 8

1. Revelle, R. R., and A. E. Maxwell, 1952. Heat flow through the floor of the eastern North Pacific Ocean, *Nature* 170: 199–202.

2. Bullard, E. C., 1952. Discussion of a paper by R. R. Revelle and A. E. Maxwell. Heat flow through the floor of the eastern North Pacific Ocean, *Nature* 170: 200.

3. Sclater, J. G., 1966. Heat flux through the ocean floor, Ph.D. thesis, Cambridge University, 124.

4. Von Herzen, R. P., 1960. Pacific Ocean heat flow measurements, their interpretation and geophysical implications, Ph.D. thesis, UCLA, 119.

5. Von Herzen, R., 1959. Heat-flow values from the South-Eastern Pacific, *Nature*, 183: 882–883; Von Herzen, R. P., and S. Uyeda, 1963. Heat flow through the eastern Pacific Ocean floor, *Journal of Geophysical Research* 68: 4219–4250; Vacquier, V., and R. P. Von Herzen, 1964. Evidence for connection between heat flow and the mid-Atlantic ridge magnetic anomaly, *Journal of Geophysical Research* 69: 1093–1101; Von Herzen, R. P., and V. Vacquier, 1966. Heat flow and magnetic profiles on the Mid-Indian Ocean Ridge, *Transactions of the Royal Society* A259: 262–270.

6. Maxwell, A. E., R. P. Von Herzen, K. J. Hsü, J. E. Andrews, T. Saito, S. F.

Percival, E. D. Milow, and R. E. Boyce, 1970. Deep-sea drilling in the South Atlantic, *Science* 168: 1047–1059.

7. Williams, D. L., R. P. Von Herzen, J. G. Sclater, and R. N. Anderson, 1974. The Galapagos spreading center: Lithospheric cooling and hydrothermal circulation, *Geophysical Journal of the Royal Astronomical Society* 38: 609–626.

8. Detrick, R. S., R. P. Von Herzen, S. T. Crough, D. Epp, and U. Fehn, 1981. Heat flow on the Hawaiian swell and lithospheric reheating, *Nature* 292: 142–143; Von Herzen, R. P., M. J. Cordery, R. S. Detrick, and C. Fang, 1989. Heat flow and the thermal origin of hot spot swells: The Hawaiian swell revisited, *Journal of Geophysical Research* 94: 13783–13799.

9. Hess, H. H., 1962. History of ocean basins. In *Petrologic Studies: A Volume in Honor of A.F. Buddington*. A. E. J. Engel, H. L. James, and B. F. Leonard, eds., Denver: The Geological Society of America, pp. 599–620.

10. *Editor's note:* see McKenzie, this volume.

11. Bullard, E. C., A. E. Maxwell, and R. Revelle, 1956. Heat flow through the deep sea floor, *Advances in Geophysics* 3: 153–181.

12. Von Herzen, 1959, note 5; Bullard, Maxwell and Revelle, 1956; note 11.

13. Nason, R. D., and W. H. Lee, 1962. Preliminary heat flow profile across the Atlantic, *Nature* 196: 975.

14. Von Herzen and Uyeda, 1963, note 5.

15. Von Herzen, R. P., 1963. Geothermal heat flow in the Gulfs of California and Aden, *Science* 140: 1207–1208.

16. Von Herzen, 1960, note 4.

17. Menard, H. W., 1964. *Marine Geology of the Pacific*. New York: McGraw-Hill, p. 271.

18. Hess, 1962, note 9; Hess' sources include Menard, 1958, note 18; Bullard, Maxwell, and Revelle, 1956, note 11; Von Herzen, 1959, note 5; Heezen, B. C., 1960. The rift in the ocean floor, *Scientific American* 203: 98–110; Runcorn, S. K., 1959. Rock magnetism, *Science* 129: 1002–1011; Irving, E., 1959. Paleomagnetic pole positions, *Geophysical Journal of the Royal Astronomical Society* 2: 51–77.

19. Heezen, 1960, note 18.

20. Vacquier and Von Herzen, 1964, note 5.

21. Von Herzen and Vacquier, 1966, note 5.

22. Sclater, J. G., 1966. Heat flow in the northwest Indian Ocean and Red Sea, *Philosophical Transactions of the Royal Society* A259: 271–278.

23. Vine, F. J., and D. H. Matthews, 1963. Magnetic anomalies over oceanic ridges, *Nature* 199: 947–949; *Editor's note:* which McKenzie, Le Pichon, Pitman, and Heirtzler, respectively, did.

24. Laughton, A. S., 1966. The Gulf of Aden, *Philosophical Transaction of the Royal Society* (London) A259: 150–171; Sclater, 1966, note 22; Von Herzen and Vacquier, 1966, note 5.

25. Heirtzler, J. R., X. Le Pichon, and J. G. Baron, 1966. Magnetic anomalies over the Reykjanes Ridge, *Deep-Sea Research* 13: 427–443.

26. Popper, K. R., 1963. *Conjectures and Refutations*. London: Routledge.

27. Heirtzler, J. R., G. O. Dickson, E. M. Herron, W. C. Pitman III, and X. Le Pichon, 1968. Marine magnetic anomalies, geomagnetic reversals, and motions of the ocean floor and continents, *Journal of Geophysical Research* 73: 2119–2136.

28. McKenzie, D. P., and R. L. Parker, 1967. The North Pacific: An example of tectonics on a sphere, *Nature* 216: 1276–1280.

29. McKenzie, D. P., and D. L. Sclater, 1971. The evolution of the Indian Ocean since the Late Cretaceous, *Geophysical Journal of the Royal Astronomical Society* 25: 437–528.

30. Sclater, J. G., and J. Francheteau, 1970. The implications of terrestrial heat flow observations on current tectonic and geochemical models of the crust and upper mantle of the earth, *Geophysical Journal of the Royal Astronomical Society* 20: 509–537; Sclater, J. G., R. N. Anderson, and M. L. Bell, 1971. The elevation of ridges and the evolution of the central eastern Pacific, *Journal of Geophysical Research* 76: 7888–7915.

31. Hess, 1962, note 9; Wilson, J. T., 1965. A new class of faults and their bearing on continental drift, *Nature* 207: 343–347.

32. Kuhn, T. S., 1962. *The Structure of Scientific Revolutions*. Chicago: University of Chicago Press.

33. Oreskes, N., 1999. *The Rejection of Continental Drift: Theory and Method in American Earth Sciences*. New York: Oxford University Press.

34. Wilson, 1965, note 31.

35. Sykes, L. R., 1967. Mechanism of earthquakes and nature of faulting on the mid-oceanic ridges, *Journal of Geophysical Research* 72: 2131–2153; McKenzie and Parker, 1967, note 28; Morgan, W. J., 1968. Rises, trenches, great faults, and crustal blocks, *Journal of Geophysical Research* 73: 1959–1982; Isacks, B., J. Oliver, and L. R. Sykes, 1968. Seismology and the new global tectonics, *Journal of Geophysical Research* 73: 5855–5899.

36. Le Pichon, X., 1968. Sea floor spreading and continental drift, *Journal of Geophysical Research* 73: 3661–3697.

37. Bullard, Maxwell, and Revelle, 1956, note 11; Von Herzen and Uyeda, 1963, note 5.

38. Von Herzen, 1960 note 4.

39. Von Herzen and Uyeda, 1963, note 5.

40. Von Herzen and Uyeda, 1963, note 5.

41. Sclater, 1966, note 3.

42. Hess, 1962, note 9.

43. Hess, 1962, note 9; Wilson, 1965, note 31.

44. Langseth, M. G., X. Le Pichon, and M. Ewing, 1966. Crustal structure of the mid-ocean ridges, 5, Heat flow through the Atlantic Ocean floor and convection currents, *Journal of Geophysical Research* 71: 5321–5355; *Editor's note:* see Le Pichon, this volume.

45. McKenzie, D. P., 1967. Some remarks on heat flow and gravity anomalies, *Journal of Geophysical Research* 72: 6261–6273.

46. Vogt, P. R., and N. A. Ostenso, 1967. Steady state crustal spreading, *Nature* 215: 810–817; Langseth, Le Pichon, and Ewing, 1966, note 44.

47. McKenzie and Parker, 1967, note 28; Morgan, 1968, note 35; Isacks, Oliver, and Sykes, 1968, note 35; Le Pichon, 1968, note 36.

48. Sleep, N. H., 1969. Heat flow, gravity, and sea-floor spreading, *Journal of Geophysical Research* 74: 542–549; Sclater, Anderson, and Bell, 1971, note 30; Sclater and Francheteau, 1970, note 30.

49. Lister, C. R. B., 1972. On the thermal balance of a mid-ocean ridge, *Geophysical Journal of the Royal Society* 26: 515–535.

50. Sclater, J. G, J. Crowe, and R. N. Anderson, 1976. On the reliability of oceanic heat flow averages, *Journal of Geophysical Research* 81: 2997–3006.

51. Williams, Von Herzen, Sclater, and Anderson, 1974, note 7.

52. Hess, 1962, note 9.

53. Runcorn, 1959, note 18; Irving, 1959, note 18.

54. Jeffreys, H., 1929. *The Earth*. 2nd ed. Cambridge: Cambridge University Press.

55. Girdler, R. W., and G. Peter, 1960. An example of the importance of natural remanent magnetizations in the interpretation of magnetic anomalies, *Geophysical Prospecting* 8: 474–483.

56. Hess, 1962, note 9.

57. Hess, 1962, note 9.

58. Cox, A., R. R. Doell, and G. B. Dalrymple, 1963. Geomagnetic polarity epochs and Pleistocene geochronometry, *Nature* 198: 1049–1051; McDougall, I., and D. H. Tarling, 1963. Dating of polarity zones in the Hawaiian Islands, *Nature* 200: 54–56.

59. Vine and Matthews, 1963, note 23.

60. Lear, J., 1967. Canada's unappreciated role as a scientific innovator, *Saturday Review*, September 2: 45–50.

61. Hess, 1962, note 9; Wilson, 1965, note 31.

62. Langseth, Le Pichon, and Ewing, 1966, note 44.

63. McKenzie, 1967, note 45.

CHAPTER 9

1. Byerly, P., 1928. The nature of the first motion in the Chilean earthquake of November 11, 1922, *American Journal of Science* 16: 232–236.

2. Bolt, B. A., 1960. The revision of earthquake epicenters, focal depths, and origin-times using a high-speed computer, *Geophysical Journal of the Royal Astronomical Society* 3: 433–440.

3. *Editor's note:* the published paper used a deep earthquake in the Celebes Sea.

4. Flinn, E. A., 1960. Local earthquake location with an electronic computer, *Bulletin of the Seismological Society of America* 50: 467–470; Brazee, R. J., and R. Gunst, 1960. Hypocenter location of earthquakes by computer methods,

Geological Society of America Bulletin 71: 2051; Nordquist, J. M., 1962. A special purpose program for earthquake location with an electronic computer, *Bulletin of the Seismological Society of America* 54: 431–437; Engdahl, E. R., and R. Gunst, 1966. Use of a high speed computer for the preliminary determination of earthquake hypocenters, *Bulletin of the Seismological Society of America* 56: 325–336.

5. Sykes, L. R., 1963. Seismicity of the South Pacific Ocean, *Journal of Geophysical Research* 68: 5999–6006; Sykes, L. R., 1966. The seismicity and deep structure of island arcs, *Journal of Geophysical Research* 71: 2981–3006.

6. *Editor's note:* there is some confusion about terminology. Early workers referred to the "worldwide network of standard seismograph stations," but over time, the acronym WWSSN became widely used, interpreted as standing for the "world wide standard seismograph network." See, for example, Oliver, Jack, and Leonard Murphy, 1971. WWNSS: Seismology's global network of observing stations, *Science* 174: 254–261; vs. Presgrave, Bruce, Russell Needham, and John Minsch, 1985. Seismograph stations codes and coordinates, *U.S. Geological Survey Open File Report* 85–714, National Earthquake Information Center. The crucial point is that the stations used standardized instrumentation, facilitating global data integration.

7. Bolt, 1960, note 2.

8. Hodgson, J. H., ed., 1959. *The Mechanics of Faulting, with Special References to the Fault-Plane Work.* Ottawa, *Dominion Observatory Publication* 20; Stauder, W., 1964. A comparison of multiple solutions of focal mechanisms, *Bulletin of the Seismological Society of America* 54: 927–937.

9. Ritsema, A. R., 1964. Some reliable fault plane solutions. *Pure and Applied Geophysics* 59: 58–74; Stauder, W., 1964. A comparison of multiple solutions of focal mechanisms, *Bulletin of the Seismological Society of America* 54: 927–937.

10. Sykes, L. F., and M. Landisman, 1964. The seismicity of East Africa, the Gulf of Aden, and the Arabian and Red Seas, *Bulletin of the Seismological Society of America* 54: 1927–1940; Sykes, L. R., 1967. Mechanism of earthquakes and nature of faulting on the mid-oceanic ridges, *Journal of Geophysical Research* 72: 2131–2153.

11. *Editor's note:* see, for example, Barazangi, M., and J. Dorman, 1969. World seismicity maps computed from ESSA Coast and Geodetic Survey, Epicenter Data, 1961–1967, *Bulletin of the Seismological Society of America* 59: 369–380; and Atwater, this volume, for a discussion of how important this work was.

12. Byerly, 1928, note 1.

13. Richter, C. F., 1958. *Elementary Seismology.* San Francisco: W. H. Freeman.

14. Benioff, H., 1954. Orogenesis and deep crustal structure: Additional evidence from seismology, *Bulletin of the Seismological Society of America* 65: 385–440.

15. Bullen, K. E., 1947. *An Introduction to the Theory of Seismology.* Cambridge: Cambridge University Press.

16. Sykes, 1967, note 10.

17. Johnson, L. E., and E. Gilbert, 1972. Inversion and inferences for teleseismic ray data. In *Method in Computational Physics*, B. A. Bolt, ed. (12). New York: Academic Press, pp. 231–266.

18. Bolt, B. A., 1973. A proposal for the global calibration of group earthquake locations, *Geophysical Journal of the Royal Astronomical Society* 33: 249–256.

19. Brillinger, D., A. Udias, and B. A. Bolt, 1980. A probability model for regional focal mechanism solutions, *Bulletin of the Seismological Society of America* 70: 149–170.

20. Bolt, B. A., 1976. *Nuclear Explosions and Earthquakes: The Parted Veil.* New York: W. H. Freeman.

21. Vine, F. J., 1966. Spreading of the ocean floor: New evidence, *Science* 154: 1405–1415.

22. Isacks, B., J. Oliver, and L. R. Sykes, 1968. Seismology and the new global tectonics, *Journal of Geophysical Research* 73: 5855–5900.

CHAPTER 10

1. Menard, H., 1986. *The Ocean of Truth.* Princeton: Princeton University Press; Allegre, C., 1988. *The Behavior of the Earth.* Cambridge, Mass.: Harvard University Press; Le Grand, H. E., 1988. *Drifting Continents and Shifting Theories.* Cambridge: Cambridge University Press; Stewart, J., 1990. *Drifting Continents and Colliding Paradigms.* Bloomington: Indiana University Press; Oreskes, N., 1999. *The Rejection of Continental Drift: Theory and Method in American Earth Science.* Oxford: Oxford University Press.

2. Oliver, J., 1991. *The Incomplete Guide to the Art of Discovery.* New York: Columbia University Press; Oliver, J., 1996. *Shocks and Rocks; Seismology in the Plate Tectonics Revolution.* History of Geophysics 6, American Geophysical Union.

3. *Editor's note:* see also Bolt, this volume.

4. For example: William, Glen, 1982. *The Road to Jaramillo: Critical Years in the Plate Tectonics Revolution.* Stanford, Calif.: Stanford University Press.

5. Sykes, L. R., 1966. Mechanism of earthquakes and nature of faulting on the mid-oceanic ridges, *Journal of Geophysical Research* 71: 2981.

6. Coats, R. R., 1962. Magma type and crustal structure in the Aleutian arc. In *Crust of the Pacific Basin*, G. A. MacDonald and H. Kuno, eds. American Geophysical Union Geophysical Monograph 6, Washington, D.C., pp. 92–109.

7. Oliver, J., and B. Isacks, 1967. Deep earthquake zones, anomalous structures in the upper mantle, and the lithosphere, *Journal of Geophysical Research* 72: 4259.

8. *Editor's note:* these terms were first introduced in 1918 by American geodesist Joseph Barrell in the context of isostasy; see Barrell, J., 1927. On continental fragmentation, and the geologic bearing of the moon's surficial features, *American Journal of Science* 213: 283–314.

9. *Editor's note:* and by Dan McKenzie and Robert Parker at the Scripps Institution of Oceanography; see their essays in this volume.

10. Isacks, B., J. Oliver, and L. Sykes, 1968. Seismology and the new global tectonics, *Journal of Geophysical Research* 73: 5855–5899.

11. Stauder, W., 1968. Mechanism of the Rat Island earthquake sequence of February 4, 1965, with relation to island arcs and sea-floor spreading, *Journal of Geophysical Research* 73: 3847–3858; Stauder, W., 1968. Tensional character of earthquake foci beneath the Aleutian Trench with relation to sea-floor spreading, *Journal of Geophysical Research* 73: 7693–7701.

12. McKenzie, D. P., and R. L. Parker, 1967. The North Pacific: An example of tectonics on a sphere, *Nature* 216: 1276–1280.

13. Molnar, P., and J. Oliver, 1969. Lateral variation in attenuation in the upper mantle and discontinuities in the lithosphere, *Journal of Geophysical Research* 74: 2648–2682.

14. *Editor's note:* on American geologists' belief in inductive science earlier in the 20th century, see Oreskes, 1999, note 1, pp. 123–156.

15. Romm, J., 1994. A new forerunner for continental drift, *Nature* 367: 207. On the early history of the idea that continents were once connected, see Carozzi, A., 1970. New historical data on the origin of the theory of continental drift, *Geological Society of America Bulletin* 81: 283–285; and Carozzi, A. V., 1969. A propos de l'origine de la théories des dérives continentales: Francis Bacon (1620), François Placet (1668), A. Von Humboldt (1801), and A. Snider (1858), *Comptes Rendus et Séances de la Société des Physique et de la Histoire Naturelle* n.s. 4:171–179.

16. Oreskes, 1999, and Le Grand, 1988, note 1.

CHAPTER 11

1. Cox, A., 1973. *Plate Tectonics and Geomagnetic Reversals.* San Francisco: W. H. Freeman.

2. Menard, H.W., 1986. *The Ocean of Truth.* Princeton: Princeton University Press.

3. Kuhn, T. S. 1970. *The Structure of Scientific Revolutions.* Chicago: University of Chicago Press.

4. Fairbrother, N., 1954. *Children in the House.* London: Hogarth; Fairbrother, N., 1954. *An English Year.* New York: Knopf; Fairbrother, N., 1960. *The Cheerful Day.* London: Hogarth and New York: Knopf.

5. Bullard, E. C. 1957. Gerald Ponsonby Lenox-Conyngham 1866–1956. *Biographical Memoirs of the Fellows of the Royal Society* 3: 129–140. Bullard, E. C. 1967. Maurice Neville Hill 1919–1966. *Biographical Memoirs of the Fellows of the Royal Society* 13: 193–203. McKenzie, D., 1987. Edward Crisp Bullard 1907–1980. *Biographical Memoirs of the Fellows of the Royal Society* 33, 67–98.

6. McKenzie, D. P. 1977. Plate tectonics and its relationship to the evolution of ideas in the geological sciences, *Daedalus* 106: 97–124.

7. Vine, F. J., and D. H. Matthews, 1963. Magnetic anomalies over ocean ridges, *Nature* 199: 947–949. Bullard, E. C., J. E. Everett, and Smith, A. C. 1965. The fit of the continents around the Atlantic, *Philosophical Transactions of the Royal Society* A258: 41–51.

8. Williams, C. A., and D. McKenzie, 1971. The evolution of the northeast Atlantic, *Nature* 232: 168–173.

9. Wilson, J. T., 1965. A new class of faults and their bearing on continental drift, *Nature* 207: 343–347. Vine, F. J. and J. T. Wilson, 1965. Magnetic anomalies of a young oceanic ridge off Vancouver Island, *Science* 150: 485–489.

10. Phinney, R. A., ed.,1968. *The History of the Earth's Crust—A Symposium.* Princeton: Princeton University Press.

11. Vine, F. J., 1968. Magnetic anomalies associated with mid-ocean ridges. In *The History of the Earth's Crust,* R. A. Phinney, ed. Princeton: Princeton University Press, pp. 73–89. Sykes, L. R., 1968. Seismological evidence for transform faults, sea-floor spreading, and continental drift. In *The History of the Earth's Crust,* R. A. Phinney, ed. Princteon: Princeton University Press, pp. 120–150.

12. McKenzie, D. P., 1967. Some remarks on heat flow and gravity anomalies, *Journal of Geophysical Research* 72: 6261–6273.

13. Langseth, M. G., X. Le Pichon, and M. Ewing, 1966. Crustal structure of mid-ocean ridges, 5, Heat flow through the Atlantic Ocean floor and convection currents, *Journal of Geophysical Research,* 71: 5321–5355.

14. Batchelor, G. K. and R. Hide. 1988. Adrian Edmund Gill 1937–1986, *Biographical Memoirs of the Fellows of the Royal Society* 34: 221–258.

15. Menard, 1986, note 2.

16. Kuhn, 1970, note 3.

17. Bullard, Everett, and Smith, 1965, note 7.

18. McKenzie, D. P., 1968. The influence of the boundary conditions and rotation on convection in the Earth's mantle, *Geophysical Journal of the Royal Astronomical Society* 15: 457–500.

19. McKenzie, D. P., and R. L. Parker, 1967. The north Pacific: An example of tectonics on a sphere, *Nature* 216: 1276–1280.

20. Morgan, J. W., 1968. Rises, trenches, great faults, and crustal blocks, *Journal of Geophysical Research* 73: 1959–1982.

21. Menard, 1986, note 2.

22. Morgan, J. W., 1971. Convection plumes in the lower mantle, *Nature* 230: 42–43.

23. Read, W. T., 1953. *Dislocations in Crystals.* New York: McGraw-Hill.

24. Le Pichon, X., 1968. Sea floor spreading and continental drift, *Journal of Geophysical Research* 73: 3661–3697; Isacks, B., J. Oliver, and L. R. Sykes, 1968. Seismology and the new global tectonics, *Journal of Geophysical Research* 73: 5855–5899; Heirtzler, J. R., G. O. Dickson, E. M. Herron, W. C. Pitman III, and X. Le Pichon, 1968. Marine magnetic anomalies, geomagnetic field reversals, and motions of the ocean floor and continents, *Journal of Geophysical Research* 73: 2119–2136. Pitman, W. C., III, E. M. Herron, and J. R. Heirtzler, 1968. Magnetic anomalies in the Pacific and sea floor spreading, *Journal of Geophysical Research* 73: 2069–2085.

25. Menard, 1986, note 2.

26. Morgan, J.W., 1975. *Heat Flow and Vertical Movements of the Crust*. In *Petroleum and Global Tectonics,* A. G. Fischer and S. Judson, eds. Princeton: Princeton University Press, pp 23–43.

27. McKenzie, D. P., and W. J. Morgan, 1969. Evolution of triple junctions, *Nature* 224: 125–133. Atwater, T., 1970. Implications of plate tectonics for the Cenozoic tectonic evolution of western North America, *Bulletin of the Geological Society of America* 81: 3513–3536.

28. Bullard, Everett, and Smith, 1965, note 7.

29. McKenzie, 1977, note 6.

30. Isacks, Oliver, and Sykes, 1968, note 24.

31. McKenzie, D., and J. G. Sclater, 1971. The evolution of the Indian Ocean since the late Cretaceous, *Geophysical Journal of the Royal Astronomical Society* 25: 437–528.

32. Hey, R. N., 1977. A new class of pseudofaults and their bearing on plate tectonics: A propagating rift model, *Earth and Planetary Science Letters* 37: 321–325.

CHAPTER 12

1. McKenzie, D. P., and R. L. Parker, 1967. The North Pacific: An example of tectonics on a sphere, *Nature* 216: 1276–1280.

2. Vine, F. J., and D. H. Matthews, 1963. Magnetic anomalies over oceanic ridges, *Nature* 199: 947–949; Hess, H. H., 1962. History of ocean basins. In *Petrologic Studies: A Volume in Honor of A.F. Buddington,* A. E. J. Engel, H. L. James, and B. F. Leonard, eds. Denver: The Geological Society of America, pp. 599–620.

3. The state of knowledge on the subject at the end of the 1960s is lucidly summarized in Teddy Bullard's 1967 Bakerian lecture: 1968. Reversals of the earth's magnetic field, *Philosophical Transactions of the Royal Society* A263: 481–524.

4. Wilson, J. T., 1965. A new class of faults and their bearing on continental drift, *Nature* 207: 343–344.

5. Sykes, L. R., 1967. Mechanism of earthquakes and the nature of faulting on the mid-oceanic ridges, *Journal of Geophysical Research* 72: 2131–2153.

6. Bullard, E. C., J. E. Everett, and A. G. Smith, 1965. The fit of the continents around the Atlantic, *Philosophical Transactions of the Royal Society* A258: 41–51.

7. Carey saw the tectonic activity of the earth as a by-product of a general expansion of the earth through geologic time, a process for which there was no plausible physical explanation. See Carey, S. W., 1976. *The Expanding Earth.* New York: Elsevier Scientific.

8. Morgan, W. J., 1968. Rises, trenches, great faults, and crustal blocks, *Journal of Geophysical Research* 73: 1959–1982.

9. Parker, R. L., 1968. Electromagnetic induction in a thin strip, *Geophysical Journal of the Royal Astronomical Society* 14: 1968.

10. Bullard, E. C., and R. L. Parker, 1970. Electromagnetic induction in the oceans. In *The Sea, Ideas and Observations on Progress in the Study of the Seas.* New York: Wiley-Interscience, Part 1: 695–730.

11. GMT is a powerful map drawing and data display program that was written by Paul Wessel and Walter Smith while they were graduate students at Lamont. GMT has become indispensable to modern geophysics, but it is accessed through a distressingly Byzantine command structure, modeled after the Unix command line. See the website http://www.soest.hawaii.edu/gmt/.

12. Bullard, Everett, and Smith, 1965, note 6.

13. McKenzie, D. P., and R. L. Parker, 1974. Plate tectonics in 'omega' space, *Earth and Planetary Science Letters* 22: 285–293; Klitgord, K. D., S. P. Huestis, J. D. Mudie, and R. L. Parker, 1975. An analysis of near-bottom magnetic anomalies: Sea-floor spreading and the magnetized layer, *Geophysical Journal of the Royal Astronomical Society* 43: 387–424.

14. Parker, R. L., and D. W. Oldenburg, 1973. Thermal model of ocean ridges, *Nature: Physical Science* 242: 137–139.

15. Parker, R. L., 1994. *Geophysical Inverse Theory.* Princeton: Princeton University Press, p. 386.

CHAPTER 13

1. Le Pichon, X., 1984. La naissance de la tectonique des plaques, *La Recherche* 153: 414–422; Le Pichon, X., 1986. The birth of plate tectonics, *Lamont Year Book,* 53–61; Le Pichon, X., 1991. Introduction to the publication of the extended outline of Jason Morgan's April 17, 1967 American Geophysical Union Paper on "Rises, Trenches, Great Faults, and Crustal Blocks," *Tectonophysics* 187: 1–22.

2. Ewing, M., and B. C. Heezen, 1956. Some problems of Antarctic submarine geology, *American Geophysical Union, Geophysical Monograph* 1: 75–81.

3. Rothé, J. P., 1954. La zone séismique médiane Indo-Atlantique, *Proceedings of the Royal Society* A222: 387–397. *Editor's note:* the association of earthquakes with the mid-Atlantic ridge was earlier recognized by Heck, N. H., 1938. The role of earthquakes and the seismic method in submarine biology, *Proceedings of the American Philosophical Society* 79(1): 97–108.

4. Ewing, M., and F. Press, 1955. Geophysical contrasts between continents and ocean basins, *Geological Society America Special Paper* 62: 1.

5. Carey, S. W., 1958. A tectonic approach to continental drift. In *Continental Drift, A Symposium,* S. W. Carey, ed. Hobart: University of Tasmania, pp. 177–355.

6. Sullivan, W., 1974. *Continents in Motion, the New Earth Debate.* New York: McGraw-Hill.

7. Sullivan, 1974, note 6.

8. Holmes, Arthur, 1925. Radioactivity and the Earth's thermal history, *Geological Magazine* 62: 504–544. *Editor's note:* it is often stated that Hess originated the idea of the ocean floor as "a kind of endless traveling belt" but in fact this precise phrase was used by Arthur Holmes in his 1945 text (Holmes, Arthur, 1945. *Principles of Physical Geology.* New York: Ronald, pp. 507–508). What Hess

did was to revive an idea that American geologists had largely ignored, and of which American geophysicists were largely ignorant.

9. Hess. H. H., 1962. History of the ocean basins. In *Petrologic Studies,* Buddington Memorial Volume. A. F. J. Engel, H. L. James, and B. F. Leonard, eds. Denver: The Geological Society of America, pp. 599–620; Dietz, R. S., 1962. Continent and ocean evolution by spreading of the sea-floor, *Nature* 190: 854–857.

10. *Editor's note:* historian Alan Allwardt supports this idea, arguing that Hess presented his theory as "geopoetry" in order to deflect possible criticism. Hess knew how badly Wegener had fared with nearly the same idea and sought to minimize negative reactions by presenting his case diffidently. See Allwardt, A. S., 1990. The role of Arthur Holmes and Harry Hess in the development of modern global tectonics, Ph.D. dissertation, University of California, Santa Cruz.

11. *Editor's note:* See Opdyke, this volume.

12. Paleomagneticians knew that the magnetic poles had switched their polarity several times in the last few million years: What is magnetic north was then south.

13. The rift along which these anomalies were formed had since disappeared within the oceanic trench that had existed in earlier geological times along the North American western margin.

14. Vine, F. J., and D. H. Matthews, 1963. Magnetic anomalies over oceanic ridges, *Nature* 199: 947–949; Morley, L. W., and A. Larochelle, 1964. Paleomagnetism as a means of dating geological events, *Royal Society of Canada Special Publication* 9: 40–51.

15. See Glen, W., 1982. *The Road to Jaramillo, Critical Years of the Revolution in Earth Science.* Stanford, Calif.: Stanford University Press.

16. *Editor's note:* transform faults form along ridge crests where the ridge is fractured perpendicular to its axis. While the dominant motion of the plates on either side of the ridge is extensional – the plates are being pulled apart – along these fracture zones the two plates slip past each other, side by side.

17. Wilson, J. T., 1965. A new class of faults and their bearing on continental drift, *Nature* 207: 343–347.

18. Bullard, E. C., J. E. Everett, and A. G. Smith, 1965. The fit of the continents around the Atlantic. In *A Symposium on Continental Drift,* P. M. S. Blackett, E. C. Bullard, and S. K. Runcorn, eds. *Philosophical Transactions of the Royal Society* A258: 41–51.

19. Carey, 1958, note 5.

20. *Editor's note:* see Vine, and Morley, this volume, for somewhat different views.

21. *Editor's note:* see Oliver and Parker, this volume.

22. Le Pichon, X., R. Houtz, C. Drake, and J. Nafe, 1965. Crustal structure of the mid-ocean ridges, 1, Seismic refraction measurements, *Journal of Geophysical Research* 70(2): 319–339; Talwani, M., X. Le Pichon, and M. Ewing, 1965. Crustal structure of the mid-ocean ridges, 2, Computed model from gravity and seismic refraction data, *Journal of Geophysical Research* 70: 341–352.

23. Heirtzler, J. R. and X. Le Pichon, 1965. Crustal structure of the mid-ocean ridges, 3, Magnetic anomalies over the mid-Atlantic ridge, *Journal of Geophysical Research* 70: 4013–4033.

24. Ewing, M., X. Le Pichon, and J. Ewing, 1966. Crustal structure of the mid-ocean ridges, 4, Sediment distribution in the South Atlantic Ocean and the Cenozoic history of the mid-Atlantic ridge, *Journal of Geophysical Research* 71: 1611–1636.

25. It was necessary to wait for the JOIDES deep scientific drilling holes, started in 1968, to discover the major role of the very large variations, in time and in space, of the depth of dissolution of calcareous sediments. This is because the dissolution of carbonates increases very rapidly below a certain depth, beyond which only the famous abyssal red clays are deposited. If the depth of dissolution changes rapidly through time and space, the thickness of sedimentary cover may vary laterally in a very complex way.

26. It was only in 1970 that the solution to this difficult mechanical problem began to appear with the combined use of high-resolution, high-penetration commercial seismic reflection and drilling techniques.

27. Talwani, M., X. Le Pichon, and J. R. Heirtzler, 1966. East Pacific Rise: The magnetic pattern and the fracture zones, *Science* 150: 1108–1115.

28. Langseth, M., X. Le Pichon, and M. Ewing, 1966. Crustal structure of the mid-ocean ridges, 5, Heat flow through the Atlantic Ocean floor and convection currents, *Journal of Geophysical Research* 71: 5321–5355.

29. It was only after 1969 that the reason for this discrepancy was found: the measured heat flow was the conductive heat flow. It ignored the heat transported by hydrothermal circulation, which since then has been shown to be so important.

30. McKenzie, D. P., 1967. Some remarks on heat-flow and gravity anomalies, *Journal of Geophysical Research* 72: 6261–6273; *Editor's note:* see McKenzie, this volume.

31. Vening Meinesz, F. A., 1965. Origin of the crustal structure of the mid-ocean ridges, *Koninklĳke Nederlandse Akademische van Wetenschappen*, Ser. 68: 114–116.

32. Pitman, W. C., III, and J. R. Heirtzler, 1966. Magnetic anomalies over the Pacific-Antarctic Ridge, *Science* 154: 1164–1171.

33. Le Pichon, 1984, 1986, 1991, note 1.

34. Vine, J. F., 1966. Spreading of the ocean floor: New evidence, *Science* 154: 1405–1415.

35. Sykes, L. R., 1967. Mechanism of earthquakes and nature of faulting on the mid-oceanic ridges, *Journal of Geophysical Research* 72: 2131–2153.

36. Oliver, J., and B. Isacks, 1967. Deep earthquake zones, anomalous structures in the upper mantle, and the lithosphere, *Journal of Geophysical Research* 72: 4259–4275.

37. *Editor's note:* see MacDonald, this volume.

38. *Editor's note:* see McKenzie, D. P., and R. L. Parker, 1967. The North Pacific: An example of tectonics on a sphere, *Nature* 216: 1276–1280; Morgan, W. J., 1968. Rises, trenches, great faults and crustal blocks, *Journal of Geophysical Research* 73: 1959–1982.

39. Personal letter from Jason Morgan to me, July 1987.

40. Le Pichon, 1991, note 1.

41. Personal letter from Jason Morgan to me, July 1987.

42. Morgan, 1968, note 38.

43. *Editor's note:* Le Pichon details the chronology in order to establish that although Morgan's critical paper on plate tectonics was published in 1968 – after McKenzie and Parker (1967) – he had already presented his ideas to the scientific community in the spring of 1967.

44. Le Pichon, 1991, note 1.

45. Heirtzler, J. R., G. D. Dickson, E. M. Herron, W. C. Pitman, III, and X. Le Pichon, 1968. Marine magnetic anomalies, geomagnetic field reversals, and motions of the ocean floor and continents, *Journal of Geophysical Research* 73: 2119–2136.

46. Isacks, B., J. Oliver, and L. R. Sykes, 1968. Seismology and the new global tectonics, *Journal of Geophysical Research* 73: 5855–5899.

47. Bullard, Everett, and Smith, 1965, note 18.

48. McKenzie and Parker, 1967, note 38.

49. Menard, H. W., 1986. *The Ocean of Truth: A Personal History of Global Tectonics.* Princeton: Princeton University Press.

50. Cox, A., 1973. *Plate Tectonics and Geomagnetic Reversals.* San Francisco: W. H. Freeman; Bullard, Everett, and Smith, 1965, note 18.

51. *Editor's note:* see Parker, this volume.

52. McKenzie and Parker, 1967, note 38.

53. Menard, 1986, note 49.

54. Menard, 1986, note 49.

55. *Editor's note:* Similar ideas had earlier and independently been proposed by Backus, George E., 1964. Magnetic anomalies over oceanic ridges, *Nature* 201: 591–592.

56. Heezen, B. C., and M. Tharp, 1965. Tectonic fabric of Atlantic and Indian oceans and continental drift. In *Symposium on Continental Drift,* P. M. S. Blackett, E. C. Bullard, and S. K. Runcorn, eds. *Philosophical Transactions of the Royal Society* A258: 90–108; Sykes, 1967, note 35; Bullard, Everett, and Smith, 1965, note 18.

57. Heezen and Tharp, 1965, note 56.

58. Isacks, Oliver, and Sykes, 1968, note 46.

59. Heirtzler et al., 1968, note 45; Pitman, W. C., III, E. M. Herron, and J. R. Heirtzler, 1968. Magnetic anomalies in the Pacific and seafloor spreading, *Journal of Geophysical Research* 73: 2069–2085; Dickson, G. O., W. C. Pitman, III, and J. R. Heirtzler, 1968. Magnetic anomalies in the South Atlantic and ocean

floor spreading, *Journal of Geophysical Research* 73: 2087–2100; Le Pichon, X., and J. R. Heirtzler, 1968. Magnetic anomalies in the Indian Ocean and seafloor spreading, *Journal of Geophysical Research*, 76: 2101–2117.

60. *Editor's note:* Le Pichon downplays the role of the Scripps Institution of Oceanography: for alternative views, see Atwater, Mason, Morley, McKenzie, Parker, and Sclater, this volume. Morley emphasizes the importance of the Mason and Raff paleomagnetic data in getting the "whole thing" started; McKenzie and Atwater credit the general intellectual environment at Scripps.

61. Garfield, E., 1981. The 1,000 contemporary scientists most-cited 1965–1978, Part I, The basic list and introduction, *Current Contents* 41: 5–14.

62. Le Pichon, X., J. Francheteau, and J. Bonnin, 1973. *Plate Tectonics.* Amsterdam: Elsevier.

63. De Mets, C., R. G. Gordon, D. F. Argus, S. Stein, et al., 1990. Current plate motions, *Geophysical Journal International* 101: 425–478.

CHAPTER 14

1. Stereographic projections are a means to plot three-dimensional structures, such as bedding planes and fault surfaces, on a two-dimensional piece of paper. This enables geologists to see patterns that might otherwise be missed.

2. Vine, F. J., and D. H. Matthews, 1963. Magnetic anomalies over oceanic ridges, *Nature* 199: 947–949.

3. Hess, H. H., 1955. The oceanic crust, *Journal of Marine Research* 73: 423–439; Hess, H. H., 1962. History of ocean basins. In *Petrologic studies.* A. E. J. Engel, H. L. James and B. F. Leonard, eds. Denver: The Geological Society of America, pp. 599–620; Dietz, R. S., 1961. Continent and ocean basin evolution by spreading of the sea-floor, *Nature* 190: 854–857.

4. Wilson, J. T., 1965. A new class of faults and their bearing on continental drift. *Nature* 207: 343–347.

5. Dewey, J. F., and G. M. Kay, 1968. Appalachian and Caledonian evidence for drift in the North Atlantic. In *The History of the Earth's Crust.* R. A. Phinney, ed. Princeton: Princeton University Press, pp. 161–167.

6. *Editor's note:* convective heat transfer does break through the continents to form rifts, which may ultimately grow into new ocean basins. Dewey's point, however, is correct: once sea floor spreading begins, it is no longer a continent; it is a new ocean basin.

7. Sykes, L. R., 1967. Mechanism of earthquakes and the nature of faulting on the mid-ocean ridges, *Journal of Geophysical Research* 72: 2131–2153.

8. McKenzie, D. P., and R. L. Parker, 1967. The North Pacific: An example of tectonics on a sphere, *Nature* 216: 1276–1280; Morgan, W. J., 1968. Rises, trenches, great faults and crustal blocks, *Journal of Geophysical Research* 73: 1959–1982; Le Pichon, X., 1968. Sea-floor spreading and continental drift, *Jour-*

nal of Geophysical Research 73: 3661–3697; Isacks, B., J. Oliver, and L. R. Sykes, 1968. Seismology and the new global tectonics, *Journal of Geophysical Research* 73: 5855-5899.

9. Quennell, A. M., 1958. The structural and geomorphic evolution of the Dead Sea rift, *Quarterly Journal of the Geological Society of London* 114: 1–24.

10. Wellman, H. W., 1955. New Zealand Quaternary tectonics, *Geologische Rundschau* 43: 248–257.

11. Griggs, D. T., 1947. A theory of mountain building, *American Journal of Science* 237: 611–650; Holmes, A., 1944. *Principles of Physical Geology*. London: Thomas Nelson.

12. Plafker, G., 1965. Tectonic deformation associated with the 1964 Alaska earthquake, *Science* 146: 1675–1687.

13. Ampferer, O., and W. Hammer, 1911. Geologischer Querschnitt durch, die Ostalpen von Allgau zum Gardasee, Austria, *Geol. Bundesanst. Jarhrt.* 61: 531–710; Amstutz, A., 1955. Structures alpines, subductions successives dans L'Ossola, *Comptes Rendus de l'Académie des Sciences de Paris* Ser. D., 241: 967–969.

14. Wilson, J. T., 1966. Did the Atlantic close and then re-open? *Nature* 211: 676–681.

15. Le Pichon, 1968, note 8; Isacks, Oliver, and Sykes, 1968, note 8; Morgan, 1968, note 8.

16. *Editor's note:* see Opdyke, this volume.

17. Dewey, J. F., 1969. Evolution of the Appalachian/Caledonian Orogen, *Nature* 222: 124–129.

18. Atwater, T., 1970. Implications of plate tectonics for the Cenozoic tectonic evolution of western North America, *Geological Society of America Bulletin* 81: 3515–3536.

19. *Editor's note:* Peter Molnar makes a similar point, this volume.

20. Dewey, J. F., 1975. Finite plate evolution: Some implications for the evolution of rock masses and plate margins, *American Journal of Science* A275: 260–284.

21. Atwater, 1970, note 18; Smith, A. G., 1971. Alpine deformation and the oceanic areas of the Tethys, Mediterranean and Atlantic. *Geological Society of America Bulletin* 82: 2039–2070; Dewey, J. F., W. C. Pitman, W. B. F. Ryan, and J. Bonnin, 1973. Plate tectonics and the evolution of the Alpine system, *Geological Society of America Bulletin* 84: 3137–3180; Dewey, J. F., S. C. Cande, and W. C. Pitman, 1989. Tectonic evolution of the India-Eurasia convergent zone, *Eclogae Geologicae Helvetiae* 82: 717–734; Dewey, J. F., and S. H. Lamb, 1992. Active tectonics of the Andes, *Tectonophysics* 205: 79–95; Pindell, J. L., and J. F. Dewey, 1982. Permo-Triassic reconstruction of Western Pangea and the evolution of the Gulf of Mexico/Caribbean region, *Tectonics* 1: 179–211.

22. McKenzie, D. P., and W. J. Morgan, 1969. Evolution of triple junction, *Nature* 224: 125–133; Dewey, 1975, note 20.

23. *Editor's note:* scientists are less poorly paid nowadays than in the early and

mid-20th century, and have overall much more reliable and abundant funding. Perhaps the increase in oversight that Dewey laments is a consequence of having more: more money brings more micromanagement.

CHAPTER 15

1. In general, I have a very poor memory. I request forgiveness in advance for events that I misremember and from colleagues I may have misrepresented or slighted.

2. When I later asked one of my MIT professors why they bothered with us women, he said it was because a woman with an MIT education would raise great children (read "sons"). I didn't even blink. I guess I was used to it by then.

3. Ten years later I spent some geological field time in the Caucasus Mountains with three Georgian geologists, including a young woman, Manana Lordkipanze. Comparing notes, we found that she had had exactly the same experience, Soviet-style. She, too, was about to give up geology when she went to an international meeting in Moscow, heard about plate tectonics, and found her calling.

4. His presentation covered the material in Wilson, J. Tuzo, 1965. A new class of faults and their bearing on continental drift, *Nature* 207: 343–347.

5. Pitman, W. C., III, and J. R. Heirtzler, 1966. Magnetic anomalies over the Pacific-Antarctic Ridge, *Science* 154: 1164–1166. A few years ago I spent a fall sabbatical at Lamont. I took some time sorting through their voluminous data sets to plot out magnetic anomaly profiles from all the world's spreading centers. I was looking for good teaching examples to show the effects of latitude and spreading rate, but while I was at it I conducted a magnetic anomaly "beauty contest." After all these decades of ships collecting new data, that old *Eltanin*-19 crossing still won first prize. Its high latitude, medium fast spreading rate, and lack of noisy seamounts make it wonderfully clear and symmetric and exceptionally easy to read.

6. Vine, F. J., 1966. Spreading of the ocean floor; new evidence, *Science* 154: 1405–1415.

7. Actually, they must have been glad they had me after the second year. About then Uncle Sam ran out of cannon fodder for the Vietnam War and cancelled most graduate student deferments. Many of the young men in my graduate class quit and joined the Coast Guard as a preferable alternative to being drafted. Some years later, Allan Cox asked me if I felt guilty about not being eligible for the draft while all my male peers were having such a hard time with it. "Guilty?" I said, shocked. "No way! Lucky? Yes, but not guilty." This was not my war; not a war that my generation could believe in.

8. Atwater, Tanya M., and John D. Mudie, 1968. Block faulting on the Gorda Rise, *Science* 159: 729–731.

9. Barazangi, M., and J. Dorman, 1969. World seismicity map compiled from ESSA Coast and Geodetic Survey epicenter data, 1961–1967, *Seismological Society of America Bulletin* 59: 369–380.

10. Isacks, B., J. Oliver, and L. R. Sykes, 1968. Seismology and the new global tectonics, *Journal of Geophysical Research* 73: 5855–5899.

11. Morgan, W. J., 1968. Rises, trenches, great faults, and crustal blocks, *Journal of Geophysical Research* 73: 1959–1982.

12. H. W. Menard began his scientific career at the Naval Electronics Laboratory, and was appointed professor of geology at Scripps' Institute of Marine Resources in 1955, where he specialized in the topography of the Pacific sea floor.

13. Vacquier, Victor, Arthur D. Raff, and Robert E. Warren, 1961. Horizontal displacements in the floor of the northeastern Pacific Ocean, *Geological Society of America Bulletin* 72: 1251–1258; Mason, Ronald G., and Arthur D. Raff, 1961. Magnetic survey off the west coast of North America, 32°N. latitude to 42°N. latitude, *Geological Society of America Bulletin* 72: 1259–1265; Raff, Arthur D., and Ronald G. Mason, 1961. Magnetic survey off the west coast of North America, 40°N. latitude to 52°N. latitude, *Geological Society of America Bulletin* 72: 1267–1270; Peter, George, 1966. Magnetic anomalies and fracture pattern in the northeast Pacific Ocean, *Journal of Geophysical Research* 71: 5365-5374.

14. These anomalies would have given the *Eltanin*-19 profile some serious competition in the "beauty contest" if only their symmetrical halves had not been subducted under North America.

15. Long after I left Scripps, whenever Menard and I found each other at a meeting, we would make a lunch date just for the pleasure of spending a couple of hours scribbling on napkins, sharing our latest "mind candy." On my last visit with him, shortly before he died, I was pleased to see his blackboard filled with his latest ideas – possible correlations he was planning to track down.

16. Menard, H. W., and Tanya Atwater, 1968. Changes in direction of sea floor spreading, *Nature* 219: 463–467; Menard, H. W., and Tanya Atwater, 1969. Origin of fracture zone topography, *Nature* 222: 1037–1040; Atwater, Tanya, and H. W. Menard, 1970. Magnetic lineations in the northeast Pacific, *Earth and Planetary Science Letters* 7: 445–450.

17. Pitman, Walter C., III, and Dennis E. Hayes, 1968. Sea-floor spreading in the Gulf of Alaska, *Journal of Geophysical Research* 73: 6571–6580.

18. Grow, John A., and Tanya Atwater, 1970. Mid-Tertiary tectonic transition in the Aleutian arc, *Geological Society of America Bulletin* 81: 3715–3721.

19. In 1968 Peter Lonsdale gave a talk "Kula Plate not kula," in which he identified a patch of the Pacific plate which he believed had started out as a piece of the Kula plate and still is not "all gone." Lonsdale, Peter, and Debbie Smith, 1968. Kula Plate not kula, *Eos, Transactions of the American Geophysical Union* 67(44): 1199.

20. Published a few months later in McKenzie, D. P., and R. L. Parker, 1967. The north Pacific: An example of tectonics on a sphere, *Nature* 216: 1276–1280.

21. Cox, Allan, 1973. *Plate Tectonics and Geomagnetic Reversals*. San Francisco: W. H. Freeman.

22. McKenzie, D. P., and W. J. Morgan, 1969. Evolution of triple junctions, *Nature* 224: 125–133.

23. Lawson, Andrew Cowper, ed., 1908. The California earthquake of April 18, 1906. *Report of the State Earthquake Investigation Commission.* Carnegie Institution of Washington Publication 87.

24. Hill, Mason Lowell, and Thomas Wilson Dibblee, Jr., 1953. San Andreas, Garlock, and Big Pine faults, California; A study of the character, history, and tectonic significance of their displacements, *Geological Society of America Bulletin* 64: 443–458.

25. For example, Hamilton, Warren, 1969. Mesozoic California and the underflow of Pacific mantle, *Geological Society of America Bulletin* 80: 2409–2429.

26. Dickinson, William R., and Arthur Grantz, 1967. Indicated cumulative offsets along the San Andreas Fault in the California coast ranges. In *Proceedings of Conference on Geologic Problems of San Andreas Fault System.* William R. Dickinson and Arthur Grantz, eds. Stanford University Publications. *Geological Sciences* 11: 117–120.

27. Huffman, O. F., 1972. Lateral displacement of upper Miocene rocks and the neogene history of offset along the San Andreas Fault in Central California, *Geological Society of America Bulletin* 83: 2913–2946; Matthews, V., 1976. Correlation of Pinnacles and Neenach volcanic formations and their bearing on San Andreas Fault problems, *Bulletin of the American Association of Petroleum Geologists* 60: 2128–2141.

28. As summarized in Cox, Allan, et al., 1968. Radiometric time-scale for geomagnetic reversals, *Quarterly Journal of the Geological Society of London* 124: 53–66.

29. Pitman, W. C., III, E. M. Herron, and J. R. Heirtzler, 1968. Magnetic anomalies in the Pacific and sea floor spreading, *Journal of Geophysical Research* 73: 2069–2085; Dickson, G. O., W. C. Pitman III, and J. R. Heirtzler, 1968. Magnetic anomalies in the south Atlantic and ocean floor spreading, *Journal of Geophysical Research* 73: 2087–2100; Le Pichon, X., and J. R. Heirtzler, 1968. Magnetic anomalies in the Indian Ocean and sea-floor spreading, *Journal of Geophysical Research* 73: 2101–2117.

30. Heirtzler, J. R., G. O. Dickson, E. M. Herron, W. C. Pitman III, and X. Le Pichon, 1968. Marine magnetic anomalies, geomagnetic field reversals, and motions of the ocean floor and continents, *Journal of Geophysical Research* 73: 2119–2136.

31. Maxwell, A. E., R. P. Von Herzen, K. Jinghwa Hsu, J. E. Andrews, T. Saito, S. F. Percival, Jr., E. D. Milow, and R. E. Boyce, 1970. Deep sea drilling in the south Atlantic, *Science* 168: 1047–1059.

32. I have been surprised to discover how differently various colleagues feel about new discoveries. A colleague once told me that he loves to keep a new idea to himself for a while, savoring the feeling that he may be the only one in the world with that particular concept. In contrast, when something dawns for me, I can't wait to tell someone. It is as if it doesn't exist until I say it out loud.

33. Atwater, Tanya, 1970. Implications of plate tectonics for the Cenozoic tectonic evolution of western North America, *Geological Society of America Bulletin* 81: 3513–3535.

34. Actually, this ocean was too poorly known when we made the first circuit calculations so we had to insert yet another step, through Australia/India, to get from Antarctica to Africa.

35. Molnar, Peter, Tanya Atwater, Jacqueline Mammerickx, and Stuart M. Smith, 1975. Magnetic anomalies, bathymetry and the tectonic evolution of the South Pacific since the late Cretaceous, *Geophysical Journal of the Royal Astronomical Society* 40: 383–420.

36. Atwater, Tanya, and Peter Molnar, 1973. Relative motion of the Pacific and North American plates deduced from sea-floor spreading in the Atlantic, Indian, and South Pacific oceans. In *Conference on Tectonic Problems of the San Andreas Fault System, Proceedings*. Stanford University Publications. *Geological Sciences* 13: 136–148.

37. For information about GPS, visit the University NAVSTAR Consortium, 1998, UNAVCO brochure, www.unavco.ucar.edu/community/brochure/.

38. I find it quite wonderful that a system developed to track missiles and other fast-moving man-made objects is equally useful for monitoring the slow, stately drift of the plates.

39. Sagiya, T., S. Miyazaki, and T. Tada, 2000. Continuous GPS array and present-day crustal deformation of Japan. *PAGEOPH* 157: 2303–2322.

40. Thatcher, Wayne, G. R. Foulger, B. R. Julian, J. Svarc, E. Quilty, and G. W. Bawden, 1999. Present-day deformation across the Basin and Range Province, Western United States, *Science* 283: 1714–1718.

41. There is a wonderful refrain in an old folk song in which all sorts of outrageous things are being planned . . . "that is, if the weather be good."

42. *Editor's note:* for a discussion of this problem with respect to locating the oceanic trenches see Shor, E. N. 1978. *Scripps Institution of Oceanography: Probing the Oceans 1936–1976*. San Diego: Tofua, pp. 288–289.

43. Smith, Walter H. F., and David T. Sandwell, 1997. Global sea floor topography from satellite altimetry and ship depth soundings, *Science* 277: 1956–1962; Sandwell, David T., and Walter H. F. Smith, 1997. Marine gravity anomaly from Geosat and ERS 1 satellite altimetry, *Journal of Geophysical Research, B, Solid Earth and Planets* 102: 10,039–10,054.

44. Atwater, T., and J. M. Stock, 1998. Pacific-North America plate tectonics of the Neogene southwestern United States: An update, *International Geological Review* 40: 375–402. (Reprinted, 1998, in *Integrated Earth and Environmental Evolution of the Southwestern United States: The Clarence A. Hall, Jr., Volume.* W. G. Ernst, and C. A. Nelson, eds. Columbia, MD: Bellwether, pp. 393–420.)

45. For example, see Gans, P. B., et al., 2001. Rapid Eocene extension in the Robinson district, White Pine Co., Nevada: Constraints from 40Ar/39Ar dating, *Geology* 29: 475–478.

46. Wernicke, B. P., and J. K. Snow, 1998. Cenozoic tectonism in the central Basin and Range: Motion of the Sierran-Great Valley block, *International Geological Review* 40: 403–410. (Reprinted, 1998, in *Integrated Earth and Environmental Evolution of the Southwestern United States: The Clarence A. Hall, Jr.,*

Volume. W. G. Ernst and C. A. Nelson, eds., Columbia, MD: Bellwether, pp. 111–118.)

47. To see this quantitative deformation in action, visit my web site: www.geol.ucsb.edu/~atwater/ and download my animations. Enjoy.

CHAPTER 16

1. Wilson, J. Tuzo, 1965. A new class of faults and their bearing on continental drift, *Nature* 207: 343–347.

2. Atwater, Tanya, 1970. Implications of plate tectonics for the Cenozoic tectonic evolution of western North America, *Geological Society of America Bulletin* 81: 3513–3536.

3. Karig, Daniel E., 1971. Origin and development of marginal seas in the western Pacific, *Journal of Geophysical Research* 76: 2542–2561.

4. Knopf, Adolph, 1960. Analysis of some recent geosynclinal theory, *American Journal of Science (Bradley Volume)* A258: 126–136.

5. Kay, Marshall, 1951. *North American Geosynclines.* New York: Geological Society of America.

6. Dietz, Robert S., and John C. Holden, 1966. Miogeoclines in space and time, *Journal of Geology* 74: 566–583.

7. Dewey, John F., and John M. Bird, 1970. Plate tectonics and geosynclines, *Tectonophysics* 10: 625–638.

8. Wilson, 1965, note 1.

9. Dickinson, William R., and Arthur Grantz, eds., 1968. *Proceedings of Conference on Geologic Problems of San Andreas Fault System* (Stanford University Publications in the Geological Sciences, vol. 11). Stanford: School of Earth Sciences, Stanford University.

10. Wilson, J. Tuzo, 1965. Transform faults, oceanic ridges, and magnetic anomalies southwest of Vancouver Island, *Science* 150: 482–485; Dickinson, William R., Darrel S. Cowan, and Richard A. Schweickert, 1972. Test of new global tectonics [discussion], *American Association of Petroleum Geologists Bulletin* 56: 375–384.

11. Dickinson, William R., and Brian P. Wernicke, 1997. Reconciliation of San Andreas slip discrepancy by a combination of interior basin and range extension and transrotation near the coast, *Geology* 25: 663–665.

12. The International Upper Mantle Committee was an arm of the International Upper Mantle project, a program of research on the solid earth sponsored by the International Council of Scientific Unions (ICSU) and operated under the aegis of the International Union of Geodesy and Geophysics (IUGG) and the International Union of Geological Sciences (IUGS).

13. Dickinson, William R., 1969. Evolution of calc-alkaline rocks in the geosynclinal system of California and Oregon. In *Proceedings of the Andesite Conference,* Alexander R. McBirney, ed. Portland: Oregon Department of Geology and Mineral Industries Bulletin 65, pp. 151–156; idem, 1970.

Relations of andesitic volcanic chains and granitic batholith belts to the deep structures of orogenic arcs, *Geological Society of London Proceedings* 1662: 27–30.

14. Kuno, Hisashi, 1959. Origin of Cenozoic volcanic provinces of Japan and surrounding areas, *Bulletin Volcanologique* (Ser. II) 20: 37–76; Sugimura, Arata, 1960. Zonal arrangement of some geophysical and petrological features in Japan and its environs, *Journal of the Faculty of Science, University of Tokyo* (Sec. II) 12(2): 133–153; Katsui, Yoshio, 1961. Petrochemistry of the quaternary rocks of Hokkaido and surrounding areas, *Journal of the Faculty of Science, University of Hokkaido* (Ser. 4) 11: 1–58.

15. Frohlich, Cliff, 1987. Kiyoo Wadati and early research on deep focus earthquakes: Introduction to special section on deep and intermediate focus earthquakes, *Journal of Geophysical Research* 92 (B13): 13,777–13,788.

16. Dickinson, William R., and Trevor Hatherton, 1967. Andesitic volcanism and seismicity around the Pacific, *Science* 157(3790): 801–803.

17. Dickinson, William R., 1975. Potash-depth (K-h) relations in continental margin and intra-oceanic magmatic arcs, *Geology* 3: 53–56.

18. Dickinson, William R., 1968. Circum-Pacific andesite types, *Journal of Geophysical Research* 73: 2261–2269; idem, 1962. Petrogenetic significance of geosynclinal andesitic volcanism along the Pacific margin of North America, *Geological Society of America Bulletin* 73: 1241–1256.

19. Miyashiro, Akiho, 1967. Orogeny, regional metamorphism, and magmatism in the Japanese Islands, *Meddelelser fra Dansk Geologisk Forening* 17(4): 390–446.

20. Takeuchi, Hitoshi, and Seiya Uyeda, 1965. A possibility of present-day regional metamorphism, *Tectonophysics* 2: 59–68.

21. Hamilton, Warren, 1966. Origin of the volcanic rocks of eugeosynclines and island arcs. In *Continental Margins and Island Arcs*, W. H. Poole, ed. Ottawa: Geological Survey of Canada Special Paper 66–15, pp. 348–356.

22. Hamilton, Warren, 1969. The volcanic central Andes – A modern model for the Cretaceous batholiths and tectonics of western North America. In *Proceedings of the Andesite Conference*, Alexander R. McBirney, ed. Portland: Oregon Department of Geology and Mineral Industries Bulletin 65, pp. 175–184.

23. Isacks, B. L., J. Oliver, and L. R. Sykes, 1968. Seismology and the new global tectonics, *Journal of Geophysical Research* 73: 5855–5899.

24. Dickinson, William R., 1971. Plate tectonic models of geosynclines, *Earth and Planetary Science Letters* 10: 165–174; idem, 1971. Plate tectonic models for orogeny at continental margins, *Nature* 232: 41–42.

25. Dickinson, William R., 1970. The new global tectonics, *Geotimes* 15(4): 18–22; idem, 1970. Global tectonics, *Science* 168: 1250–1259.

26. Dickinson, William R., 1971. Plate tectonics in geologic history, *Science* 174: 107–113.

27. Dickinson, William R., 1998. A revolution in our time, *Geotimes* 43(11): 21–25.

28. Dickinson, William R., 1972. Evidence for plate-tectonic regimes in the rock record, *American Journal of Science* 272: 551–576; idem, 1972. Symposium: Plate tectonics [preface], *American Journal of Science* 272: 549–550.

29. Dickinson, William R., 1974. Preface. In *Tectonics and Sedimentation*, W. R. Dickinson, ed. Tulsa: Society of Economic Paleontologists and Mineralogists Special Publication 22, pp. iii–iv.

30. Dickinson, William R., 1974. Plate tectonics and sedimentation. In *Tectonics and Sedimentation*, W. R. Dickinson, ed. Tulsa: Society of Economic Paleontologists and Mineralogists Special Publication 22, pp. 1–27.

31. Dickinson, William R., 1977. Subduction tectonics in Japan, *American Geophysical Union Transactions (EOS)* 58(10): 948–952.

32. Curtis, G. H., J. F. Evernden, and J. Lipson, 1958. *Age Determination of Some Granitic Rocks in California by the Potassium-Argon Method.* San Francisco: California Division of Mines Special Report 54.

33. Irwin, William P., 1957. Franciscan group in coast ranges and its equivalents in Sacramento Valley, California, *American Association of Petroleum Geologists Bulletin* 41(10): 2284–2297.

34. Bailey, Edgar H., William P. Irwin, and David L. Jones, 1964. *Franciscan and Related Rocks, and their Significance in the Geology of Western California.* San Franciso: California Division of Mines and Geology Bulletin 183.

35. Dickinson, William R., 1965. Folded thrust contact between Franciscan rocks and Panoche group in the Diablo Range of Central California [abstract], *Geological Society of America Special Paper* 82: 248–249; idem, 1966. Table mountain serpentinite extrusion in California coast ranges, *Geological Society of America Bulletin* 77: 451–472.

36. Bailey, Edgar H., M. Clark Blake, Jr., and David L. Jones, 1970. On-land Mesozoic oceanic crust in the California coast ranges, *U.S. Geological Survey Professional Paper* 700-C: C70–C81.

37. Ojakangas, Richard W., 1964. Ph.D. dissertation, Stanford University; Ojakangas, Richard W., *Petrology and sedimentation of the Cretaceous Sacramento Valley sequence, Cache Creek, California,* 1968. Cretaceous sedimentation, Sacramento Valley, California, *Geological Society of America Bulletin* 79: 973–1008.

38. Dickinson William R., and Benjamin M. Page, 1970. *Central California Coast Ranges.* Stanford, Calif.: Cordilleran Section, Geological Society of America Annual Meeting Field Trip No. 1.

39. Hamilton, Warren, 1969. Mesozoic California and the underflow of Pacific mantle, *Geological Society of America Bulletin* 80: 2409–2430.

40. Ernst, W. Gary, 1970. Tectonic contact between the Franciscan mélange and the Great Valley sequence – Crustal expression of a late Mesozoic Benioff zone, *Journal of Geophysical Research* 75: 886–901.

41. Dickinson, William R., 1970. Relations of andesites, granites, and derivative sandstones to arc-trench tectonics, *Reviews of Geophysics and Space Physics* 8: 813–862.

42. Hsü, K. Jinghwa, 1971. Franciscan mélanges as a model for eugeosyncli-

nal sedimentation and underthrusting tectonics, *Journal of Geophysical Research* 76: 1162–1170.

43. Dickinson, William R., 1995. Forearc basins. In *Tectonics of Sedimentary Basins*, C. J. Busby and R. V. Ingersoll, eds. Cambridge: Blackwell, pp. 221–261.

44. Coats, Robert R., 1962. Magma type and crustal structure in the Aleutian arc. In *The Crust of the Pacific Basin*, G. A. MacDonald and H. Kuno, eds. Washington, D.C.: American Geophysical Union Geophysical Monograph 6, pp. 92–109.

45. Anderson, Roger N., Stephen E. DeLong, and W. M. Schwarz, 1980. Dehydration, asthenospheric convection and seismicity in subduction zones, *Journal of Geology* 88: 445–451.

46. Stille, Hans, 1955. Recent deformations of the earth's crust in the light of those of earlier epochs. In *Crust of the Earth*, Arie Poldervaart, ed. New York: Geological Society of America Special Paper 62, pp. 171–191.

47. Platt, John R., 1964. Strong inference, *Science* 146: 347–353.

48. Cassirer, Ernst, 1923. *Substance and Function*. Chicago: Open Court, 1923.

49. Coney, Peter J., 1970. Geotectonic cycle and the new global tectonics, *Geological Society of America Bulletin* 81: 739–748.

50. Dickinson, William R., 1981. Plate tectonics and the continental margin of California. In *The Geotectonic Development of California*. W. G. Ernst, ed. Englewood Cliffs: Prentice-Hall, pp. 1–28.

51. Dickinson, William R., 1973. Reconstruction of past arc-trench systems from petrotectonic assemblages in island arcs of the western Pacific. In *The Western Pacific: Island Arcs, Marginal Seas, and Geochemistry*, P. J. Coleman, ed. New York: Crane Russak, pp. 569–601; idem, 1978. Plate tectonic evolution of the North Pacific Rim, *Journal of the Physics of the Earth* 26 (Supplement): S1–S19.

52. Dickinson, William R. and William C. Luth, 1971. A model for plate tectonic evolution of mantle layers, *Science* 174: 400–404; idem, 1974. Subduction and oil migration, *Geology* 2: 421–424.

53. Dickinson, William R., and Christopher A. Suczek, 1979. Plate tectonics and sandstone composition, *American Association of Petroleum Geologists Bulletin* 63: 2164–2182; Dickinson, William R., and Renzo Valloni, 1980. Plate settings and provenance of sands in modern ocean basins, *Geology* 8: 82–86; Dickinson, William R., 1982. Compositions of sandstones in circum-Pacific subduction complexes and fore-arc basins, *American Association of Petroleum Geologists Bulletin* 66: 121–137; Dickinson, William R., 1985. Interpreting provenance relations from detrital modes of sandstones. In *Provenance of Arenites*, G. G. Zuffa, ed. NATO ASI Series. Dordrecht: Reidel, pp. 333–361.

54. Dickinson, William R., 1977. Tectono-stratigraphic evolution of subduction-controlled sedimentary assemblages. In *Island Arcs, Deep-Sea Trenches, and Back-Arc Basins*, M. Talwani and W. C. Pitman III, eds. Washington, D.C.: American Geopysical Union Maurice Ewing Series 1, pp. 33–40; idem, 1980. Plate tectonics and key petrologic associations. In *The Continental Crust and its Mineral Deposits*, D. W. Strangway, ed. Waterloo: Geological Association of Canada Special Paper 20, pp. 341–360; idem, 1988. Provenance and sediment

dispersal in relation to paleotectonics and paleogeography of sedimentary basins. In *New Perspectives in Basin Analysis,* K. L. Kleinspehn and C. Paola, eds. New York: Springer-Verlag, pp. 3–25; idem, 1992. Cordilleran sedimentary assemblages. In *The Cordilleran Orogen: Conterminous U.S.,* B. C. Burchfiel, P. W. Lipman, and M. L. Zoback, eds. *The Geology of North America,* vol. G-3, Boulder: The Geological Society of America, pp. 539–551.

55. Dickinson, William R., and Richard Shutler, Jr., 1971. Temper sands in prehistoric pottery of the Pacific Islands, *Archaeology in Oceania* 6(3): 191–203; Dickinson, William R., 1998. Petrographic temper provinces of prehistoric pottery in Oceania, *Records of the Australian Museum* 50: 263–276; Dickinson, William R., and Richard Shutler, Jr., 2000. Implications of petrographic temper analysis for oceanian prehistory, *Journal of World Prehistory* 14(3): 203–266.

56. Dickinson, William R. 1976. *Plate Tectonic Evolution of Sedimentary Basins.* Tulsa: American Association of Petroleum Geologists Continuing Education Course Note Series 1; Ingersoll, Raymond V., 1988. Tectonics of sedimentary basins, *Geological Society of America Bulletin* 100: 1704–1719; Ingersoll, Raymond V., and Cathy J. Busby, 1995. Tectonics of sedimentary basins. In *Tectonics of Sedimentary Basins,* C. J. Busby and R. V. Ingersoll, eds. Cambridge: Blackwell, pp. 1–51.

57. Dickinson, William R., 1993. Basin geodynamics, *Basin Research* 5: 195–196.

CHAPTER 17

1. Oliver, J., and B. Isacks, 1967. Deep earthquake zones, anomalous structures in the upper mantle, and the lithosphere, *Journal of Geophysical Research* 72: 4259–4275.

2. Molnar, P., and L. R. Sykes, 1969. Tectonics of the Caribbean and Middle America regions from focal mechanisms and seismicity, *Geologic Society of America Bulletin* 80: 1639–1684; Molnar, P., and J. Oliver, 1969. Lateral variations of attenuation in the upper mantle and discontinuities in the lithosphere, *Journal of Geophysical Research* 74: 2648–2682.

3. Talwani, M., X. Le Pichon, M. Ewing, G. H. Sutton, and J. L. Worzel, 1966. Comments on paper by W. Jason Morgan, "Gravity anomalies and convection currents. 2. The Puerto Rico Trench and the Mid-Atlantic Rise," *Journal of Geophysical Research* 71: 3602–3606.

4. Morgan, W. J., 1965. Gravity anomalies and convection currents. 1. A sphere and a cylinder sinking beneath the surface of a viscous fluid, *Journal of Geophysical Research* 70: 6175–6187; Morgan, W. J., 1965. Gravity anomalies and convection currents. 2. The Puerto Rico Trench and the Mid-Atlantic Rise, *Journal of Geophysical Research* 70: 6189–6204.

5. Morgan, W. J., 1968. Rises, trenches, great faults, and crustal blocks, *Journal of Geophysical Research* 73: 1959–1982.

6. Isacks, B., J. Oliver, and L. R. Sykes, 1968. Seismology and the new global tectonics, *Journal of Geophysical Research* 73: 5855–5899.

7. Oliver and Isacks, 1967, note 1; Sykes, L. R., 1967. Mechanism of earthquakes and the nature of faulting on the mid-ocean ridges, *Journal of Geophysical Research* 72: 2131–2153.

8. Oliver, J. E., 1993. *The Incomplete Guide to the Art of Discovery.* New York: Columbia University Press.

9. Isacks and Molnar, 1969, note 6. From seismograms recorded at numerous stations surrounding an earthquake, we can determine two perpendicular planes: one that is parallel to the plane of the fault that ruptured during the earthquake, and one that is perpendicular to the direction that one side of the fault slid with respect to the other. Sykes (see note 7), for instance, showed that earthquakes on fracture zones that offset mid-ocean ridges were not only consistent with strike-slip faulting, but also with transform faulting.

10. Benioff, H., 1954. Orogenesis and deep crustal structure: Additional evidence from seismology, *Geological Society of America Bulletin* 65: 385–400.

11. In addition to the orientations of the two planes, three other parameters are also estimated from a fault-plane solution of an earthquake: the P- and T-axes, which bisect the two planes, give the approximate orientations of maximum and minimum principal compressive stresses; the B-axis, defined by the intersection of the two planes, is approximately parallel to the intermediate principal stress.

12. Sykes, 1967, note 7.

13. Wilson, J. T., 1965. A new class of faults and their bearing on continental drift, *Nature*, 207: 343–347.

14. Morgan, W. J.,1971. Convection plumes in the lower mantle, *Nature* 230: 42–43.

15. Below the mantle, a layer of rock some 2,700 kilomoeters thick, lies the liquid "core" of the earth, which consists largely of iron and whose movements generate the earth's magnetic field.

16. Molnar, P., and T. Atwater, 1973. Relative motion of hotspots in the mantle, *Nature* 246: 288–291.

17. For example, Cande, S. C., C. A. Raymond, J. Stock, and W. F. Haxby, 1995. Geophysics of the Pitman fracture zone and the Pacific-Antarctic plate motions during the Cenozoic, *Science* 270: 947–953.

18. Morgan, 1968, note 5.

19. McKenzie, D. P., 1970. Plate tectonics of the Mediterranean region, *Nature* 226: 239–243; McKenzie, D. P., 1972. Active tectonics of the Mediterranean system, *Geophysical Journal of the Royal Astronomical Society* 30: 109–172.

20. Molnar, P., T. J. Fitch, and F. T. Wu, 1973. Fault plane solutions of shallow earthquakes and contemporary tectonics of Asia, *Earth and Planetary Science Letters* 16: 101–112.

21. Kostrov, V. V.,1974. Seismic moment, and energy of earthquakes, and the seismic flow of rock (in Russian), *Izvestiya Fiziki Zemlyi, Akademiya Nauk USSR* 1: 23–44.

22. McKenzie, D., and R. L. Parker, 1967. The North Pacific: An example of tectonics on a sphere, *Nature* 216: 1276–1280. The slip vector is the orientation (not strictly a vector) of relative movement of one side of a fault relative to the others. For earthquakes at plate boundaries, the slip vector is parallel to the direction that one plate moves with respect to the other.

23. For example, Aggarwal, Y. P., L. R. Sykes, J. Armbruster, and M. L. Sbar, 1973. Premonitory changes in seismic velocities and prediction of earthquakes, *Nature* 241: 101–106; Nur, A., 1972. Dilatancy, pore fluids, and premonitory variations in t_S/t_P travel times, *Bulletin of the Seismological Society of America* 62: 1217–1223; Scholz, C. H., L. R. Sykes, and Y. P. Aggarwal, 1973. Earthquake prediction: A physical basis, *Science* 181: 803–810.

24. Aggarwal et al., 1973, note 23; Whitcomb, J. H., J. D. Garmany, and D. L. Anderson, 1973. Earthquake prediction: Variation of seismic velocities before the San Fernando earthquake, *Science* 180: 632–641.

25. Molnar P., and P. Tapponnier, 1975. Cenozoic tectonics of Asia: Effects of a continental collision, *Science* 189: 419–426.

26. Bullard, E. C., 1936. Gravity measurements in East Africa, *Philosophical Transactions of the Royal Society* A235: 445–531.

27. Molnar and Tapponier, 1975, note 25.

28. Tapponnier, P., and P. Molnar, 1976. Slip-line field theory and large-scale continental tectonics, *Nature* 264: 319–324.

29. For example, Dewey, J. F., and J. M. Bird, 1971. Origin and emplacement of the ophiolite suite: Appalachian ophiolites in Newfoundland, *Journal of Geophysical Research* 76: 3179–3206.

30. Atwater, T., 1970. Implications of plate tectonics for the Cenozoic evolution of western North America, *Geological Society of America Bulletin* 81: 3513–3536.

31. Atwater, T., 1989. Plate tectonic history of the northeast Pacific and western North America. In *The Geology of North America*, vol. N, *The Eastern Pacific Ocean and Hawaii*, E. L. Winterer, D. M. Hussong, and R. W. Decker, eds. Boulder, Colo.: The Geological Society of America, pp. 21–72.

32. McKenzie, 1970, note 19; McKenzie, 1972, note 19.

33. McKenzie, D., 1977. Can plate tectonics describe continental deformation? In *International Symposium on the Structural Geology of the Mediterranean Basins: Split, (Yugoslavia), 25–29 October 1976*, B. Biju-Duval and L. Montadert, eds. Paris: Editions Technip, pp. 189–196.

34. For example, Avouac, J.-P., and P. Tapponnier, 1993. Kinematic model of active deformation in Central Asia, *Geophysical Research Letters* 20: 895–898; Leloup, P. H., R. Lacassin, P. Tapponnier, U. Schärer, Zhong Dalai, Liu Xiaohan, Zhang Liangshang, Ji Shaocheng, and Phan Trong Trinh, 1995. The Ailao Shan–Red River shear zone (Yunnan, China), Tertiary transform bound-

ary of Indochina. *Tectonophysics* 251: 3–84; Matte, Ph., P. Tapponnier, N. Arnaud, L. Bourjot, J. P. Avouac, Ph. Vidal, Liu Qing, Pan Yusheng, and Wang Yi, 1996. Tectonics of western Tibet, from the Tarim to the Indus, *Earth and Planetary Science Letters* 142: 311–330; Meyer, B., P. Tapponnier, L. Bourjot, F. Métevier, Y. Gaudemer, G. Peltzer, Guo Shunmin, and Chen Zhitai, 1998. Crustal thickening in Gansu-Qinghai, lithospheric mantle subduction, and oblique strike-slip controlled growth of the Tibet plateau, *Geophysics Journal International* 135: 1–47; Tapponnier, P., R. Lacassin, P. H. Leloup, U. Schärer, Zhong Dalai, Wu Haiwei, Liu Xiaohan, Ji Shaocheng, Zhang Lianshang, and Zhong Jiayou, 1990. The Ailao Shan/Red River metamorphic belt: Tertiary left-lateral shear between Indochina and South China, *Nature* 343: 431–437; Tapponnier, P., B. Meyer, J. P. Avouac, G. Peltzer, Y. Gaudemer, Guo Shunmin, Xiang Hongfa, Yin Kelun, Chen Zhitai, Cai Shuahua, and Dai Huagang, 1990. Active thrusting and folding in the Qilian Shan, and decoupling between upper crust and mantle in northeastern Tibet, *Earth and Planetary Science Letters* 97: 382–403; Wittlinger, G., P. Tapponnier, G. Poupinet, Jiang Mei, Shi Danian, G. Herquel, and F. Masson, 1998. Tomographic evidence for localized lithospheric shear along the Altyn Tagh Fault. *Science* 282: 74–76.

35. Cobbold, P. R., and P. Davy, 1988. Indentation tectonics in nature and experiment. 2. Central Asia, *Bulletin of the Geological Institute of Uppsala* 14: 143–162; England, P., and P. Molnar, 1990. Right-lateral shear and rotation as the explanation for strike-slip faulting in eastern Tibet, *Nature* 344: 140–142.

36. Menard claimed that he became the expert of the marine geology of the Pacific Ocean basin when one day he went downstairs at Scripps, looked at all of the echo-sounding records, and an hour later had seen all of the data. He and his colleagues then wrote most of the literature on this subject over the next 15 years.

37. For example, Molnar, P., and D. Gray, 1979. Subduction of continental lithosphere: Some constraints and uncertainties, *Geology* 7: 58–62.

38. Chopin, C., 1984. Coesite and pure pyrope in high-grade blueschists of the western Alps: A first record and some consequences, *Contributions to Mineralogy and Petrology* 86: 107–118.

39. For example, Brace, W. F., and D. L. Kohlstedt, 1980. Limits on lithospheric stress imposed by laboratory experiments, *Journal of Geophysical Research* 85: 6248–6252.

40. Brace, W. F., and J. D. Byerlee, 1966. Stick-slip as a mechanism for earthquakes, *Science* 153: 990–992; Brace, W. F., and J. D. Byerlee, 1970. California earthquakes: Why only shallow focus? *Science* 168: 1573–1576.

41. Byerlee, J. D., 1978. Friction in rocks, *Pure and Applied Geophysics* 116: 615–626.

42. Goetze, C., 1978. The mechanisms of creep in olivine, *Philosophical Transactions of the Royal Society* A288: 99–119; Kohlstedt, D. L., and C. Goetze, 1974. Low-stress high-temperature creep in olivine single crystals, *Journal of Geophysical Research* 79: 2045–2051.

43. Brace and Kohlstedt, 1980, note 39.

44. It helps to know that the Mayfair station serves a neighborhood that was once very posh, but no longer is.

45. Bird, P., and K. Piper, 1980. Plane stress finite element models of tectonic flow in southern California, *Physics of the Earth and Planetary Interiors* 21: 58–175; England, P., and D. McKenzie, 1982. A thin viscous sheet model for continental deformation, *Geophysical Journal of the Royal Astronomical Society* 70: 295–321; England, P., and D. McKenzie, 1983. Correction to: A thin viscous sheet model for continental deformation, *Geophysical Journal of the Royal Astronomical Society* 73: 523–532; Houseman, G. A., and P. C. England, 1986. Finite strain calculations of continental deformation, 1, Methods and general results for convergent zones, *Journal of Geophysical Research* 91: 3651–3663.

46. England and McKenzie, 1982, note 45.

47. For example, Anderson, R. E., 1971. Thin-skinned distension in Tertiary rocks of southeastern Nevada, *Geological Society of America Bulletin* 82: 43–58; Davis, G. A., J. L. Anderson, E. G. Frost, and T. J. Shackleford, 1980. Mylonitization and detachment faulting in the Whipple-Buckskin-Rawhide Mountains terrain, southeastern California and western Arizona, *Cordilleran Metamorphic Complexes, Geological Society of America Memoirs* 153: 79–130.

48. Jackson, J. A., and N. J. White, 1989. Normal faulting in the upper crust: Observations from regions of active extension, *Journal of Structural Geology* 11: 15–36.

49. Block, L., and L. H. Royden, 1990. Core complex geometries and regional scale flow in the lower crust, *Tectonics* 9: 557–567; Buck, W. R., 1988. Flexural rotation of normal faults, *Tectonics* 7: 959–974; Hamilton, W. B., 1988. Detachment faulting in the Death Valley region, California and Nevada. In *Geologic and Hydrologic Investigations of a Potential Nuclear Waste Disposal Site at Yucca Mountain, Southern Nevada*, M. D. Carr and J. C. Yount, eds. U.S. Geological Survey Bulletin 1790: 51–85; Jackson, J., and D. McKenzie, 1983. The geometrical evolution of normal fault systems, *Journal of Structural Geology* 5: 471–482; Kruse, S., M. McNutt, J. Phipps-Morgan, L. Royden, and B. Wernicke, 1991. Lithospheric extension near Lake Mead, Nevada: A model for ductile flow in the lower crust, *Journal of Geophysical Research* 96: 4435–4456; Spencer, J. E., 1984. The role of tectonic denudation in the warping and uplift of low-angle normal faults, *Geology* 12: 95–98; Wernicke, B. P., and G. J. Axen, 1988. On the role of isostasy in the evolution of normal fault systems, *Geology* 16: 848–851.

50. Isostasy is Archimedes' principle applied to the earth. As occurs with icebergs, the mass above sea level is compensated by lighter material at depth. For icebergs, most of the ice lies below sea level, and in many mountain ranges compensation occurs by thick crust underlying the high terrain.

51. Jones, C. H., 1987. Is extension in Death Valley accommodated by thinning of the mantle lithosphere beneath the Sierra Nevada, California? *Tectonics* 6: 449–473; Jones, C. H., and R. A. Phinney, 1998. Seismic structure of the lithosphere from teleseismic converted arrivals observed at small arrays in the southern Sierra Nevada and vicinity, California, *Journal of Geophysical Research*

103: 10,865–10,090; Jones, C. H., H. Kanamori, and S.W. Roecker, 1994. Missing roots and mantle "drips": Regional P_n and teleseismic arrival times in the southern Sierra Nevada and vicinity, California, *Journal of Geophysical Research* 99: 4567–4601.

52. Koestler, A., 1959. *The Sleepwalkers: A History of Man's Changing Vision of the Universe*. London: Penguin. Koestler expressed this so well that I feel the debt to quote him. On page 399, he mentioned "the Fifth book of *Harmonice Mundi*, which contains, almost hidden among the luxuriant growth of fantasy, Kepler's Third Law of planetary motion." Then on page 401, he wrote, "Not the least achievement of Newton was to spot the Three Laws in Kepler's writings, hidden away as they were like forget-me-nots in a tropical flowerbed."

53. Bally, A. W., P. L. Gordy, and G. A. Stewart, 1966. Structure, seismic data, and orogenic evolutions of southern Canadian Rocky Mountains, *Bulletin of Canadian Petroleum Geology* 14: 337–381.

54. For example, Dahlstrom, C. D. A., 1970. Structural geology in the eastern margin of the Canadian Rocky Mountains, *Bulletin of Canadian Petroleum Geology* 84: 1407–1422; Suppe, J., 1983. Geometry and kinematics of fault-bend folding, *American Journal of Science* 90: 648–721; Suppe, J., 1985. *Principles of Structural Geology*. Englewood Cliffs, N.J.: Prentice-Hall; Suppe, J., and D. A. Medwedeff, 1990. Geometry and kinematics of fault-propagation folding, *Eclogae geologicae Helvetiae* 83(3): 409–454. Numerous Russian geophysicists told me a joke popular in the Soviet Union in the 1970s: if you have n geologists, you will get $n + 1$ geological interpretations.

55. Sieh, K., 1978. Prehistoric large earthquakes produced by the San Andreas fault at Pallett Creek, California, *Journal of Geophysical Research* 83: 3907–3939.

56. Bibby, H. M., A. J. Haines, and R. I. Walcott, 1986. Geodetic strain and the present day plate boundary through New Zealand. In *Recent Crustal Movements of the Pacific Region, The Royal Society of New Zealand Bulletin* 24, W. I. Reilly and B. E. Hartford, eds. pp. 427–438; Walcott, R. I., 1984. The kinematics of the plate boundary zone through New Zealand: A comparison of short- and long-term deformations, *Geophysical Journal of the Royal Astronomical Society* 79: 613–633; Haines, A. J., 1982. Calculating velocity fields across plate boundaries from observed shear rates, *Geophysical Journal of the Royal Astronomical Society* 68: 203–209.

57. Haines, A. J., and W. E. Holt, 1993. A procedure for obtaining the complete horizontal motions within zones of distributed deformation from the inversion of strain rate data, *Journal of Geophysical Research* 98: 12,057–12,082; Holt, W. E., J. F. Ni, T. C. Wallace, and A. J. Haines, 1991. The active tectonics of the eastern Himalayan syntaxis and surrounding regions, *Journal of Geophysical Research* 96: 14,595–14,632.

58. For example, England, P., and P. Molnar, 1997. The field of crustal velocity in Asia calculated from Quaternary rates of slip on faults. *Geophysical Journal International* 130: 551–582; Holt, W. E., N. Chamot-Rooke, X. Le Pichon, A. J.

Haines, B. Shen-Tu, and J. Ren, 2000. Velocity field in Asia inferred from Quaternary fault slip rates and global positioning system observations, *Journal of Geophysical Research* 105: 19,185–19,209.

59. England, P., and P. Molnar, 1997. Active deformation of Asia: From kinematics to dynamics, *Science* 278: 647–650.

60. For example, Bazhenov, M. L., and V. S. Burtman, 1981. Formation of the Pamir-Punjab syntaxis: Implications from paleomagnetic investigations of Lower Cretaceous and Paleogene rocks of the Pamirs. In *Contemporary Scientific Researches in Himalaya*, A. K. Sinha, Bishen Singh Mahendra Pal Sing, eds. Dehra Dun, India, pp. 71–81; Beck, M. E., 1976. Discordant paleomagnetic pole positions as evidence of regional shear in the western cordillera of North America, *American Journal of Science* 276: 674–648; Luyendyk, B. P., M. J. Kamerling, and R. R. Terres, 1980. Geometric model for Neogene crustal rotations in southern California. *Geological Society of America Bulletin* 91: 211–217; Walcott, R. I., D. A. Christoffel, and T. C. Mumme, 1981. Bending within the axial tectonic belt of New Zealand in the last 9 Myr from paleomagnetic data, *Earth and Planetary Science Letters* 52: 427–434.

61. McKenzie, D., and J. Jackson, 1983. The relationship between strain rates, crustal thickening, paleomagnetism, finite strain and fault movements within a deforming zone, *Earth and Planetary Science Letters* 65: 182–202.

62. McKenzie, D., and J. Jackson, 1986. A block model of distributed deformation by faulting, *Journal of the Geological Society London* 143: 349–353.

63. Pratt, J. H., 1855. On the attraction of the Himalaya mountains, and of elevated regions beyond them, upon the plumb-line in India, *Philosophical Transactions of the Royal Society* 145: 53–100; Airy, G. B., 1855. On the computation of the effect of the attraction of mountain-masses, as disturbing the apparent astronomical latitude of stations of geodetic surveys, *Philosophical Transactions of the Royal Society* 145: 101–104.

64. Smithson, S. B., J. A. Brewer, S. Kaufman, J. E. Oliver, and C. A. Hurich, 1979. Structure of the Laramide Wind River uplift, Wyoming, from COCORP deep reflection data and from gravity, *Journal of Geophysical Research* 84: 5955–5972; Cook, F. A., D. Albaugh, L. Brown, S. Kaufman, J. Oliver, and R. Hatcher, Jr., 1979. Thin-skinned tectonics in the crystalline southern Appalachians: COCORP seismic reflection profiling of the Blue Ridge and Piedmont, *Geology* 7: 563–567.

65. Aki., K., A. Christoffersson, and E. S. Husebye, 1977. Determination of the three-dimensional seismic structure of the lithosphere, *Journal of Geophysical Research* 82: 277–296.

66. Roecker, S. W., 1982. Velocity structure of the Pamir-Hindu Kush region: Possible evidence for subducted crust, *Journal of Geophysical Research* 87: 945–959.

67. Chopin, 1984, note 38.

68. Hess, H. H., 1962. History of the ocean basins. In *Petrologic Studies: A Volume in Honor of A.F. Buddington*, A. E. J. Engel, H. L. James, and B. F.

Leonard, eds. Denver: The Geological Society of America, pp. 599–620; Hess, H. H., 1964. Seismic anisotropy of the uppermost mantle under oceans, *Nature* 203: 629–631.

69. Silver, P. G., and W. W. Chan, 1991. Shear-wave spitting and subcontinental mantle deformation, *Journal of Geophysical Research* 96: 16,429–16,454; Vinnik, L. P., V. Farra, and B. Romanowicz, 1989. Azimuthal anisotropy in the earth from observations of SKS at Geoscope and NARS broadband stations, *Bulletin of the Seismological Society of America* 79: 1542–1558.

70. Savage, M. K., 1999. Seismic anisotropy and mantle deformation: What have we learnt from shear wave splitting, *Reviews of Geophysics* 27: 64–106; Silver, P. G., 1996. Seismic anisotropy beneath the continents: Probing the depths of geology, *Annual Review of Earth and Planetary Sciences* 24: 385–432.

71. Suppe, 1983, note 54.

72. Langston, C. A., and D. V. Helmberger, 1975. A procedure for modeling shallow dislocation sources, *Geophysical Journal of the Royal Astronomical Society* 42: 117–130; Jackson, J. A. and T. J. Fitch, 1981. Basement faulting and the focal depths of the larger earthquakes in the Zagros Mountains (Iran), *Geophysical Journal of the Royal Astronomical Society* 64: 561–586.

73. Vine, F. J., 1977. The case of the midwife Fred or the confessions of a confirmed drifter (abstract), *Eos, Transactions of the American Geophysical Union* 58: 366.

74. Hess,1962, note 68.

75. Vine, F. J., and D. H. Matthews, 1963. Magnetic anomalies over oceanic ridges, *Nature* 199: 947–949.

76. Pitman, W. C., III, and J. R. Heirtzler, 1966. Magnetic anomalies over the Pacific-Antarctic ridge, *Science* 154: 1164–1171; Glen, W., 1982. *The Road to Jaramillo: Critical Years of the Revolution in Earth Sciences.* Stanford, Calif.: Stanford University Press, pp. 336–337: Vine, F. J., 1966. Spreading of the ocean floor: New evidence, *Science* 154: 1405–1415; Vine, F. J., and J. T. Wilson, 1965. Magnetic anomalies over a young ocean ridge off Vancouver Island, *Science* 150: 485–489.

77. Glen, 1982, note 76.

78. Vine, 1977, note 73.

79. For example, Fitch, T. J.,1972. Plate convergence, transcurrent faults, and internal deformation adjacent to southeast Asia and the western Pacific, *Journal of Geophysical Research* 77: 4432–4460; McCaffrey, R., 1992. Oblique plate convergence, slip vectors, and forearc deformation, *Journal of Geophysical Research* 97: 8905–8915.

80. For example, Le Grand, H. E., 1988. *Drifting Continents and Shifting Theories.* Cambridge: Cambridge University Press.

81. Stewart, J. A., 1990. *Drifting Continents and Colliding Paradigms, Perspectives on the Geoscience Revolution.* Bloomington: Indiana University Press, p. 141.

82. Lliboutry, L., 1998. The birth and development of the concept of glacio-isostasy, and its modelling up to 1974, *GeoResearch Forum* 3–4: 1–16.

83. Oreskes, N., 1999. *The Rejection of Continental Drift: Theory and Method in American Earth Science.* New York: Oxford University Press.

84. Oreskes, 1999, note 83.

85. For example, Hager, B. H., 1984. Subducted slabs of lithosphere and the geoid: Constraints on mantle rheology and flow, *Journal of Geophysical Research* 89: 6003–6015; Parsons, B., and S. Daly, 1983. The relationship between surface topography, gravity anomalies, and temperature structure of convection, *Journal of Geophysical Research* 88: 1129–1144; Richards, M. A., and B. H. Hager, 1984. Geoid anomalies in a dynamic earth, *Journal of Geophysical Research* 89: 5987–6002.

86. Argand, E.,1924. *La Tectonique de l'Asie, 13th International Geological Congress, Reports of Sessions* 13(1): 171–372.

87. Oreskes, 1999, note 83; Le Grand, 1988, note 80; Molnar, P., 1988. Continental tectonics in the aftermath of plate tectonics, *Nature* 335: 131–137.

88. Wilson, J. T., 1968. Revolution in earth science, *Geotimes* December: 10–16.

89. Le Grand, 1988, note 80, p. 2.

90. For example, Glen, W., 1975. *Continental Drift and Plate Tectonics.* Columbus: Charles E. Merrill; Hallam, A., 1973. *A Revolution in the Earth Sciences, From Continental Drift to Plate Tectonics.* London: Oxford University Press; Le Grand, 1988, note 80; Marvin, U. B., 1973. *Continental Drift: The Evolution of a Concept.* Washington, D.C.: Smithsonian Institution.

91. Glen, 1975, note 90, p. 186.

92. Le Grand, 1988, note 80.

93. Wilson, J. T., 1976. Preface to *Continents Adrift and Continents Aground: Readings from Scientific American.* San Francisco: W. H. Freeman; Wilson, 1968, note 88.

94. Menard, H. W., 1986. *The Ocean of Truth, A Personal History of Global Tectonics.* Princeton: Princeton University Press.

95. Kuhn, T. S., 1996. *The Structure of Scientific Revolutions.* 3rd ed. Chicago: University of Chicago Press, pp. 92–94.

96. Brinton, C., 1965. *The Anatomy of Revolution.* New York: Vintage.

97. Vine, 1966, note 76; Sykes, 1967, note 7; Wilson, 1965, note 13; Hess, 1962, note 68.

98. Oreskes, 1999, note 83.

99. Doyle, W., 1989. *The Oxford History of the French Revolution.* Oxford: Oxford University Press, pp. 405–406.

100. Doyle, 1989, note 99, pp. 407–413.

101. Schama, S., 1989. *Citizen: A Chronicle of the French Revolution* New York: Knopf, pp. 854–855.

102. Udintsev, G. B., 1995. The development of plate tectonics in Russia: An historical view, *Terra Nova* 7: 603–606.

103. Churchill, W. S., 1957. *The Age of Revolution.* New York: Dodd, Mead, p. 383.

104. Wilson, 1968, note 88.

105. Braudel, F., 1975. *The Mediterranean and the Mediterranean World in the Age of Philip II.* vol. 1. London: Fontana Press, p. 147.

106. Schama, 1989, note 101, p. xiii.

107. Sleep, N. H., 1971. Thermal effects of the formation of Atlantic continental margin by continental break up, *Geophysical Journal of the Royal Astronomical Society* 24: 325–350.

108. McKenzie, D., and M. J. Bickle, 1988. The volume and composition of melt generated by extension of the lithosphere, *Journal of Petrology* 29: 625–679.

109. Bally et al., 1966, note 53.

110. Kuhn, 1996, note 95.

111. Holton, G., 1973. *Thematic Origins of Scientific Thought, Kepler to Einstein.* Cambridge, Mass.: Harvard University Press.

112. Press, Frank, 1976. Personal communication; Frank Press was the head of the Department of Earth and Planetary Sciences at MIT and had hired John Edmond a few years earlier.

113. Heilbron, J. L., 1998. Beware scientists writing history, "The Atom in the History of Human Thought" (book review), *Nature* 395: 339–340.

CHAPTER 18

1. Kious, W. J., and R. I. Tilling, 1997. *This Dynamic Earth: The Story of Plate Tectonics.* Washington, D.C.: U.S. Geological Survey, U.S. Government Printing Office. http://pubs.usgs.gov/publications/text/dynamic.html

2. Smith, W. H. F., and D. T. Sandwell, 1997. Global sea floor topography from satellite altimetry and ship depth soundings, *Science* 277: 1956–1961.

3. Mammerickx, J., S. M. Smith, I. L. Taylor, and T. E. Chase, 1975. *Topography of the South Pacific.* La Jolla, Calif.: Institute of Marine Resources, University of California at San Diego.

4. Turcotte, D. L., and G. Schubert, 1982. *Geodynamics: Applications of Continuum Physics to Geological Problems.* New York: Wiley; Press, F., and R. Siever, 1994. *Understanding Earth.* San Francisco: W. H. Freeman.

5. Cox, A., 1973. *Plate Tectonics and Geomagnetic Reversals.* San Francisco: W. H. Freeman.

6. Raff, A. D., and R. G. Mason, 1961. Magnetic survey off the west coast of North America, 40°N latitude to 50°N latitude, *Geological Society of America Bulletin* 72: 1267–1270; Vine, F. J., 1966. Spreading of the ocean floor: New evidence, *Science* 154: 1405–1415; Heirtzler, J. R., G. O. Dickson, E. M. Herron, W. C. Pitman and X. Le Pichon, 1968. Marine magnetic anomalies, geomagnetic field reversals, and motions of the ocean floor and continents, *Journal of Geophysical Research* 73: 2119–2136; Atwater, T., and J. Severinghaus, 1989. *Tectonic Map of the Northeast Pacific Ocean.* Boulder, Colo.: Geological Society of America; Cande, S. C., J. L. LaBrecque, R. L. Larson, W. C. Pitman, X. Golovchenko, and W. F. Haxby, 1989. *Magnetic Lineations of the World's Ocean Basins.* Tulsa, Okla.: American Association of Petroleum Geologists.

7. Foster, J. S., and L. D. Welch, 2000. The evolving battlefield, *Physics Today* December: 31–35.

8. Cloud, J., 2000. Crossing the Olentangy River: The figure of the Earth and Military-Industrial-Academic Complex, 1947–1972. *Studies in the History and Philosophy of Modern Physics* 31: 371–404; Day, D. A., J. M. Logsdon, and B. Latell, 1998. *Eye in the Sky: The Story of the CORONA Spy Satellites.* Washington, D.C.: Smithsonian; Chesbrough, G. L., 1991. Letter from Oceanographer of the Navy to David Sandwell acknowledging the request for declassification and noting the reasons for classification of Geosat altimeter data, December 23.

9. MEDEA, 1995. *Scientific Utility of Naval Environmental Data.* McClean, Va.: MEDEA Office.

10. Kaula, W. M., 1963. Determination of the Earth's gravitational field, *Reviews of Geophysics* 1:507–551.

11. Haxby, W. F. 1987. *Gravity Field of the World's Oceans.* Boulder, Colo.: National Geophysical Data Center, NOAA.

12. Barazangi, M., and J. Dorman, 1969. World seismicity map compiled from ESSA Coast and Geodetic Survey epicenter data, 1961–1977, *Seismological Society of America Bulletin* 59: 369–380; Engdahl, E. R., R. van der Hilst and R. Buland, 1998. Global teleseismic earthquake relocation with improved travel times and procedures for depth determination, *Bulletin of the Seismological Society of America* 88: 722–743.

13. Wadati, K., 1928. Shallow and deep earthquakes, *Geophysical Magazine* 1: 162–202; Benioff, H., 1954. Orogenesis and deep crustal structure: Additional evidence from seismology, *Geological Society of America Bulletin Special Paper* 62: 61–75.

14. Kirby, S. H., S. Stein, E. A. Okal, and D. C. Rubie, 1996. Metastable mantle phase transformations and deep earthquakes in subducting oceanic lithosphere, *Reviews of Geophysics* 34: 261–306. These earthquakes occur in the transition zone 200 to 450 miles (300 to 720 kilometers) deep of rapidly increasing seismic velocity that also corresponds to phase transitions in the mantle. In particular, the olivine to spinel mineral phase transformation is associated with a significant volume decrease. As a relatively cold subducting slab falls through this phase transition depth, the phase transformation is delayed in a triangular wedge between the still cold interior of the slab and the outer edges of the slab. This difference in contraction results in a large stress. Why this stress is relieved in sudden events – earthquakes – is not well understood.

15. Pitman, W. C., and J. R. Heirtzler, 1966. Magnetic anomalies over the Pacific-Antarctic ridge, *Science* 154: 1164–1171.

16. Menard, H. W., 1986. *The Ocean of Truth.* Princeton: Princeton University Press.

17. Parsons, B., and J. G. Sclater, 1977. An analysis of the variation of ocean floor bathymetry and heat flow with age, *Journal of Geophysical Research* 82: 803–827.

18. Brace, K. L., 1977. *Preliminary Ocean-Area Geoid from GEOS-III Satellite Radar Altimetry.* St. Louis, Mo.: Defense Mapping Agency.

19. Tapley, B. D., G. H. Born, and M. E. Parke, 1982. The SEASAT altimeter data and its accuracy assessment, *Journal of Geophysical Research* 87: 3179–3188.

20. Apel, J., 1987. *The Navy Geosat Mission*. Laurel, Md.: Johns Hopkins Applied Physics Laboratory.

21. Chesbrough, 1991, note 8.

22. Minster, J. B., D. Sandwell and D. McAdoo, 1991. *Scientific Rationale for Declassification of Geosat Altimeter Data*, unpublished white paper; Chesbrough, G. L., 1992. Letter from Oceanographer of the Navy to David Sandwell acknowledging the declassification of all Geosat data south of 30° south latitude, June 10.

23. Medea, 1995, note 9.

24. Medea, 1995, note 9.

26. Connerney, J. E. P., M. H. Acuna, P. J. Wasilewski, N. F. Ness, H. Reme, C. Mazelle, D. Vignes, R. P. Lin, D. L. Mitchell and P. A. Cloutier, 1999. Magnetic lineations in the ancient crust of Mars, *Science* 284: 794–798.

26. Jordan, T. H., and J. B. Minster, 1988. Beyond plate tectonics: Looking at plate deformation with space geodesy. In *The Impact of VLBI on Astrophysics and Geophysics: Proceedings of the 129th Symposium of the International Astronomical Union held in Cambridge*, M. J. Ried and J. M. Moran, eds. Boston: Kluwer Academic, pp. 341–350.

27. Herring, T. A., I. I. Shapiro, T. A. Clark, C. Ma, J. W. Ryan, B. R. Schupler, C. A. Knight, G. Landquist, D. B. Shaffer, N. R. Vandenberg, B. E. Corey, H. F. Hinteregger, A. E. E. Rogers, J. C. Webber, A. R. Whitney, G. Elgered, B. O. Ronnang, and J. L. Davis, 1986. Geodesy by radio interferometry: Evidence for contemporary plate motion, *Journal of Geophysical Research* 91: 8341–8347.

FURTHER READING

Allegre, Claude, 1988. *The Behavior of the Earth*. Cambridge, Mass.: Harvard University Press.

Blackett, P. M. S., Sir Edward Bullard, and S. K. Runcorn, 1965. *A Symposium on Continental Drift, Philosophical Transactions of the Royal Society A258*. London: The Royal Society.

Cox, Allan, 1973. *Plate Tectonics and Geomagnetic Reversals*. San Francisco: W. H. Freeman.

Davies, Geoffrey F., 1999. *Dynamic Earth: Plates, Plumes and Mantle Convection*. Cambridge: Cambridge University Press.

Frankel, Henry, 1987. The continental drift debate. In *Resolution of Scientific Controversies: Case Studies in the Resolution and Closure of Disputes in Science and Technology*, H. T. Engelhardt, Jr., and A. L. Caplan, eds. Cambridge: Cambridge University Press, pp. 203–248.

Glen, William, 1982. *The Road to Jaramillo: Critical Years of the Revolution in Earth Science*. Stanford, Calif.: Stanford University Press.

Greene, Mott T., 1982. *Geology in the Nineteenth Century: Changing Views of a Changing World*. Ithaca: Cornell University Press.

Laudan, Rachel, 1980. The method of multiple working hypotheses and the discovery of plate tectonic theory. In *Scientific Discovery: Case Studies, Boston Studies in the Philosophy of Science 60*. Dordrecht: Reidel, pp. 331–343.

Le Grand, H. E., 1988. *Drifting Continents and Shifting Theories*. Cambridge: Cambridge University Press: Cambridge.

Le Pichon, Xavier, Jean Francheteau, and Jean Bonnin, 1973. *Plate Tectonics*. Amsterdam: Elsevier Scientific.

Marvin, Ursula B., 1973. *Continental Drift: The Evolution of a Concept*. Washington, D.C.: Smithsonian Institution Press.

Menard, H. W., 1986. *The Ocean of Truth: A Personal History of Global Tectonics*. Princeton: Princeton University Press.

Oreskes, Naomi, 1999. *The Rejection of Continental Drift: Theory and Method in American Earth Science*. New York: Oxford University Press.

Phinney, R. A. ed., 1968. *The History of the Earth's Crust*. Princeton: Princeton University Press.

Stewart, J. A., 1990. *Drifting Continents and Colliding Paradigms: Perspectives on the Geoscience Revolution*. Bloomington: Indiana University Press.

Strahler, Arthur N., 1998. *Plate Tectonics*. Cambridge, Mass.: Geobooks.

Wegener, Alfred, 1929. *The Origin of Continents and Oceans*, translated from the 4th rev. German ed. by John Biram. New York: Dover.

ABOUT THE CONTRIBUTORS

Tanya Atwater received her Ph.D. from the Scripps Institution of Oceanography in 1972, and taught at MIT before joining the faculty at the University of California, Santa Barbara. Her research in tectonics has taken her to the bottoms of the oceans and the tops of mountains on many continents. She is especially well known for her work on the plate tectonic history of western North America and the origins of the San Andreas Fault system. At Santa Barbara, she teaches geology and tectonics at all levels, and is deeply involved in working with the media, museums, and K-12 teachers to bring Earth awareness to all. She is the winner of the Newcomb Prize of the American Association for the Advancement of Science, and was elected in 1997 to the U.S. National Academy of Sciences.

Bruce Bolt is Professor of Seismology, Emeritus, at the University of California, Berkeley, where he was Director of the Seismographic Stations from 1963 to 1993. Bolt began his scientific career as an applied mathematician, and moved into seismology while on sabbatical leave at the Lamont Geological Observatory, where he developed a computer program for locating the sources of earthquakes that was used worldwide for many years. He is the author of numerous scientific articles and books, including "Earthquakes" (W. H. Freeman, New York 1999), and consults internationally earthquake hazards.

John Dewey is Professor of Geology at the University of California, Davis, having previously taught at Manchester, Cambridge, Durham, and Oxford Universities in the United Kingdom, and the State University of New York, Albany, in the United States. He has published over 140 scientific papers, and is a Fellow of the Royal Society, the U.S. National Academy of Science, and the European Academy of Sciences. His research has been principally focused on tectonics and structural geology, from the hand sample to the continental scale. His philosophy is that "the truth resides in the rocks," and that little can be achieved in the earth sciences without a good geological map.

Bill Dickinson is Emeritus Professor of Geosciences at the University of Arizona. After 33 years on the faculties of Stanford and Arizona, he retired in 1991 to pursue geologic research full time. His field and petrographic research focuses on sedimentary basins, sandstones, and geoarchaeology in a tectonic context, with time divided over the years between western North America and islands of the tropical Pacific Ocean. His recent contributions include analyses of the tectonic evolution of Mexico, and the history of sea level change in the Pacific during the very recent geological past.

Xavier Le Pichon is titular of the "Chaire de Géodynamique" at the Collège de France in Paris. In 1968, while terminating a five-year stay in the Lamont Geological Observatory, he published the first global quantitative model of plate motion, showing that it could explain most of the world's seismicity. In 1973, after his return to France, he published, with Jean Francheteau and Jean Bonnin, the first book on plate tectonics. He was one of the initiators of the submersible exploration of mid-ocean ridges and trenches, and in recent years has used geodetic data as a new tool for plate tectonic analysis of geodynamic problems.

Gordon MacDonald retired in 2001 as Director of the International Institute for System Analysis (IIASA), an international organization devoted to the study of major contemporary social issues. His 50-year career in geology and geophysics began with studies on metamorphic terrains in New England followed by analysis of the stability at high temperature and pressure of minerals, employing direct laboratory synthesis and thermodynamic analysis. Subsequently his interest shifted to policy related issues. In the early 1970s, he served on the White House Council on Enviornmental Quality. In this position Professor MacDonald played a major role in setting the legislative, administrative, and regulatory framework for how the United States deals with environmental issues. In recent years, MacDonald continues his work on both the technical and policy issues of climate change.

Ron Mason has been associated with Imperial College, London, for most of his working life, and is currently a senior fellow in the Department of Earth Science and Engineering. He has been a research geophysicist at the Scripps Institution of Oceanography, and at the Institute of Geophysics of the University of Hawaii. His research has focused on various aspects of Earth's magnetic field and its relation to recent crustal movement.

Dan McKenzie is a Royal Society Research Professor in the Department of Earth Sciences at Cambridge University, England. He first became interested in geology as a schoolboy reading Charles Lyell's *Principles of Geology*. However, as an undergraduate at Cambridge University, he found the undergraduate courses in geology disappointing, and so took a degree in physics. He then became a graduate student of Sir Edward Bullard, whose old office he now occupies, and wrote a Ph.D. on the shape of the Earth. He worked on plate tectonics for six years after receiving his Ph.D. in 1966, and has since worked in many areas of the earth sciences. At present his principal interests are in geochemistry and petrology, and in planetary geology.

Peter Molnar is a half-time Professor in the Department of Geological Sciences and a fellow of the Cooperative Institute for Research in Environmental Science (CIRES) at the University of Colorado in Boulder. Following a Ph.D. in seismology from Columbia University, he held postdoctoral positions there (at Lamont-Doherty Geological Observatory) and at the Institute of Geophysics and Planetary Physics, Scripps Institution of Oceanography. He then spent four months in the Soviet Union studying, among other subjects, Soviet methods of earthquake prediction. In 1974 he began a 27-year association with MIT, culminating in 15 years of freedom as a Senior Research Associate when teaching exhausted him. He currently is trying to renew himself by learning meteorology and oceanography to study climate change. He enjoys walking in the mountains and letting their beauty fuel his subconscious.

Lawrence Morley, a physics and geology major from the University of Toronto, worked with Fairchild Aerial Surveys and the Gulf Research and Development Company to pioneer the commercial application of airborne magnetometry in oil and mineral exploration. In 1953, he founded the Geophysics Division of the Geologic Survey of Canada, which over the next 30 years managed the detailed aeromagnetic survey of all of Canada and its continental shelves. In 1971, he promoted the establishment of the Canada Centre for Remote Sensing, and served as its director for ten years. After leaving government service in 1982, he joined the physics department in Toronto, and became the first director of the Institute for Space and Terrestrial Science, a university-government-industry consortium.

Jack Oliver, currently Professor Emeritus at Cornell University and the 1998 recipient of the Penrose Medal of the Geological Society of America, is known in the world of earth science for a variety of activities, achievements, and discoveries, mostly in the fields of seismology, tectonics, and earth exploration. He was a founding member of Columbia's Lamont-Doherty Earth Observatory, initiated in 1949, and pioneered the installation of seismographs at a variety of locations—in deep mines, on the sea floor, on the moon. The data generated played a key role in the development of plate tectonics by demonstrating the existence of down-going crustal slabs in subduction zones along continental margins and oceanic island arcs. Since the 1970s, Oliver has focused his efforts on COCORP—Consortium for Continental Reflection Profiling—a research project to use seismic reflection profiling to understand the deep continental crust.

Neil Opdyke received his Ph.D. from the University of Durham, United Kingdom, in 1958, and his D.Sc. from the University of Newcastle-on-Tyne in 1982. From 1963 to 1981, he was a research associate at the Lamont-Doherty Geological Observatory at Columbia University, and then Professor of Geology at the University of Florida from 1981 to the present. He is the author of over 150 scientific papers, and a member of many distinguished scientific societies, including the U.S. National Academy of Sciences and the American Academy of Arts and Sciences. He holds the Fleming Medal of the American Geophysical Union, the Woolard Award of the Geological Society of America, and the Stillwell Award of the Geological Society of Australia.

Robert Parker is a Professor of Geophysics at the Institute of Geophysics and Planetary Physics in Scripps Institution of Oceanography at the University of California, San Diego, where he has been since 1969. Parker's primary professional interest is mathematical geophysics, concentrating on inverse theory, which aims to develop techniques enabling geophysicists to draw conclusions from incomplete and inaccurate observations (*Geophysical Inverse Theory*, Princeton, 1994). He also studies the Earth's magnetic field, particularly with a view to understanding its variation in space and time. He is an enthusiastic recreational cyclist, having ridden more than 5,000 miles for each of the last 10 years. Parker is a Fellow of the Royal Society and the American Geophysical Union, and in 1998 was awarded the Gold Medal of the Royal Astronomical Society.

Walter Pitman received his Ph.D. from Columbia University in 1967. His research has been in plate tectonics, sequence stratigraphy, and global sea level change, and he is currently doing research on Arctic climatology. He was elected to the U.S. National Academy of Sciences in 2000.

David Sandwell is a Professor of Geophysics at the Scripps Institution of Oceanography, University of California, San Diego. The focus of his research is on the use of spaceborne radar to measure the physical properties of Earth and Venus. He pioneered the use of radar altimetry for revealing the topography and crustal structure of the deep oceans and has used these data to guide seagoing investigations. He was an investigator on the Magellan radar mapping of Venus and proposed that arcuate trenches on Venus were once active subduction zones. Although he serves on a number of committees to guide the development of Earth observations, he would prefer to stay at Scripps to work with students, do research, and go surfing.

Between Cambridge University, where he completed his Ph.D. in 1965, and the Scripps Institution of Oceanography, where he helped run the marine heat flow research program from 1965 to 1972, **John Sclater** met most of the major figures in plate tectonics. In 1972, he joined the faculty at the Massachusetts Institute of Technology, spending summers at Woods Hole. He left there for the University of Texas, Austin, in 1983, and returned to Scripps in 1991. As a seagoing scientist, he especially remembers sharing a cabin with Fred Vine on the *RRS Discovery*, in 1963, sailing with Bill Menard on the R/V *Horizon* in 1967, and taking Dan McKenzie to sea in 1968 on the R/V *Argo*.

Fred Vine is a Professorial Fellow and a former Dean of the School of Environmental Sciences at the University of East Anglia, Norwich, England. Having started his career in marine geophysics, in 1968 he began long-term and detailed studies of the structure and physical properties of the Troodos mountains ophiolite, in Southern Cyprus, an ancient slice of the ocean floor. The magnetization of rocks and the nature of the Earth's magnetic field have been common themes in much of his other research, and he has also been involved with measuring the electrical conductivity of continental rocks under simulated crustal conditions. Together with Phil Kearey, of Bristol University, he is currently working on the third edition of *Global Tectonics* (Blackwell, 1990, 1996).

ABOUT THE EDITORS

Naomi Oreskes is Associate Professor in the Department of History and Program in Science Studies at the University of California, San Diego, and is affiliated with the Institute of Geophysics and Planetary Physics (IGPP) at the Scripps Institution of Oceanography. Having started her career as a field geologist in Australia, her research now focuses on how scientists choose their research topics, how they decide what methods will provide good answers, and how they recognize a good answer when they see it. A 1994 recipient of the NSF Young Investigator Award, she has served as a consultant to the U.S. Environmental Protection Agency and the U.S. Nuclear Waste Technical Review Board on the validation of computer models. She is the author of *The Rejection of Continental Drift: Theory and Method in American Earth Science* (Oxford University Press, 1999).

Homer Le Grand completed degrees in history, chemistry, and history of science in the United States before moving to Australia in 1975. He is currently Dean of Arts at Monash University in Melbourne, and teaches in the Faculty of Arts and the Faculty of Science. He began researching and writing in the history of the earth sciences in 1980. Since completing *Drifting Continents and Shifting Theories* (Cambridge University Press, 1988), he has focused on the development of, and arguments over, the terrane concept in tectonics. He is currently collaborating with historian Bill Glen on a project entitled *Plate Tectonics Goes Ashore: Arguing Accreted Terranes.*

INDEX